Immunology of the GI Tract - Recent Advances

Edited by Luis Rodrigo

Published in London, United Kingdom

Immunology of the GI Tract - Recent Advances
http://dx.doi.org/10.5772/intechopen.101293
Edited by Luis Rodrigo

Contributors
Alireza Kazempour, Mwangala Nalisa, Thifhelimbilu E. Luvhengo, Jitendra Kumar, Sonia Sangwan, Murli Dhar Mitra, Haribrahma Singh, Priya Sharma, Somprakas Basu, Farhanul Huda, Sudhir Kumar Singh, Navin Kumar, Manisha Naithani, Jyoti Sharma, Isabel Comino, Ángela Ruiz-Carnicer, Verónica Segura, Carolina Sousa, Justus Reunanen, Surbhi Mishra, Juha Saarnio, Raffaella Canali, Marianna Roselli, Alberto Finamore, Andrea Ghiselli, Chiara Devirgiliis, Giacomo Rossi, Gratiela Gradisteanu Pircalabioru, Octavian Savu, Grigore Mihaescu, Corneliu Ovidiu Vrancianu, Mariana-Carmen Chifiriuc, Luis Rodrigo, Darmadi Darmadi, Riska Habriel Ruslie, Huseyin C. Yalcin, Mahmoud Khatib A. Al-Ruweidi, Nada Khater, Haya Rashid Alkaabi, Maram Hasan, Haya R. Al-Kaabi, Mohammed Murtaza

First published in London, United Kingdom, 2022 by IntechOpen
IntechOpen is the global imprint of INTECHOPEN LIMITED, registered in England and Wales, registration number: 11086078, 5 Princes Gate Court, London, SW7 2QJ, United Kingdom

British Library Cataloguing-in-Publication Data
A catalogue record for this book is available from the British Library

Additional hard and PDF copies can be obtained from orders@intechopen.com

Immunology of the GI Tract - Recent Advances
Edited by Luis Rodrigo
p. cm.
Print ISBN 978-1-80356-086-1
Online ISBN 978-1-80356-087-8
eBook (PDF) ISBN 978-1-80356-088-5

For EU product safety concerns:
IN TECH d.o.o., Prolaz Marije Krucifikse Kozulić 3, 51000 Rijeka, Croatia,
info@intechopen.com or visit our website at intechopen.com.

Meet the editor

Luis Rodrigo has been a Full Professor of Medicine at the University of Oviedo, Spain, since 2010, and Emeritus Professor since 2014. He has specialised in gastroenterology since 1972, achieving his Ph.D. in 1975 and being appointed Titular Professor in Medicine at the University of Oviedo in 1983. He has been head of the Gastroenterology Service of HUCA in Oviedo since 1975. He is the author of eight books on the treatment of gastroenterological and other digestive and liver diseases, has contributed 58 chapters to books on gastroenterology, and is the main author or co-author of 440 scientific papers in English and 282 in Spanish.

Contents

Preface

I am delighted to present this book, *Immunology of the GI Tract - Recent Advances*, an interesting and relatively little-known topic in which notable advances have been made in recent decades.

The 13 chapters, written by experienced and knowledgeable international authors, are arranged in four sections. In Section 1, concerning structure and function, after the Introductory Chapter, Prof. Rossi in chapter 2 and Thifhelimbilu E. Luvhengo and Mwangala Nalisa in chapter 3 detail the cellular components and diverse functions of the GI tract.

Contributions to Section 2 by Dr. Pircalabiouru et al. and Dr. Canali et al. describe the composition of the human microbiome and the role of different types of diet in its maintenance. A chapter by Dr. Darmadi and Riska Habriel Ruslie is devoted to Helicobacter Pylori infection, which appears with great frequency worldwide, especially in developing countries and has important clinical implications such as the development of gastroduodenal ulcers and gastric cancer. Prevention and precise eradication with a combination of special antibiotics and IBPs in high doses are discussed.

Section 3 examines the relationship of intestinal microbiota with various diseases. Ph.D. Student Kazempour deals with their influence on the development of autoimmune and neurological diseases, Dr. Basu et al. discuss their possible relationship with the development of malignant tumors of the gall bladder, Dr. Isabel Comino et al. look at the influence of microbiota on celiac disease and its treatment, and Dr. Reunanen et al. consider their importance in relation to the development of inflammatory bowel disease and its associated complications.

The three chapters of Section 4 develop the theme of the importance of immunology in relation to autoimmune diseases of the digestive system. Dr. Kumar et al. describe the role of molecular changes in the presentation and development of these dieseases. In the other two chapters, Dr. Yalcin et al. describe in detail the appearance, evolution, and complications of five autoimmune diseases of the digestive tract: achalasia, eosinophilic esophagitis, autoimmune atrophic gastritis, celiac disease and inflammatory bowel disease.I would like to cordially thank all the authors for their excellent and comprehensive contributions, without which the completion of this book would not have been possible.

Finally, I want to express my sincere thanks to Ms. Karla Skuliber for her continuous collaboration and editorial assistance, and to IntechOpen Books for their excellent editorial work.

Luis Rodrigo, MD
Professor of Medicine,
University of Oviedo,
Oviedo, Asturias, Spain

Section 1

Immune System - Structure and Function

Chapter 1

Introductory Chapter: Structure and Functions of the Small Intestine

Luis Rodrigo

1. Introduction

The small intestine is a tubular structure that connects the stomach to the colon. The intestinal mucosa is the innermost layer that is in contact with the intestinal lumen.

It is constituted by a glandular epithelium supported on the lamina propria, below which, the muscular layer is located. The folds on its surface, called conniving valves or Kerckring folds, that are formed by mucosa and submucosa, while the villi that project on its surface, are covered only by the mucosal layer, being its height greater in the proximal sections of the duodenum and jejunum to decrease progressively toward the ileum. This design of the intestinal mucosa, forming folds alternating with villi, is aimed at obtaining the maximum possible nutrient absorption surface. At the base of the villi lie the crypts of Lieberkühn, which form glandular structures that extend to the muscularis mucosa [1].

2. Microscopic structure

In the intestinal epithelium, both at the level of the villi and in the crypts, five types of cells are identified, which are the following:

1. Enterocytes: These are the epithelial cells responsible for absorbing nutrients. Their surface is covered with microvilli, which further increase the absorption surface, containing numerous enzymes for their functions.

2. Goblet cells: They are responsible for the secretion of mucus that acts as a lubricant and protector.

3. Paneth cells: They have phagocytic and lysozyme elimination capacity, playing an important role in the regulation of the microbiota and in the immune defense mechanisms.

4. Entero-endocrine cells: They secrete different hormones with important functions on intestinal motility and secretion, such as cholecystokinin (CCK), secretin, neurotensin, peptide YY, ghrelin, and gastric inhibitory peptide (GIP).

5. M cells: They are located in the domes of the lymphoid aggregates and have a relevant function, which consists of transporting various tumors, both food and bacteria, toward the underlying lymphoid tissue.

The lamina propria on which the epithelial cells are located is made up of a connective component and gut-associated lymphoid tissue. Its main function is to participate in the defense against microorganisms and other pathogens.

3. Intestinal absorption

There are three main mechanisms that are carried out mainly through the apical membrane of the enterocytes.

a. Passive diffusion: It does not need carrier molecules and does not consume energy. It is produced by a concentration gradient from the intestinal lumen.

b. Facilitated diffusion: There is also a concentration gradient, but a specific carrier protein is also involved, which facilitates the passage of the nutrient through the membrane of the enterocyte.

c. Active transport: A carrier protein intervenes, which requires a cellular energy supply. In this way, the substance can be absorbed although there is no greater intraluminal gradient.

Prior to intestinal absorption, a process of chemical and mechanical digestion of foods takes place, which begins in the mouth, through chewing and salivary secretion, and continues in the stomach, by mixing food with gastric juice rich in pepsin. In this way, food is converted into substances capable of crossing the epithelial barrier and passing into the blood and lymphatic circulation [2].

At the level of the brush border formed by the intestinal microvilli of the enterocytes, some enzymes are located, which are not released into the intestinal lumen, carrying out there, various specific hydrolytic functions against disaccharides, peptides, and other nutrients.

4. Microbiota

The organization of the small intestine is in the form of a model of strata formed through the creation of the intestinal barrier in which two layers are located, one "external" (consisting of the microbiota, the mucous layer, and the intestinal epithelium) that acts as an anatomical and, therefore, physical barrier, preventing their adhesion and other more "internal," mainly formed by the GALT (gut-associated lymphoid tissue), responsible for the production of the immune response and tolerance mechanisms. The correct interrelation between both layers contributes to maintaining proper functioning of the small intestine, actively ensuring its intestinal permeability.

The microbiota contributes to the digestion and recovery of energy from dietary waste and the production of vitamins and hormones. It prevents the growth of pathogenic bacteria, increasing its protection and local defense. It also contributes to trophism, favoring the production of mucin, as well as the proliferation and differentiation of the intestinal epithelium. It also exerts immunological functions, participating in the development and maturation of the immune system [3].

5. Mucosal immune system

Peyer's patches located at the level of the submucosa are sites of controlled uptake of antigens and activation of naive T and B lymphocytes. They are made up of several aggregates of B cells (lymphoid follicles) surrounded by rings of T cells, and also by antigen-presenting cells (APC), usually macrophages.

The epithelium that covers the dome of the lymphoid aggregates contains M cells, with the capacity to transport antigens from the intestinal lumen to the underlying lymphoid tissue, together with dendritic cells located in the lamina propria, whose formation is induced by various factors produced by the epithelial cells through stimulation of Toll-like receptors (TLR).

The products captured and processed by the APCs are presented to the naive T lymphocytes, initiating clonal expansion to collaborator or helper lymphocytes (Th1 or Th2) or regulatory T lymphocytes (Th3, Tr1, or CD25+/CD4+).

Following the process of antigenic stimulation, lymphocytes migrate into the mesenteric lymph nodes where further antigenic exposure and clonal expansion occur, then passing into the systemic circulation, returning to other mucosal surfaces, forming the so-called associated lymphoid tissue to the mucous membranes [4].

The humoral response occurs with the binding of the antigen to the IgM membrane of the B lymphocyte. The action signal generated by this binding stimulates a clonal expansion and as a consequence. The induction by antigen-specific T lymphocytes, or mediated by various factors, an additional differentiation takes place, which includes several processes of reorganization of the immunoglobulin chains, mainly of the IgA type. Activated B lymphocytes undergo a terminal differentiation process and become as plasma cells [5].

Secretory IgA (sIgA) is a powerful protector of the intestinal mucosa against bacterial invasions, constituting the first line of defense against multiple external infections including various viruses and other pathogens.

Author details

Luis Rodrigo
Professor of Medicine, Department of Internal Medicine, University of Oviedo, Oviedo, Spain

*Address all correspondence to: lrodrigosaez@gmail.com

References

[1] Chun HH, Seol SY, Choi MG, Cho JY, editors. Small Intestine Diseases. 1st ed. Singapore: Springer-Verlag; 2022

[2] Henderson JM. Gastrointestinal Pathophysiology. Philadelphia: Lippincott, Williams and Wilkins; 1996

[3] Koulaouzidis A, Marlicz W. Diagnosis and treatment of small bowel disorders. Journal of Clinical Medicine (Special Issues). 2021

[4] Fisch EM, Burns B. Physiology of Small Bowel. Treasure Islands (FL): Stat Pearls Publishing; 2022

[5] Volk N, Lacy B. Anatomy and physiology of the small bowel. Gastrointestinal Endoscopy Clinics of North America. 2017;**1**:1-13

Chapter 2

New Acquisitions Regarding Structure and Function of Intestinal Mucosal Barrier

Giacomo Rossi

Abstract

The purpose of this chapter is to illustrate the role of the intestinal barrier in keeping separate, but also communicating, the "world above" represented by the resident microbial flora (microbiota) and the "world below" (the immune system associated with the gastrointestinal tract or GALT). Description will be given for how it is possible that the intestinal microbiota, in the course of *dysbiosis*, can alter the junctional complex that unites the enterocytes, and how the probiotic bacteria (and their metabolites) to restore a homeostasis in the gastrointestinal tract. The fundamental role of enterocyte mitochondria will be highlighted, where being archaic methylotrophic bacteria have retained the ability to "interpret" the bacterial signals (*eubiotic* or *dysbiotic*) derived from the intestinal lumen. In this perspective, everything starts from an altered mitochondrial functioning, deriving from a condition of *dysbiosis*, which alters the tightness of the TJs, opening up to bacterial translocation and bacterial products. Probiotics and their metabolites act by restoring mitochondrial activity and function and the enteric barrier functionality. The author will exemplify this "story" with *in vitro* and *in vivo* tests, deriving from original studies on different animal models (mouse, dog, and cat) including humans (patients with IBD and with HIV-related enteropathy).

Keywords: tight junctions, mitochondria, microbiota, Toll-like receptors, innate immunity, oral tolerance, probiotics

1. Introduction

The gastrointestinal system is, together with the skin and the respiratory system, the habitat most exposed to the external environment. Every day, thousands of microorganisms and compounds derived from digestion come into contact with it. This condition requires a complex defense system capable of separating the intestinal contents from the host tissues, regulating the absorption of nutrients and allowing the interaction between the resident microbial flora and the mucosal immune system, inhibiting the translocation of pathogens in the underlying tissues. All these functions are performed by the intestinal barrier.

The intestinal barrier is a functional unit, organized as a multi-layered system, in which it is possible to recognize two main parts: a physical surface barrier, which prevents bacterial adhesion and regulates the paracellular diffusion towards the underlying host tissues and a deeper functional barrier, which is able to discriminate between commensal and pathogenic microorganisms, organizing the immunological tolerance towards the commensal bacteria and the immune response towards the pathogens [1].

The intestinal epithelium is organized into a monolayer of cells with a thickness of only 20 μm and is composed of 5 different cell types: enterocytes (IECs), mucus-producing goblet cells (GCs), endocrine cells, "M" cells, "G" cells, and defensin-producing Paneth cells, all of which differentiate from intestinal epithelial Lgr5+ stem cells [2–4]. Lgr5+ cells are crypt base columnar (CBCSs) stem cells, a population of rapidly dividing cells at the crypt base expressing leucine-rich-repeat containing G-protein coupled receptor 5 (Lgr5), giving rise to all terminally differentiated intestinal epithelial cell (IEC) types [5] CBCSs divide into progenitor cells which move upward within the crypt into the transit amplifying zone [6]. It is here that the cells differentiate further and travel to the villus where their functions are required. At the villus tip, senescent IECs slough off through *anoikis*, a specific type of programmed cell death for anchorage-dependent cells, and make room for newly formed cells to take their place [6]. Paneth cells are the exception as these cells are long-lived secretory cells that migrate to the crypt base and reside between $Lgr5^{+\ CBCSs}$ where they produce and secrete antimicrobial peptides and stem cell factors such as epidermal growth factor (EGF), and other factors that sustain the stem cell niche [7].

IECs are the most represented cell type. They act as a physical barrier that inhibits the translocation of the luminal content into the innermost tissues; IECs form a seamless structure. In fact, they are connected by particular inter-cellular binding structures called adherent junctions (AJs) and tight junctions (TJs), characterized by trans-membrane proteins that interact with adjacent cells and with intracellular proteins, intimately connected with the enterocyte cytoskeleton. The fundamental elements on which the integrity of the "*intestinal barrier*" depends are, therefore, the IECs and the intercellular junctions.

1.1 Intestinal epithelial cells (IECs) and their metabolism

The main function of enterocytes is the absorption of nutrients, and this function is performed by the mature or "*absorptive*" IECs, which are differentiated from the intestinal stem cells, CBCSs, residing at the bottom of the crypt. Nutrients such as glucose and amino acids are transported and absorbed by various transporters embedded on the membranes of these enterocytes. Metabolism occurs in each cell along the crypt-villus axis (CVA). The intestinal epithelial cells are the most vigorous, self-renewing cells, regenerating from the crypt bottom to the villus tip in only 3–5 days. Intestinal epithelial cells continuously migrate and mature along the CVA; the energy metabolism in intestinal epithelial cells increases from the bottom of the crypt to the top of the villi. Moreover, the expression of proteins related to the metabolism of glucose, most amino acids, and fatty acids increases in intestinal epithelial cells during maturation along the CVA, while the expression of proteins related to glutamine metabolism decreases from crypt to villus tip. The expression of proteins involved in the citrate cycle is also increased in IECs during maturation along CVA [8].

L-Glutamate is one of the most abundant amino acids in alimentary proteins, but its concentration in blood is among the lowest. This is largely because L-glutamate is extensively oxidized in small intestine epithelial cells during its transcellular journey from the lumen to the bloodstream and after its uptake from the bloodstream. This oxidative capacity coincides with a high energy demand of the epithelium, which is in rapid renewal and responsible for the nutrient absorption process. L-Glutamate is a precursor for glutathione and *N*-acetylglutamate in enterocytes. Glutathione is involved in the enterocyte redox state and in the detoxification process. *N*-acetylglutamate is an activator of carbamoyl phosphate synthetase 1, which is implicated in L-citrulline production by enterocytes. Furthermore, L-glutamate is a precursor in enterocytes for several other amino acids, including L-alanine, L-aspartate, L-ornithine, and L-proline. Thus, L-glutamate can serve both locally inside enterocytes and through the production of other amino acids in an inter-organ metabolic perspective. In colonocytes, L-glutamate also serves as a fuel but is provided from the bloodstream. Alimentary and endogenous proteins that escape digestion enter the large intestine and are broken down by colonic bacterial flora, which then release L-glutamate into the lumen. L-Glutamate can then serve in the colon lumen as a precursor for butyrate and acetate in bacteria. L-Glutamate, in addition to fiber and digestion-resistant starch, can thus serve as a luminally derived fuel precursor for colonocytes (**Figure 1**) [9].

Glutamine is the principal energy source for IECs, and during acute illnesses, patients experience nutritional depletion that is correlated to low plasma and low mucosal glutamine concentrations. Such deficiencies are common among hospitalized dogs and cats or human patients and are associated with an increased risk of developing infectious complications, organ failure, and death [10, 11]. A number of clinical studies reveal a significant benefit of glutamine use on mortality, length of hospital stay [12, 13], and infectious morbidity in critical illnesses, as well as in dog or cat parvovirus infection [11, 12, 14]. Patients receiving high-dose parenteral (rather than orally) glutamine presented the highest beneficial effects, and it is estimated that high doses of parenteral Gln (>0.50 g/kg/day) are the best treatment for humans and animals, demonstrating a greater potential to benefit [15]. However,

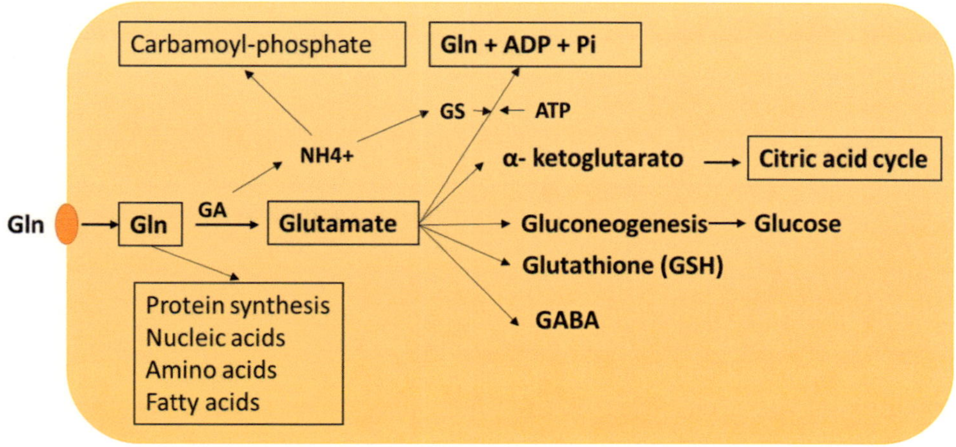

Figure 1.
Metabolic role of glutamine at the cellular level. In the catabolic phase, glutamine is transformed into glutamate and ammonium ions, thanks to the mitochondrial enzyme glutaminase (GA), while in the anabolic phase, at the level of most tissues, glutamine can be synthesized starting from glutamate and ammonia, in the presence of ATP, thanks to the enzyme glutamine synthetase (GS). Ammonia can be converted into Carbamoyl-phosphate, while Glutamate can form a-Ketoglutarate but also Glucose in the liver and kidney, while it is the basis for the synthesis of Glutathione in most cells, and of GABA (Gamma aminobutyric acid) at the neuronal level.

the role of glutamine in the maintenance of normal gut and immune system function may be even more important for critically ill animals [16]. Glutamine is now considered by many investigators to be a conditionally essential nutrient during protein-calorie malnutrition, required in quantities that are greater than those that can be synthesized by the body. Based on this hypothesis and preclinical studies performed in dogs [17] the commercial veterinary critical care rations often recommended for cats and dogs with some severe enteropathies and cancer are routinely supplemented with glutamine. Glutamine supplementation has also been suggested as a way to promote more rapid resolution of acute side effects of the oral mucosa in dogs receiving oronasal radiotherapy and to maintain gut immunity and integrity in patients receiving radiotherapy or chemotherapy [18].

Recently, glutamine parenteral supplementation evidenced restoration of interdigestive migrating contraction in an experimental canine model of postoperative ileus [19]; in this research is hypothesized that the benefit derives from glutamine's ability to maintain glutathione concentration and thereby counteract the deleterious effects from surgical injury, inflammation, and oxidative stress. Similarly, parenteral administration of L-alanyl-L-glutamine [20] in dogs prevented the immune suppression induced by high-dose methylprednisolone sodium succinate, and experimental studies in the current literature indicate that glutamine use may prevent the occurrence of lung injury, tissue metabolic dysfunction, and reduce mortality after injury [21]. Glutamine's beneficial effects on critical illnesses or during IBD, may result from two principal ways: (a) the direct effect on IECs metabolism that helps to maintain the integrity of the epithelial barrier, preventing bacteria translocation; and (b) enhanced heat shock proteins (HSP) expression [22, 23] by enterocytes, and leucocytes [10, 24]. Heat shock proteins are a group of proteins essential to cellular survival under stressful conditions. The stress-inducible HSP60, HSP70 and HSP72 are inducible forms of the stress protein, which may confer cellular protection [10]. The cellular functions of intracellular HSP70 and HSP72 are responsible for limiting protein aggregation, facilitating protein refolding, and chaperoning proteins; an intra-mitochondrial concentration of these proteins is associated with an increase of mitochondria wellness, metabolic activity and ATP production for IECs (see below). IECs-specific deletion of the mitochondrial chaperone protein heat-shock protein 60 (HSP60) led to mitochondrial dysfunction, impairment of cell proliferation and loss of stemness of intestinal stem cells [25]. Additionally, mitochondrial dysfunction impaired the ability of the CBCSs to produce ATP, leading to altered CBCSs self-renewal and differentiation. Furthermore, L-glutamine potentiation of HSP72 is associated with increased gut epithelial resistance to apoptotic injury, and reduced HSP72 may be associated with increased caspase activity in glutamine-deficient [26]. In fact, glutamine induces autophagy under stressed conditions, and prevents apoptosis under heat stress through its regulation of the mTOR and p38 MAP kinase pathways [27]. Glutathione (GSH) metabolism is also closely related to the apoptotic processes of epithelial and immune cells. The increase of intracellular GSH is sufficient to reduce Fas-triggered increase in apoptotic cells. Over expression of Bcl-2, an anti-apoptotic protein, causes redistribution of glutathione to the nucleus, thereby altering nuclear redox and blocking caspase activity [28, 29].

Also, the amount and type of dietary fiber influence the end-products of fermentation and thus fuel availability to intestinal tissue in a *specie* depending manner. The metabolic fuel usage was studied in intestinal cells isolated from dogs consuming a commercial diet to examine preferential fuel usage and the effect of diet on canine enterocytes and canine colonocytes, respectively, indicating that glutamate/glutamine is preferentially used by enterocytes, while butyrate (found in food and produced as an intestinal fermentation by-product of dietary fiber by gut bacteria) followed by glutamine is preferentially used by isolated canine colonocytes [30].

IBD has been suggested to involve a state of energy-deficiency, whereby oxidative metabolism is altered within IECs [31, 32]. Butyrate undergoes catabolic degradation through β-oxidation in the mitochondrial matrix of colonocytes, providing over 70% of the energy demand of the colonic epithelium [33]. Butyrate metabolism was demonstrated to be impaired in an animal model of colitis [34], and numerous studies have reported impaired metabolism in the intestinal mucosa of patients with IBD [35, 36]. Similarly, intestinal mucosal inflammation results when butyrate oxidation is inhibited in experimental animals [33]. Santhanam et al. [37] showed that the mitochondrial acetoacetyl CoA thiolase, which catalyzes the critical last step in butyrate oxidation, was significantly impaired in the colonic mucosa of patients with ulcerative colitis. Furthermore, they conclude that an increase in mitochondrial ROS may trigger this enzymatic defect [37]. Thus, defective β-oxidation in the mitochondria has deleterious effects beyond energy requirements. Likewise, a dysfunctional gut microbiome or a poor diet may also result in a decrease of butyrate metabolism in the colonic epithelium. Enhanced production of butyrate may potentially benefit the colonic epithelial cells by stimulating an enhancement in cellular homeostasis, including antioxidant and anti-inflammatory roles as well as protective gut-barrier functions.

1.2 Role of mitochondria in IECs homeostasis and barrier integrity

The integrity of the intestinal epithelium, tight junction maintenance, and β-oxidation are key cellular processes within the intestinal epithelium that are not only dependent upon properly functioning mitochondria but are also known to be altered in animal models of intestinal inflammation and in humans with IBD.

Control of intestinal epithelial stemness is crucial for tissue homeostasis. Disturbances in epithelial function are implicated in inflammatory and neoplastic diseases of the gastrointestinal tract. Mitochondrial function plays a critical role in maintaining intestinal stemness and homeostasis. Using murine IECs, Berger et al. [25] demonstrated that loss of mitochondrial chaperone HSP60, activates the mitochondrial unfolded protein response (MT-UPR) and results in mitochondrial dysfunction [25].

During IBD, a destruction of the intestinal epithelial barrier, an increased gut permeability, and an influx of immune cells through the intestinal mucosa are observed. Given that, most cellular functions as well as maintenance of the epithelial barrier are energy-dependent, it is logical to assume that mitochondrial dysfunction may play a key role in both the onset and recurrence of disease. Indeed, several studies have demonstrated evidence of mitochondrial stress and impaired functions, such as oxidative stress and impaired ATP production, within the intestinal epithelium of patients with IBD and mice undergoing experimental colitis [38].

Recently, we have observed that mitochondria dysfunction has a central role in human detrimental intestinal barrier effects of chronic HIV infection [39].

Mitochondria are membrane-bound organelles that maintain cellular energy production through oxidative phosphorylation [40], and contain a circular small genome that encodes only 13 proteins [41]. Despite the limited coding-capacity of the mtDNA, mitochondria regulate vital cellular functions aside from energy production, such as the generation of ROS and reactive nitrogen species (RNS), the induction of programmed cell death, and the transduction of stress and metabolic signals [42]. The current literature would support a key correlation between mitochondrial function and intestinal barrier dysfunction/inflammation. Nonetheless, it is important to understand how any alteration in the multifaceted functionality of the mitochondrion may contribute to the initiation and propagation of an inflammatory insult (**Figure 2**).

Supporting the importance of mitochondrial form and function, enterocytes isolated from patients with IBD have been reported to exhibit swollen mitochondria with irregular cristae [43, 44]. Abnormal mitochondrial structure is also seen in IECs from mice subjected to experimental models of colitis [45]. Similar observations are made on canine IECs during IBD or lymphangiectasia [46].

These morphological changes are suggestive of cellular stress and bioenergetic failure. Indeed, patients with IBD have reduced ATP levels within the intestine [33, 47]. As would be expected, morphological changes in mitochondria have been shown to result in deficiencies in the β-oxidation of short-chain fatty acids (SCFA) [48]. The intestinal mucosa of IBD patients has been demonstrated to be in a state of energy deficiency characterized by low ATP levels and low energy charge potential [33, 49], calling into question the functionality of this organelle during disease. To further prove this, in a recent study, it was demonstrated that mtDNA released into the serum in IBD patients was recognized as a damage-associated molecular pattern (DAMP) potentially by toll-like receptor 9 (TRL9), and could provide a biomarker of inflammation [50].

Thus, defects in intestinal epithelial homeostasis result in an inadequate intestinal barrier defense, which may allow luminal antigens and/or microbes to interact with or violate the intestinal epithelium and consequently cause inflammation [51]. However, the role of mitochondrial dysfunction during IECs differentiation needs to be further

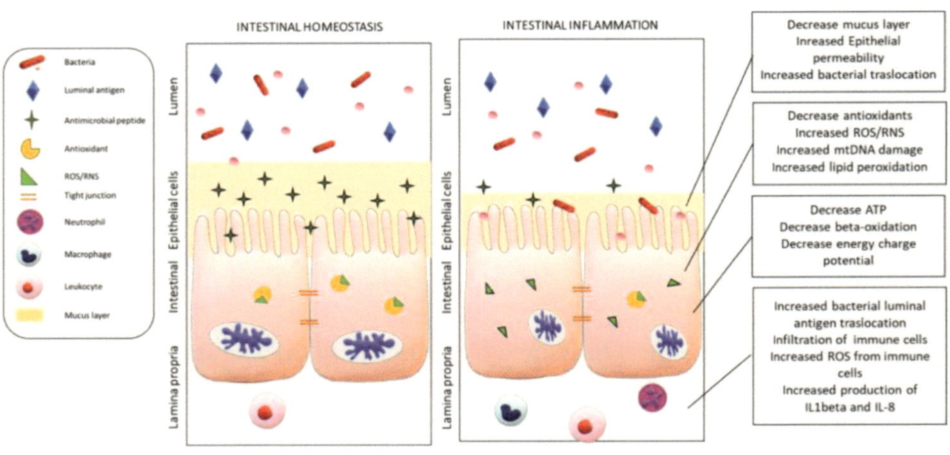

Figure 2.
A condition of eubiosis involves the correct synthesis/absorption of glutamine and glutathione by the enterocytes. Furthermore, the presence of "healthy" bacterial species producing NEFAs in the correct proportion, with an excess of butyrate, preserves the mitochondria from oxidative damage from ROS. A condition of dysbiosis increases mitochondrial damage, critically reducing the number of mitochondria but above all modifying their morphology and permeability. A critical reduction in mitochondria leads to a decrease in the production of ATP by the enterocyte (due to a reduction in the Krebs cycle and beta-oxidation). A reduction in energy leads to a lower "hold" of the intercellular junctional complexes and an increase in bacterial translocation through the intestinal epithelium, which becomes more permeable. At the submucosal level, this condition increases inflammation and the recall of leukocytes, further worsening the condition of the mucosal barrier.

evaluated in order to understand the role it may play in the development of intestinal inflammation. Recently, Bär et al. [52] demonstrated that altered mitochondrial oxidative phosphorylation activity influences intestinal inflammation in mice models of experimental colitis. The study suggests that increased regeneration of the intestinal epithelium (by means of increased mitochondrial function) is a key factor in combating intestinal inflammation. Mucosal healing also results in improved mitochondrial structure in the IECs of patients with ulcerative colitis [53].

1.3 Mitochondria cross-talking with intestinal microbiota maintaining intestinal barrier integrity

Maintenance of TJ integrity is an energy-dependent process, and it is not surprising that disruption of the barrier by toxins, pathogens, or noxious stimuli can be initiated by damaged mitochondria [39, 54, 55].

Mitochondria in animals, as well as chloroplasts in plant cells, are old- primitive bacteria that have lost the ability to live a "*free*" life by entering into a complex system of cooperation, the eukaryotic cell, and leaving some fundamental functions to the nucleus and other cellular organelles. The fact that mitochondria are ancestral bacteria makes them particularly sensitive to metabolic "motifs" produced by other bacteria. New research shows bidirectional communication exists between the gut microbiota and mitochondria [56, 57].

Certain insults, such as NSAID exposure, are known to disrupt the structure and function of the mitochondria, and at least transiently, increase gut permeability [58–60]. Additionally, it has been reported that some patients with Crohn's disease develop immune reactivity against components of their gut microbiome [61]. Consistent with these reports, Nazli et al. [44] demonstrated that treating a cell monolayer with dinitrophenol (an oxidative phosphorylation uncoupler) resulted in cellular internalization of a non-invasive strain of *Escherichia coli*. From this, the authors hypothesized that under metabolic stress resulting from mitochondrial dysfunction, the enteric epithelium loses its ability to distinguish between commensals and pathogens, and as a result, begins internalizing commensal organisms, which can lead to an exacerbated intestinal inflammatory response [44]. Studies do suggest that both mitochondrial dysfunction [62] and increased gut permeability [63] affect the overall competence of the intestinal epithelial barrier, but the stimuli that initiate either process are not known. Nonetheless, these studies reinforce the implication of epithelial mitochondrial dysfunction as a predisposing factor for an increase in gut epithelial permeability and a loss of gut barrier function, resulting in intestinal inflammation. The intestinal lumen and epithelium are continuously exposed to noxious stimuli, such as ingested nutrients, local microbes or infections, gastric acid production, and periods of ischemia/reperfusion that have the potential to stimulate the generation of oxygen and nitrogen radicals [64–66]. Additionally, the infiltration of leukocytes, monocytes, and neutrophils during inflammation can further enhance intestinal ROS production through both respiratory burst enzymes and prostaglandin and leukotriene metabolism [67]. Several studies have demonstrated increased ROS/RNS levels within the intestinal epithelium of animals and patients with spontaneous and experimentally induced IBD [68–70].

Typical gut bacterial families found in healthy dogs and cats include Bacteroidaceae, Clostridiaceae, Prevotellaceae, Eubacteriaceae, Ruminococcaceae, Bifidobacteriaceae, Lactobacillaceae, Enterobacteriaceae, Saccharomycetaceae, and Methanobacteriaceae [71].

The gut microbiota are key to host metabolism as they aid in the digestion and absorption of food, neutralize drugs and carcinogens, synthesize choline [72], secondary bile acids [73, 74], folate, vitamin K2 and short chain fatty acids (SCFA). Additionally, the gut microbiota protects the host against pathogenic infection, stimulating and maturing the immune system [75] and epithelial cells [76] and regulating oxidative stress [77].

Bacterial metabolites, including short-chain fatty acids (SCFAs) and hydrogen sulfide (H_2S), serve as messengers to enteric/colonic epithelial and immune cells, impacting their metabolism, epigenetic modifications, and gene expression. SCFAs are currently the most studied bacterial metabolites and are beneficial to intestinal and colon homeostasis. The three major SCFAs, acetate, propionate, and butyrate, are produced in the colon by bacterial fermentation of carbohydrates and are an important source of energy for colon epithelial cells. SCFAs are ligands for free fatty acid receptors 2 and 3, which modulate glucose metabolism and mitochondrial fatty acid β-oxidation (FAO). Additionally, SCFAs regulate PGC1α, a transcriptional coactivator that is a central inducer of mitochondrial biogenesis in cells [78]. These responses to SCFAs result, at the organelle level, in increased glucose uptake, FAO, oxidative phosphorylation, and mitochondrial biogenesis. In terms of intestinal homeostasis, these responses to SCFAs in colon epithelial cells facilitate the development of a tolerant mucosal immune system, promote epithelial barrier integrity, promote "physiologic hypoxia", and suppress colitis [7]. In addition, steady-state inflammasome machinery activation in the colon is mediated by SCFAs, which produces basal IL-18 levels, regulates the microbiome composition, and dampens overt inflammatory responses.

Butyrate, a by-product of the microbial fermentation of SCFAs, is one of the key molecules of mitochondria/gut microbiota cross-talk; butyrate may influence mitochondrial-endoplasmic reticulum (ER) contact signaling pathways. A body of recent evidence reveals that the microbiome impacts the host by communicating with its intracellular relatives, the mitochondria. This perspective mode of chemical communication between bacteria and mitochondria may help us understand complex and dynamic environment-microbiome-host interactions regarding their vital impacts on health and diseases. Communications between bacteria and mitochondrial are mediated by chemical signals from intestinal bacteria. In one case, a cluster of bacterial metabolites including betaine, methionine, and homocysteine initiate a signaling cascade that triggers the nuclear receptor 5A nuclear receptor and activates hedgehog signaling to regulate mitochondrial fission-fusion balance in intestinal cells [79]. This bacteria-mitochondria communication ultimately regulates fat storage homeostasis in the host [80]. Additionally, a slime polysaccharide named *colanic acid*, a major biofilm component of *E. coli*, secreted from intestinal bacteria, after entering the host cytoplasm via endocytosis, increases the fragmentation of intestinal mitochondria in a dependent fashion to the Dynamin Related protein-1 (Drp-1), a cytosolic guanosine triphosphate (GTPase) protein-key player of mitochondria fission, as well as enhances Mitochondrial Unfolded-Protein Response (UPR^{mt}) in response to mitochondrial stress. These signaling effects of bacterial colanic acid on mitochondrial dynamics and UPR^{mt} consequently lead to lifespan extension and protection against age-associated pathologies, like germline tumor progression and toxic amyloid-beta accumulation, in the host [81]. Besides SCFA, secondary bile acids produced by the gut microbiota also play an important role in regulating mitochondrial energy metabolism. Anaerobic bacteria of the genera *Bacteroides, Eubacterium*, and *Clostridium hiranonis* degrade 5–10% of the primary bile acids, forming secondary bile acids [71, 82, 83] Secondary bile acids interact with mitochondria by modulating transcription factors related to lipid and carbohydrate metabolism, including farnesoid X receptor (*FXR*) and G-coupled membrane protein 5 (*TGR5*) [84]. *FXR* is a target of NAD-dependent protein deacetylase *silent regulator 1 (SIRT1)* [85] and regulates the steroid response element binding protein-1c, carbohydrate response element binding protein, and Peroxisome proliferator-activated receptor alpha (*PPAR*-α), which stimulates fatty acid uptake and oxidation [74]. There is increasing evidence that secondary bile acid metabolism might also directly modify *SIRT1* expression as well as mitochondrial biogenesis, inflammation, and intestinal barrier function in different types of cells (**Figure 3**) [86, 87].

Together, these results consistently show that mitochondria undergo chemical communication with bacteria, a process modulating metabolic and senescent states of eukaryotic cells. The impact of microbiota on mitochondrial functions has been further supported by studies intending to manipulate gut microbiota through the use of probiotics. One example is the probiotic *Escherichia coli* Nissle 1917 (EcN) with proven effectiveness in the treatment of inflammatory intestinal disorders and acute diarrhea. Outer membrane vesicles (OMVs) released by the probiotic EcN and the commensal ECOR63 are taken up by intestinal epithelial cells, and modulate the epithelial barrier integrity through several mechanisms, mediated by the restoring of the mitochondria [88]. Administration of the probiotic *Lactobacillus rhamnosus* CNCMI–4317 induced a series of modulating factors that modified the oxidative phosphorylation (OXPHOS) capacity of mitochondria [89]. Certain intestinal bacteria such as *Eubacterium hallii* and *Anaerostipes caccae* have the capacity to transform the byproduct of anaerobic glycolysis lactate into SCFA during glucose depletion thus

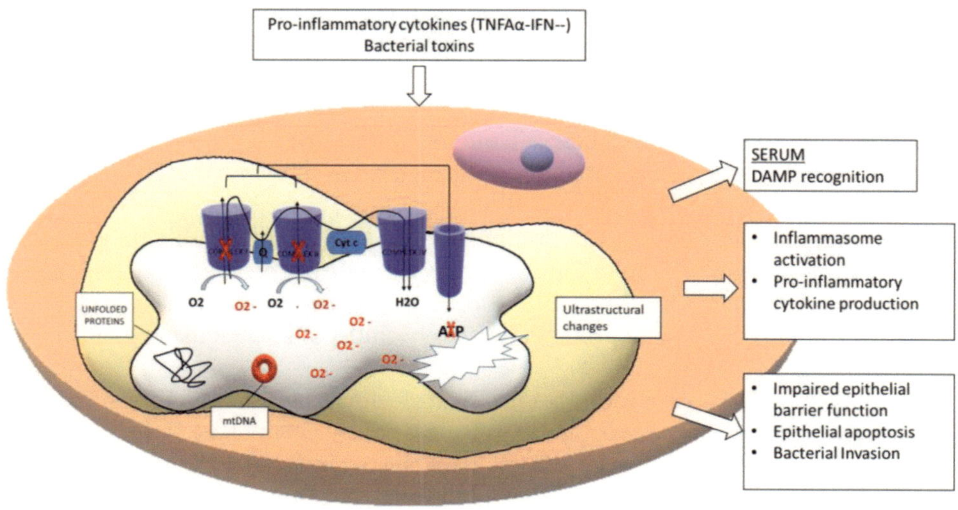

Figure 3.
Effect of mitochondrial morpho-functional alterations at the basis of the "leaky gut" as observed in the course of IBD. Dysbiosis, bacterial toxins and free radicals linked to a reduced intake of glutamine, involve the activation of signals of the extrinsic and intrinsic pathways of apoptosis, which pass through the structural and functional alterations of the mitochondria (i.e., increased permeability, translocation of HSPs and of the APAF-Cytochrome C complex, loss of Ca ++ etc.). A reduction in mitochondria, resulting in a reduction in ATP, causes a decreased activity of ETC complexes, accumulation of mtROS, accumulation of misfolded or unfolded proteins in the matrix, and ultrastructural changes such as cresting. Subsequent loss of epithelial barrier integrity, epithelial cell apoptosis, and bacterial invasion have been demonstrated following mitochondrial dysfunction in the epithelium. mtDNA is released in the serum of IBD patients and acts as a DAMP for the activation of immune cells. Furthermore, damaged mitochondria can signal the activation of the inflammasome, leading to the production of pro-inflammatory cytokines and increasing leukocyte infiltration of the intestinal mucosa.

creating an alternative energy source for the host, while bypassing OXPHOS [90, 91]. Finally, probiotic mixture Slab51™ administration for a period of two or six months, restores mitochondria inducing HSP60 and 70 mitochondrial internalization and increasing number and size of mitochondria in intestinal cells of IBD, and suffering dogs, and HIV chronically affected patients [39, 92, 93].

Unlike the beneficial effects commensal bacteria and certain probiotics have on energy metabolism, pathogens such as *Salmonella* and *E. coli* [94] can produce negative effects on the host mitochondria energy metabolism by degrading sulfur amino acids to produce hydrogen sulfide (H_2S) in the large intestines. H_2S is an important mediator of many physiological and pathological processes. High amounts of H_2S can inhibit a key component of the mitochondrial respiratory chain by penetrating cell membranes and inhibiting COX activity and energy production [56]. Pathobionts can also produce NO, which may affect host mitochondrial activity and favor bacterial infection [95]. Beaumont et al. [96] concluded that exposure of high levels of H_2S to HT-29 human cells showed not only reduced mitochondrial oxygen consumption but also an increase in the expression of inflammatory genes such as IL-6, which was increased following a high protein diet. Mottawea et al. [56] recently demonstrated that a proliferation of pathobionts, many of which are known to be potent H_2S producers, down regulated mitochondrial proteins. Additionally, H_2S induces genotoxic damage to the epithelium, inhibits metabolism of SCFAs, and induces breaks in the mucus barrier, allowing exposure of luminal contents to the underlying tissue [7, 97].

1.4 The system of intercellular junctions

In the intestinal epithelium there are two main types of junctions: adherent junctions (AJs) and tight junctions (TJs). Both types are formed from the proteins of the classes of cadherins, claudins and occludins, present in different concentrations and control the paracellular permeability through the intercellular spaces. In epithelial barriers, TJs and AJs are well defined and distributed: the TJs are present in the apical part, while the AJs are located in the basolateral part, below the TJs (**Figure 4**). Both are connected to the actin cell cytoskeleton.

The tight junctions seal the paracellular space and for their assembly they need adherent joints. As can be seen in **Figure 4**, they are multi-protein complexes made up of integral membrane proteins (claudins, occludins and junctional adhesion molecules), peripheral membrane proteins (zonula occludens) and regulatory molecules such as kinases.

Claudins (18–27 kDa) are proteins with 2 extracellular loops and a C-terminal cytoplasmic domain. They constitute a large gene family in which 24 isoforms have been identified that determine the selectivity of the paracellular pathway in terms of tissue, charge and size. They are expressed in a tissue-specific way and a mutation or deletion of one of the members of this family can have significant effects on the function of the epithelial barrier [98, 99].

The data obtained in some in vitro studies indicate that the claudins -1, -3, -4, -5, -8, -11, -14, and -19 play a determining role in the selectivity of the paracellular barrier. The permeability of the ions through the TJs is regulated by the claudins -4, -8 and -14 which are involved in the cationic barrier, while other claudins such as -2, -7 and -13 form the paracellular pores for cations and anions. In the gastrointestinal tract claudins -2, -3, -4, -7, -8, -12, and -15 are expressed, but the levels of expression and their subcellular localization are different in the different intestinal segments [98].

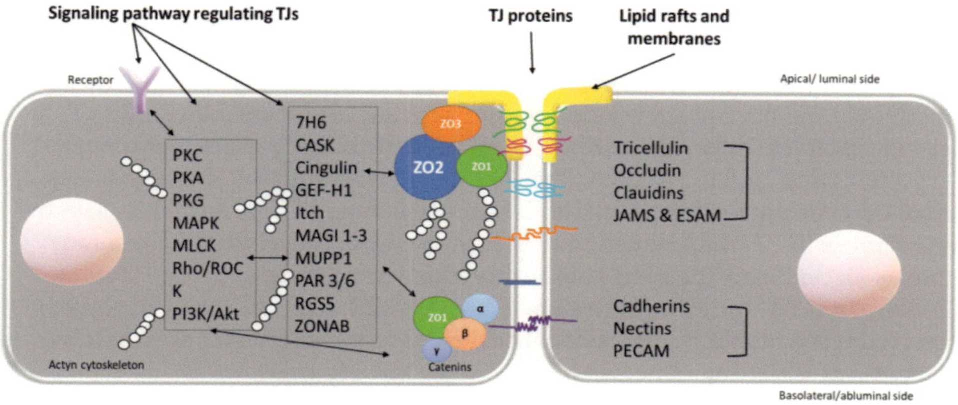

Figure 4.
Molecular structure of tight junctions. When the intestinal barrier is intact, the paracellular space between two enterocytes is sealed by TJs which are made up of a series of transmembrane proteins that include occludin, claudins, and the junctional adhesion molecule-1 (JAM-1). Thanks to TJs, the intestinal barrier is perfectly able to keep the luminal environment separate from the underlying immune system. Claudins adhere to each other in a homotypic as well as a heterotypic manner. ZO-1, -2, and -3 bind the cytoplasmic tail of occludin and link the TJ to the actin cytoskeleton. Proteins of the ZO family can shuttle to the nucleus to influence transcriptional processes in cellular proliferation and differentiation. The ZO-proteins have also been shown to interact with claudins and provide molecular scaffolds for TJ assembly. In the composition of TJ we also find cingulin, a protein of 140 kDa, which is associated with the cell cytoskeleton of actomyosin. Tyrosine phosphorylated Par3 / 6 regulates tight junction assembly and promotes cell polarity via intracellular signaling. Localization of TJ-associated 7H6 antigen along the cell border of vascular endothelial cells has been shown to be related to paracellular barrier function. The ZO-1 and ZO-2 scaffold proteins form dimers and bind to claudins, thereby contributing to the targeting and polymerization of claudins at tight junctions. Dimerization involves the SH3/GUK domain of ZO-1 / ZO-2. Also, ZO-1 and ZO-2 interact with the underlying actin cytoskeleton and act as a scaffold at tight junctions. The apical polarity protein complexes, including the Crumbs and Par complexes, localize to tight junctions.

Occludins (65 kDa) are proteins with 4 transmembrane domains and 2 extracellular loops and exist in 2 isoforms. The C-terminal domain, located in the cytoplasm, binds directly to ZO-1 (zonula occludens) which in turn binds the apical part of the actin. This portion of occludin is rich in sites of phosphorylation (thyroxine, serine and threonine) which can be modified by kinases and phosphatases. The non-phosphorylated occludin is distributed in the basolateral membrane and in the cytoplasmic vesicles, while the phosphorylated occludin is localized in the TJ and determines a reduced paracellular permeability [100]. Alterations (chronic inflammations or hyperplasias) have been observed in occludin deficient mice in all those districts characterized by the presence of TJs, suggesting more complex functions to be attributed to occludin, whose role is not yet fully known [101]. The interaction of occludins, claudins, JAMs, and tricellulin between cells and with ZOs maintains the integrity of the tight junction and controls the passage of molecules through the paracellular space.

Junctional adhesion molecules (JAM) (32 kDa, 3 isoforms) contain a transmembrane segment and an extracellular domain. They are proteins involved in the adhesion between the barrier cells and between the barrier and the blood cells and can form homophilic and heterophilic interactions with different ligands including integrins. They can also interact with partners such as ZO-1 and the protease-activated receptor PAR-3 [98].

Peripheral membrane proteins associated with zonula occludens (ZO) are crucial for the assembly and maintenance of TJs because they have multiple domains for interaction with other proteins, including integral membrane proteins and actin. On the intracellular side of the membrane, the carboxy-terminal ends of claudin, occludin and actin interact with the proteins ZO-1 (220 kDa), ZO-2 (160 kDa) and ZO-3 (130 kDa). These proteins belong to the membrane-associated guanylate kinase (MAGuK) superfamily and have an enzymatically inactive guanylate kinase domain. The TJ multiprotein complex, hitherto described, is linked to the actin cytoskeleton through the ZO proteins that bind to the integral membrane proteins with the N-terminal domain and to the actin cytoskeleton with the C-terminal domain. The protein that plays the central role is ZO-1 which directly and indirectly connects the integral membrane proteins (occludin and claudin) to the other cytoplasmic proteins of the TJs and to the actin cytoskeleton. It has been shown that, like occludins, ZO-2 and ZO-3 cannot interact directly with actin filaments since their C-terminal domains show similarities only towards ZO-1. Therefore, the binding to the actin cytoskeleton is limited to ZO-1 which has the potential to organize the structural components and to modulate the paracellular pathway [102].

There are many other proteins involved in TJ: tricellulin, the coxsackie and adenovirus receptor (CAR), the selective adhesion molecule for endothelial cells (ESAM), JAM4, AF-6/afadine, PAR3, MUPP-1, cingulin, PILT (protein subsequently incorporated into TJ) and JEAP (junction-enriched and -associated protein). All this gives the idea of the complex organization of TJs [98].

Until a few years ago, tight junctions and adherent junctions were seen as discrete and independent complexes. However, new evidence has emerged that highlights their interdependence. From these studies, it is clear that there are both physical and biochemical connections between adherent and tight junctions. The ZO-1 complex physically connects the two junctional complexes through its interactions with the binding proteins of actin, α-catenin and afadin. These interactions promote the maturation of the AJs and the subsequent assembly of the TJs [103].

1.5 Mechanisms of passage of different molecules through the intestinal epithelial barrier

The intestinal epithelium is a single layer of cells that covers the intestinal mucosa, separating it from the lumen and has two critical functions: first, it acts as a barrier to prevent the passage of harmful intraluminal entities including antigens, foreign microorganisms and their toxins. Its second function is to act as a selective filter that allows the translocation of essential nutrients, electrolytes and water, from the intestinal lumen to the circulatory stream. Enterocytes have a high transport activity because they have ion channels, transporters and pumps in the apical and basolateral membranes. Net fluid absorption in the intestine is the result of the balance between absorption and secretion. This transport is carried out selectively via two routes: the paracellular route and the transcellular route. The paracellular pathway allows 85% of the total passive trans-epithelial flow of molecules through the space between two adjacent epithelial cells and is regulated by TJs, which have pores of different sizes, limiting and selecting the passage of the molecules. This pathway constitutes an effective barrier for the passage of luminal antigens and is decisive for establishing intestinal permeability [104].

Transcellular transport involves the transportation of solutes through the enterocyte membrane. There are several mechanisms that mediate the passage of molecules through the transcellular pathway. Small-sized lipophilic and hydrophilic compounds are spread, by passive transport, through the lipid double layer of the enterocyte membrane. Furthermore, epithelial permeability is conditioned by active transport, mediated by transporters and by various mechanisms of endocytosis, transcytosis and exocytosis for ions, amino acids or some antigens. Large molecules, such as proteins and bacterial products, are captured by cells through the mechanism of endocytosis and are actively transported through the cytoplasm, by transcytosis process, for further processing and presentations, as part of the intestinal immune response (**Figure 5**) [105].

1.6 Mechanisms of damage and rupture of the intestinal epithelial barrier

The intestinal barrier is a dynamic system in which various factors intervene and the increase in the passage of substances due to the increased permeability does not necessarily imply its dysfunction. The progressive increase in intestinal permeability during the development of a pathological process implies an imbalance of the various factors that maintain the barrier function; the immune system being the main candidate to exert a greater effect on it, given the association between inflammation and barrier dysfunction in various digestive diseases. Under normal conditions, the increase in permeability is insufficient to cause a state of "intestinal disease" since the epithelial barrier has the ability to restore itself once the inducer stimulus has ceased. However, in certain pathological conditions, this self-regulating ability can be lost and this condition can facilitate an increase in permeability, facilitating chronic intestinal inflammation. Although the etiology of inflammatory bowel disease (IBD) is unknown, it has been observed that IBD patients have greater intestinal permeability than healthy subjects. It has been identified that this is due to the structural alterations of the TJ proteins, mainly due to the reduction of the expression of claudin-3, 4, 5 and 8 and of occludin, as well as an increased expression of claudin-2 and the phosphorylation of the myosin-light-chain (MLC); this phosphorylation is catalyzed by the specific myosin-light-chain kinase (MLCK), which is activated when

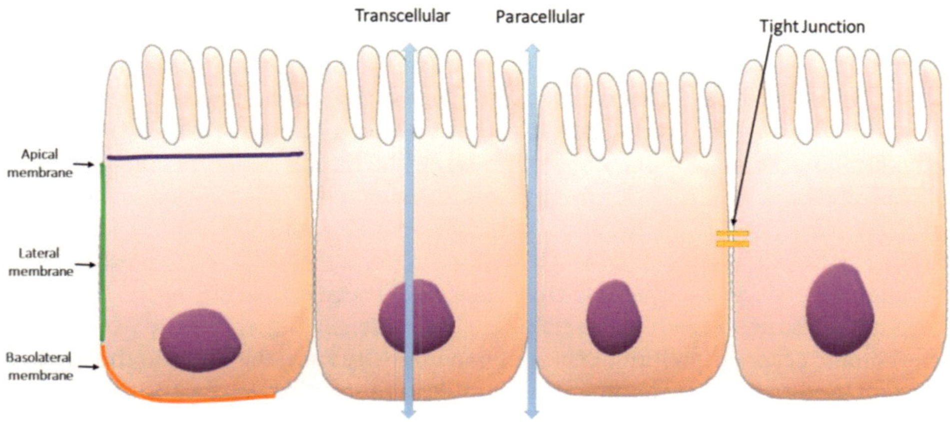

Figure 5.
Schematic representation of epithelia and transport pathways across a monolayer, and prototypic arrangement of junctions in polarized epithelial cells. The apical junction complex is formed by the tight junction, adherens junction and the most apically located desmosome. Gap junctions and additional desmosomes associate beneath the apical junction complex along the remainder of the lateral cell membranes. Hemidesmosomes interact with the basal lamina at the base of the cells. Intermediate filaments dock into desmosomes and hemidesmosomes whereas actin filaments attach to both tight and adherens junctions. Transcellular permeability is associated with the movement of solutes or water through intestinal epithelial cells. Paracellular permeability is associated with movement in the intercellular space between epithelial cells and is regulated by tight junctions located at the junction of the apical-lateral membranes.

it binds to calcium and calmodulin, forming a complex (Ca ++−calmodulin-MLCK) which facilitates the contraction of the cytoskeleton and the opening of the junctions [106–108]. The exaggerated inflammatory response would presumably be the cause of these alterations, given the increase in IFN-γ and TNF-α in these patients [109] and the in vitro effect that these cytokines have on the epithelial barrier. In the final analysis, as mentioned above, the alterations of the intercellular junctional complex during enteropathy are linked to an altered mitochondrial function with an energy deficit of the epithelium.

IECs culture and enteroid models have provided important mechanistic insight, suggesting that decreased mitochondrial function in epithelial cells drives a loss in barrier integrity and subsequent bacterial invasion of the underlying intestinal tissue. Loss of barrier function can manifest from epithelial cell death or leakiness of paracellular epithelial cell-cell junctions. DSS-induced colitis is associated with epithelial barrier dysfunction and mechanistic studies using Caco-2 cell monolayers demonstrated that mitochondrial reactive oxygen species (mtROS) play a key role in the loss of barrier integrity during DSS via stimulating the redistribution of Occludin and ZO-1 from intercellular junctions into intracellular compartments, causing leakiness of the tight junctions without altering cell viability [110]. Many forms of ROS have been implicated in disrupting tight junctions through the rearrangement of the actin cytoskeleton to decrease its interaction with tight junction proteins Occludin and ZO-1 and interaction with myosin heavy chain [111]. Additionally, hydrogen peroxide alters phosphorylation of Occludin, disrupting the tight junction, and phosphorylation of β-catenin, disrupting the adherens junction due to the redistribution of E-cadherin preventing interaction with β-catenin [111]. Indeed, dysfunctional mitochondria and accumulation of mtROS during deficiency of the autophagy mechanism induced epithelial barrier defects and the transcellular passage of bacteria that perpetuated intestinal inflammation [112].

In healthy dogs, similarly to the results of Ohta et al. [113], we describe a characteristical pattern of expression of AJ proteins along the small and large intestine [106]. Occludin-specific labeling is uniformly expressed throughout the epithelium of the small and large intestine, with the most intense labeling at the epithelial cell AJC, with fainter labeling observed along the basolateral membranes. Concerning the overall intensity of E-cadherin expression, we observe a decrease from the luminal epithelium to the distal crypts. At the luminal epithelium, E-cadherin labeling is uniform along the length of the intercellular junction, while the expression becomes polarized toward the AJC in the distal glands/crypts. At cellular levels, E-cadherin-specific labeling is restricted to the AJC and basolateral membranes of intestinal epithelial cells. Moreover, there is little evidence of specific labeling outside the epithelium. Claudin-2 readily detectable in the duodenal epithelium and glands and in the colonic crypt epithelium, decreasing in intensity from the distal to the proximal crypt, and remaining minimally detectable at the luminal surface of the colon. Interestingly, the expression pattern of AJC proteins in healthy dogs of our study, is very similar to the AJC proteins distribution, associated with clinical improvement, in IBD suffering dogs, after an oral probiotic treatment of 60 days [106] instead, a different pattern of AJC protein expression was observed in a homogeneous group of IBD affected dogs, apparently improved after a canonical association of metronidazole and prednisone therapy. In this classically treated group, claudin-2 expression was severely increased in the large intestine, particularly at the level of the proximal crypt and luminal epithelium. On the contrary, in the same group of dogs occludin was significantly lower, with a weak to absent expression in the luminal epithelium and in the small intestinal glands. No discernible difference in the distribution or staining intensity of E-cadherin was observed between normal and all IBD affected dogs. This greater deviation from the physiological conditions in the expression of Occludin in the small intestine and Claudin-2 in the colon of IBD suffering dogs, treated with a classical therapeutic protocol, resembles that previously described in samples from the colon of dogs with colitis [114]. In our experience, the effects of a multi strain, live and highly concentrated probiotic association, restored the epithelial barrier integrity, also from a morphological point of view, increasing the number and average size of IECs mitochondria [92]. In our studies, this restoration suggests a potential anti-inflammatory effect of probiotics, on the moment that in treated dogs, decreased mucosal CD3+ T-lymphocytes, and increased FoxP3+ and TGF-β+ positive cells were observed 30 days after the end of probiotic administration. More specifically, the probiotic treated dogs showed increases in CD3+/FoxP3+ cells in the intestinal mucosa, while dogs treated with prednisone and metronidazole displayed an overall decrease in all inflammatory cell populations that was accompanied by a decrease of FoxP3+ lymphocytes and TGF-β expressing cells.

The combination of different factors, genetic, environmental and defects in the barrier function, it is what ultimately predisposes the patient to an abnormal immune response and a greater susceptibility to developing intestinal inflammation. In fact, the appearance of IBD has been linked to the presence of mutated proteins such as X-box binding protein 1 (XBP1) or mutations in the NOD-2 gene related to lower IL-10 production or inadequate immune tolerance to antigens and luminal microbial products [115, 116].

TNF-α and IFN-γ have been extensively studied for their effects on the tight junction barrier in the gut. The effect of TNF-α on the intestinal barrier has been associated with IBD [117]; graft-versus-host disease [118], and celiac disease (CD) [119]. In patients with Crohn's disease (CrD) anti-TNF-α treatment is able to correct barrier disruption seen in the colon [117].

The mechanism of TNF-α barrier disruption has been shown to be mediated by MLCK. MLCK activation alone has been shown to decrease tight junction permeability both in vitro and in vivo [120, 121]. IFN-γ increases intestinal permeability through changes in expression and localization of tight junction proteins as well as rearrangement of the cytoskeleton [122].

Toll-like receptors (TLRs) are a class of transmembrane PRRs that are important for microbial recognition and control of immune responses. TLR2 is one member of the TLR family, which recognizes conserved patterns on both Gram-negative and Gram-positive bacteria. TLR2 is expressed on many cell types through the intestine including epithelial cells [123]. Stimulation of TLR2 in vitro increased trans epithelial electrical resistance through protein kinase C (PKC = a group of enzymes activated by signals such as increases in the concentration of diacylglycerol or calcium ions, and involved in several signal transduction cascades) activation and translocation of ZO-1 to the tight junction complex [123]. Proteinase activated receptors (PARs) are a family of g-protein-couple-receptors that are activated by proteolytic cleavage of their N-terminus revealing a tether ligand. PAR2 is found on both the apical and baso-lateral sides of enterocytes [124]. Stimulation of basolateral PAR2 results in increased permeability through redistribution of ZO-1, occludins, and F-actin [125]. Stimulation of PAR1 has also been shown to increase intestinal permeability [126].

In humans, a large number of chronic inflammatory diseases (CID) have been described to have alterations in intestinal permeability, including IBD [127], IBS [128], type-1-diabetes (T1D) [129], etc. Under normal physiological conditions, the majority (~90%) of antigens that pass through the intestinal epithelium travel through the transcellular pathway. The transcellular pathway is regulated and leads to lysosomal degradation of antigens into small non-immunogenic peptides. The remaining ~10% of proteins cross the epithelium through the paracellular pathway as full intact proteins or partially digested peptides as tightly regulated antigen trafficking through intestinal tight junction modulation, which leads to antigenic tolerance [130].

Zonula occludens toxins (Zot), is an enterotoxin which is able to reversibly open intracellular tight junctions [131]. Zot causes polymerization of actin of targeted cells leading to disassembly of tight junction complexes through a protein kinase C (PKC) dependent mechanism [132]. Immunofluorescent studies have shown that Zot is able to interact with epithelial cells along the GI tract with the highest binding in the jejunum and distal ileum and also decreasing along the villous to crypt axis [133]. Anti-Zot antibodies led to the identification of a ~47 kDa human analog to Zot, named zonulin [134]. Ex vivo studies show endogenous human zonulin is able to increase permeability in both the jejunum and ileum [135].

Studies on human sera from CD patients, who have increased zonulin levels [134] as determined by ELISA measurement using polyclonal zonulin cross reacting anti-Zot antibodies [136], revealed that zonulin is pre-haptoglobin (Hp)-2, the pro-protein of Hp2 before enzymatic cleavage into its mature form. After this discovery, an analogue of human zonulin (Hp2) has been evidenced in dogs. Dog and human Hp2 are proteins with a 98% similarity.

It was therefore hypothesized that zonulin may disassemble TJ through epidermal growth factor (EGF) activation, since it has been described that EGF can modulate the actin cytoskeleton, similar to the effects seen with zonulin [134, 135]. In vitro studies in Caco-2 cells showed zonulin caused EGF receptor (EGFR) phosphorylation and subsequent increases in permeability, which were blocked by an EGFR inhibitor. To confirm the effect was due to zonulin and not mature Hp2, trypsin digested zonulin was tested and showed no EGFR activation. Additionally, it was shown that EGFR activation was dependent on PAR2 as demonstrated both in Caco2 cells in which the receptor was silenced, and in PAR2−/− mice [136]. Zonulin contains a PAR2 activating peptide-like sequence in its β-chain, and it had been reported previously that PAR2 is able to transactivate EGFR [137].

The signaling pathways triggered by Zot and zonulin leading to tight junction disassembly have been extensively studied and resulted being similar, passing by PAR2 binding, and increasing permeability through displacement of ZO-1 and occludin from the cell junctions [138]. The displacement of ZO-1 and occludin was shown to be secondary to PCKα-dependent phosphorylation of ZO-1, causing decreased tight junction protein-protein interactions, and of myosin-1C that, together with the cytoskeletal rearrangement, temporarily removes ZO-1 and occludin from the junctional complex. While ZO-1 displacement per se is not sufficient to cause a barrier defect [139], the combination with other intracellular signaling events affecting TJ, including occludin displacement, actin polymerization, and myosin-1C phosphorylation [132] may contribute to a more profound rearrangement of the junctional complex that ultimately causes transient TJ disassembly (**Figure 6**).

High alteration in intestinal barrier permeability was observed also during IBS syndrome in man and, recently, in dogs [140]. In both species, IBS is associated with low grade inflammatory infiltration, often rich in mast cells, in both the small and large bowel. The close association of mast cells with major intestinal functions, such

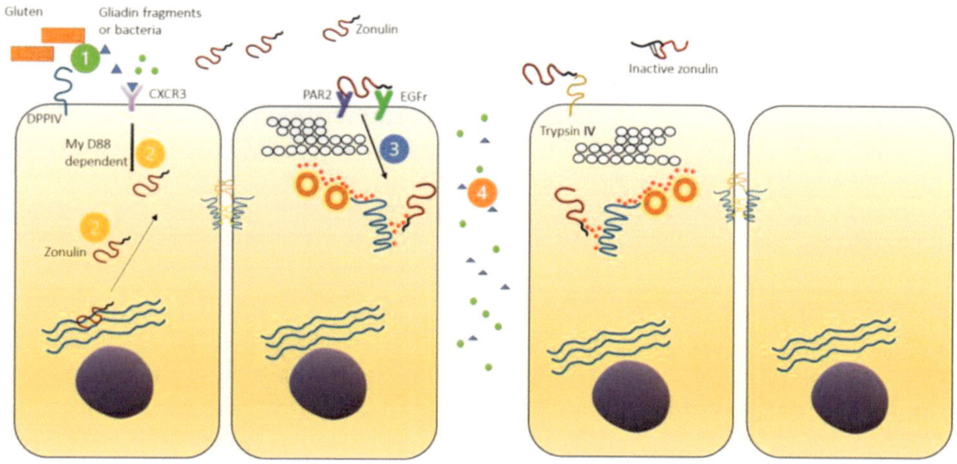

Figure 6.
Schematic representation of the gliadin and bacteria-induced release mechanism of zonulin, with the consequent increase in intestinal mucosal permeability, alteration of the barrier and increase in paracellular permeability. In phase 1, some specific peptides, such as gliadin, or deriving from other food sources or from bacteria, induce the release of zonulin mediated by the activation of the C-X-C Motif Chemokine Receptor 3 (CXCR-3 receptor or IFN-gamma induced G protein-coupled chemokines receptor 3—CD183) and dependent on MyD88 (or Myeloid differentiation primary response 88—a innate immune signal transduction adaptor) (phase 2). Zonulin transactivates EGFR (Epidermal Growth factor Receptor) via the PAR2 receptor leads to disassembly of the PCK-α-dependent (Protein kinase alfa) tight junction (phase 3). There is therefore an increase in intestinal permeability due to the opening of the intercellular junctions and the paracellular passage of "non–self-" antigens (phase 4) which diffuse into the lamina propria where they are able to interact with the immune system.

as epithelial secretion and permeability, neuroimmune interactions, visceral sensation, and peristalsis, makes it necessary to focus attention on the key roles of mast cells in the pathogenesis of IBS. Numerous evidence showed a positive relationship between the number of mucosal MCs and intestinal permeability [141], and the MC-derived tryptase was well identified as a key factor disrupts the intestinal barrier [142]. MC tryptase cleaves PAR2 on colonocytes to increase paracellular permeability by acting, as previously described, on the intercellular apical junction complex, which mainly consists of the tight junctions such as claudins, occludin, zonula occludens, junctional adhesion molecule, and the adherens junction such as E-cadherin [143]. Furthermore, PAR2 may induce the activation of extracellular signal-related kinase 1/2 (ERK1/2) and phosphorylation of MLCK, which regulates reorganization of F-actin and cytoskeleton and redistribution of tight junction, to increase epithelial permeability [105]. Other MC mediators such as interferon-γ, tumor necrosis factor-α (TNF-α), interleukin (IL)-1β, IL-4, IL-13, and prostaglandin E2 also have destructive effects on both trans- and paracellular permeability.

The most important triggers of zonulin release that have been described are bacteria, gliadin, and intestinal mast cells (MCs) tryptase. Enterotoxins and several enteric pathogens such as *E. coli*, and *Salmonella typhi* have been shown to cause a release of zonulin from the intestine when applied to the apical surface of IECs. Following the release of zonulin, that may be find and quantified in intestinal lumen (in fecal material) or in plasma, intestine showed increased permeability and disassembly of ZO-1 from the tight junction complex, permitting antigen and bacteria translocation and/or inflammatory cells passage throughout the epithelial layer. As we described in the previous paragraphs, conditions of dysbiosis, IECs absorption of bacterial/alimentary toxins, and other substances can induce IEC mitochondrial dysfunction with an increase of intracellular mitochondrial reactive oxygen species. MtROS, mostly from complex III, provides a pathway through which PAR1 and PAR2 are activated. Other sources of ROS do not participate in this induction. While PAR1 signaling ultimately involves NF-kappaB activation, inducing nuclear transcription for many pro-inflammatory molecules, PAR2 induces the activation of ERK1/2 and phosphorylation of MLCK, which regulates reorganization of F-actin and cytoskeleton and redistribution of tight junction, particularly of ZO-1 and occluding that break the integrity of the TJ complex [144], increasing the epithelial permeability [145]. This pathogenic mechanism is proposed for IBD pathogenesis. Gliadin is the other trigger that has been described to release zonulin; only when applied to the IECs apical surface, gliadin causes a release of zonulin and a subsequent increase in permeability, in both cell culture models and *ex vivo* studies of intestinal tissue. Lammers et al. described that specific non-digestible gliadin peptides are able to bind the CXCR3 receptor on the apical surface of enterocytes with subsequent MyD88-dependent zonulin release [146, 147]. The CXCR3 receptor is also overexpressed on the apical IECs surface of biopsies from celiac disease suffering patients (CD), which may explain the increased levels of zonulin detected in intestinal explants obtained from CD patients when exposed to gliadin [148].

CD suffering patients have a reorganization of actin filaments and an altered expression of occludin, claudin-3 and claudin-4, as well as ZO-1 and the adhesion protein E-cadherin [149, 150]. Generally, under physiological circumstances, there is a tight control of mucosal antigen trafficking (antigen sampling) that, in concert with specific immune cells and chemokine and cytokine mediators, leads to anergy and, therefore, to mucosal tolerance. In the pathological conditions above expressed, the inappropriate production of an increased amount of zonulin causes a functional loss of barrier function, with subsequent inappropriate and uncontrolled antigen trafficking instigating an innate immune response by the submucosal immune compartment, with production of pro-inflammatory cytokines, including IFN-γ and TNF-α that cause further opening of the paracellular pathway to the passage of antigens, creating a vicious cycle.

In conclusion, the loss of gut barrier function, through increased zonulin release from of both epithelial and endothelial barriers, as an essential step to initiate the intestinal inflammatory process. In many human and canine chronic intestinal diseases, whole bacteria or bacteria toxins, as well as gliadin or MCs tryptase are the triggers of zonulin release, leading to gut barrier dysfunction. Similar results, with increase plasma and fecal levels of Zonulin, plasma LPS and cleaved C18 cytokeratin [93] were recently described in sera of dogs with lymphangiectasia, and in cats with enteritis associated T cell lymphoma type II (EATCL II) [151] by the author [46].

An imbalanced microbiome or its inappropriate distribution along the gastrointestinal tract causes dysbiosis, mitochondrial dysfunction with an increase of intracellular mitochondrial reactive oxygen species (MtROS), and the induction of the release of zonulin leading to the passage of luminal contents across the epithelial barrier, causing the release of pro-inflammatory cytokines. The presence of cytokines eventually sustains the ulterior increased permeability, causing a massive influx of dietary and microbial antigens, leading to the activation of T-cells. Depending on the genetic background of the host, these T-cells can remain within the GI tract, causing chronic inflammation restricted to the intestinal mucosa (IBD, IBS, CD), or migrate to several different organs to cause a systemic chronic disease. Generally the main alterations in the expression of TJ proteins are the decrease in ZO

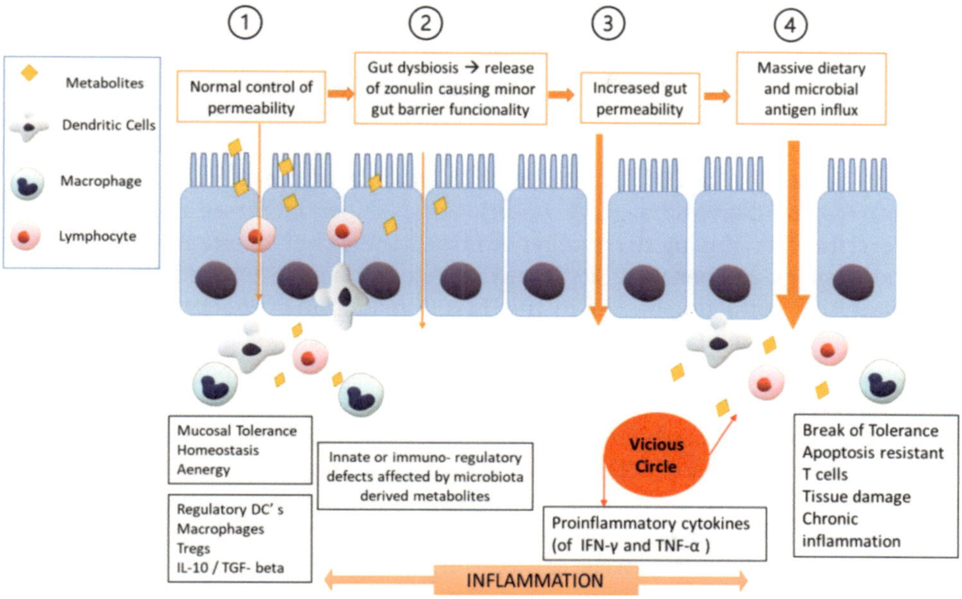

Figure 7.
Pathogenic mechanism of chronic intestinal diseases (CID), linked to the loss of impermeability and selectivity of the intestinal barrier induced by the action of TJs-released zonulin. In phase 1 it is observed that, thanks to the barrier effect, the condition of eubiosis, and the physiological traffic through the barrier of non–self-antigens, which are suitably presented to the leukocyte cells of the lamina propria (Th3, Tregs, etc.), there is the establishment of "oral tolerance" with the homeostasis of the mucosa. In phase 2 it is observed how environmental stimuli cause an imbalance of the microbiota, triggering the release of zonulin, loss of paracellular permeability, and an increase in the flow of antigens from the intestinal lumen to the lamina propria. In phase 4, the antigens in the lamina propria activate the immune system in a "pro-inflammatory" manner by causing the release of IFN-γ and TNF-α. This inflammation further exacerbates the increase in intestinal permeability and immune response, worsening and chronicizing the inflammation. This vicious circle, even more serious in genetically predisposed individuals, causes the interruption of oral tolerance to food antigens and causes the aggravation of chronic enteropathies.

and occludin, as well as an increase in claudin-2 and myosin light chain MLC phosphorylation (**Figure 7**) [152].

2. Conclusions

The gastrointestinal system is, together with the skin and the respiratory system, the habitat most exposed to the external environment, microorganisms and compounds derived from digestion. This condition requires a complex defense system capable of separating the intestinal contents from the host tissues, regulating the absorption of nutrients and allowing interaction between the resident microbial flora and the mucosal immune system, inhibiting the translocation of pathogens into the underlying tissues. All these functions are performed by the intestinal barrier, a functional unit, organized as a multi-layered system: The barrier is more superficially composed of a physical surface barrier, which prevents bacterial adhesion and regulates the paracellular diffusion towards the underlying host tissues. More in depth, we find a deeper functional barrier, which is able to discriminate between commensal and pathogenic microorganisms, organizing the immunological tolerance towards the commensal bacteria and the immune response towards the pathogens. The fundamental elements on which the integrity and functionality of the "intestinal barrier" depends are therefore the IECs and the intercellular junctions. Glutamine plays a fundamental role in the metabolism of IECs. A condition of eubiosis involves the correct synthesis/absorption of glutamine and glutathione by the IECs. Furthermore, the presence of "healthy" bacterial species producing NEFAs in the correct proportion, with an excess of butyrate, preserves the IEC's mitochondria from ROS oxidative damage. A condition of dysbiosis increases mitochondrial damage, critically reducing the number of mitochondria but, above all modifying their morphology and permeability. A critical reduction in mitochondria leads to a decrease in the production of ATP by the IECs. A reduction in energy leads to a lower "hold" of the intercellular junctional complexes and an increase in bacterial translocation through the intestinal epithelium, which becomes more permeable. At the submucosal level, this condition increases inflammation and the recall of leukocytes, further worsening the condition of the mucosal barrier. Pathogenic mechanism of chronic intestinal diseases (CID), linked to the loss of impermeability and selectivity of the intestinal barrier, are induced by the action of TJs-released zonulin. Zonulin is a protein that modulates the permeability of TJs between cells of the intestinal barrier. Zonulin has been implicated in the pathogenesis of important GI diseases (i.e. coeliac disease and diabetes), and some glycoproteins, such as the gluten protein gliadin, activate zonulin signaling, increasing intestinal barrier permeability of macromolecules and contributing to "leaky gut" conditions. Thanks to the barrier effect, the condition of eubiosis, and the physiological traffic through the barrier of non-self-antigens, which are suitably presented to the leukocyte cells of the lamina propria (Th3, Tregs, etc.), there is the establishment of "oral tolerance" with the homeostasis of the GI mucosa. When environmental stimuli cause an imbalance of the microbiota, triggering the release of zonulin, loss of paracellular permeability, and an increase in the flow of antigens from the intestinal lumen to the lamina propria, antigens activate the immune system in a "pro-inflammatory" manner by causing the release of IFN-γ and TNF-α. This inflammation further exacerbates the increase in intestinal permeability and immune response, worsening and chronicizing the inflammation. This vicious circle, even more serious in genetically predisposed individuals, causes the interruption of oral tolerance to food antigens and causes the aggravation of chronic enteropathies.

Acknowledgements

The author would like to thank Dr. Lucia Biagini for her helpful advice on various technical issues examined in this paper, and for her help in the graphic realization of the figures that exemplify the concepts expressed in the paragraphs of this chapter. The author also wishes to thank Dr. Livio Galosi, for his usual support activities in the revision of the proofs and in the editing of this and other papers written by the author. We also thank Mrs. Subeide Mari for her technical assistance for all the experimental parts of the numerous papers that have been written by the author over the years on the topics covered in this chapter.

Conflict of interest

The author declares no conflict of interest.

Author details

Giacomo Rossi
School of Biosciences and Veterinary Medicine, University of Camerino, Matelica, MC, Italy

*Address all correspondence to: giacomo.rossi@unicam.it

References

[1] Scaldaferri F, Pizzoferrato M, Gerardi V, Lopetuso L, Gasbarrini A. The gut barrier: New acquisitions and therapeutic approaches. Journal of Clinical Gastroenterology. 2012;**46**(Suppl):S12-S17

[2] Gibson PR, Anderson RP, Mariadason JM, Wilson AJ. Protective role of the epithelium of the small intestine and colon. Inflammatory Bowel Diseases. 1996;**2**(4):279-302

[3] Crosnier C, Stamataki D, Lewis J. Organizing cell renewal in the intestine: Stem cells, signals and combinatorial control. Nature Reviews Genetics. 2006;**7**(5):349-359

[4] van der Flier LG, Clevers H. Stem cells, self-renewal, and differentiation in the intestinal epithelium. Annual Review of Physiology. 2009;**71**:241-260

[5] Barker N, Clevers H. Leucine-rich repeat-containing G-protein-coupled receptors as markers of adult stem cells. Gastroenterology. 2010;**138**(5):1681-1696

[6] Clevers HC, Bevins CL. Paneth cells: Maestros of the small intestinal crypts. Annual Review of Physiology. 2013;**75**:289-311

[7] Jackson DN, Theiss AL. Gut bacteria signaling to mitochondria in intestinal inflammation and cancer. Gut Microbes. 2020;**11**(3):285-304

[8] Yang H, Xiong X, Wang X, Tan B, Li T, Yin Y. Effects of weaning on intestinal upper villus epithelial cells of piglets. PLoS One. 2016;**11**(3):e0150216

[9] Blachier F, Boutry C, Bos C, Tomé D. Metabolism and functions of L-glutamate in the epithelial cells of the small and large intestines. The American Journal of Clinical Nutrition. 2009;**90**(3):814S-821S

[10] Oehler R, Pusch E, Dungel P, Zellner M, Eliasen MM, Brabec M, et al. Glutamine depletion impairs cellular stress response in human leucocytes. The British Journal of Nutrition. 2002;**87**(Suppl. 1):S17-S21

[11] Kathrani A, Allenspach K, Fascetti AJ, Larsen JA, Hall EJ. Alterations in serum amino acid concentrations in dogs with protein-losing enteropathy. Journal of Veterinary Internal Medicine. 2018;**32**(3):1026-1032

[12] Zheng Y-M, Li F, Zhang M-M, Wu X-T. Glutamine dipeptide for parenteral nutrition in abdominal surgery: A meta-analysis of randomized controlled trials. World Journal of Gastroenterology. 2006;**12**(46):7537-7541

[13] Zheng Y, Li F, Qi B, Luo B, Sun H, Liu S, et al. Application of perioperative immunonutrition for gastrointestinal surgery: A meta-analysis of randomized controlled trials. Asia Pacific Journal of Clinical Nutrition. 2007;**16**(Suppl. 1): 253-257

[14] Andreasen AS, Pedersen-Skovsgaard T, Mortensen OH, van Hall G, Moseley PL, Pedersen BK. The effect of glutamine infusion on the inflammatory response and HSP70 during human experimental endotoxaemia. Critical Care (London, England). 2009;**13**(1):R7

[15] Déchelotte P, Hasselmann M, Cynober L, Allaouchiche B, Coëffier M, Hecketsweiler B, et al. L-alanyl-L-glutamine dipeptide-supplemented total parenteral nutrition reduces

infectious complications and glucose intolerance in critically ill patients: The French controlled, randomized, double-blind, multicenter study. Critical Care Medicine. 2006;**34**(3):598-604

[16] Chan DL. Nutritional support of the critically ill small animal patient. The Veterinary Clinics of North America. Small Animal Practice. 2020;**50**(6):1411-1422

[17] Humbert B, Nguyen P, Dumon H, Deschamps J-Y, Darmaun D. Does enteral glutamine modulate whole-body leucine kinetics in hypercatabolic dogs in a fed state? Metabolism. 2002;**51**(5):628-635

[18] Withrow SJ. Withrow and MacEwen's small animal clinical oncology. Elsevier Health Sciences. 2007;**2007**:863

[19] Ohno T, Mochiki E, Ando H, Fukasawa T, Toyomasu Y, Ogata K, et al. Glutamine decreases the duration of postoperative ileus after abdominal surgery: An experimental study of conscious dogs. Digestive Diseases and Sciences. 2009;**54**(6):1208-1213

[20] Kang J-H, Kim S-S, Yang M-P. Effect of parenteral l-alanyl-l-glutamine administration on phagocytic responses of polymorphonuclear neutrophilic leukocytes in dogs undergoing high-dose methylprednisolone sodium succinate treatment. American Journal of Veterinary Research. 2012;**73**(9):1410-1417

[21] Tang Z-F, Ling Y-B, Lin N, Hao Z, Xu R-Y. Glutamine and recombinant human growth hormone protect intestinal barrier function following portal hypertension surgery. World Journal of Gastroenterology. 2007;**13**(15):2223-2228

[22] Singleton KD, Serkova N, Banerjee A, Meng X, Gamboni-Robertson F, Wischmeyer PE. Glutamine attenuates endotoxin-induced lung metabolic dysfunction: Potential role of enhanced heat shock protein 70. Nutrition. 2005;**21**(2):214-223

[23] Morrison AL, Dinges M, Singleton KD, Odoms K, Wong HR, Wischmeyer PE. Glutamine's protection against cellular injury is dependent on heat shock factor-1. American Journal of Physiology. Cell Physiology. 2006;**290**(6):C1625-C1632

[24] Ganter MT, Ware LB, Howard M, Roux J, Gartland B, Matthay MA, et al. Extracellular heat shock protein 72 is a marker of the stress protein response in acute lung injury. American Journal of Physiology. Lung Cellular and Molecular Physiology. 2006;**291**(3):L354-L361

[25] Berger E, Rath E, Yuan D, Waldschmitt N, Khaloian S, Allgäuer M, et al. Mitochondrial function controls intestinal epithelial stemness and proliferation. Nature Communications. 2016;**27**(7):13171

[26] Ropeleski MJ, Riehm J, Baer KA, Musch MW, Chang EB. Anti-apoptotic effects of L-glutamine-mediated transcriptional modulation of the heat shock protein 72 during heat shock. Gastroenterology. 2005;**129**(1):170-184

[27] Sakiyama T, Musch MW, Ropeleski MJ, Tsubouchi H, Chang EB. Glutamine increases autophagy under Basal and stressed conditions in intestinal epithelial cells. Gastroenterology. 2009;**136**(3):924-932

[28] Chang W-K, Yang KD, Chuang H, Jan J-T, Shaio M-F. Glutamine protects activated human T cells from apoptosis by up-regulating glutathione and Bcl-2 levels. Clinical Immunology. 2002;**104**(2):151-160

[29] Yang H, Söderholm JD, Larsson J, Permert J, Lindgren J,

Wirén M. Bidirectional supply of glutamine maintains enterocyte ATP content in the in vitro using chamber model. International Journal of Colorectal Disease. 2000;**15**(5-6):291-296

[30] Beaulieu AD, Drackley JK, Overton TR, Emmert LS. Isolated canine and murine intestinal cells exhibit a different pattern of fuel utilization for oxidative metabolism. Journal of Animal Science. 2002;**80**(5):1223-1232

[31] Fukushima K, Fiocchi C. Paradoxical decrease of mitochondrial DNA deletions in epithelial cells of active ulcerative colitis patients. American Journal of Physiology. Gastrointestinal and Liver Physiology. 2004;**286**(5):G804-G813

[32] Saitoh T, Fujita N, Jang MH, Uematsu S, Yang B-G, Satoh T, et al. Loss of the autophagy protein Atg16L1 enhances endotoxin-induced IL-1beta production. Nature. 2008;**456**(7219):264-268

[33] Roediger WEW. The colonic epithelium in ulcerative colitis: An energy-deficiency disease? The Lancet. 1980;**316**(8197):712-715

[34] Ahmad MS, Krishnan S, Ramakrishna BS, Mathan M, Pulimood AB, Murthy SN. Butyrate and glucose metabolism by colonocytes in experimental colitis in mice. Gut. 2000;**46**(4):493-499

[35] Ramakrishna BS, Roberts-Thomson IC, Pannall PR, Roediger WE. Impaired sulphation of phenol by the colonic mucosa in quiescent and active ulcerative colitis. Gut. 1991;**32**(1): 46-49

[36] Chapman MA, Grahn MF, Boyle MA, Hutton M, Rogers J, Williams NS. Butyrate oxidation is impaired in the colonic mucosa of sufferers of quiescent ulcerative colitis. Gut. 1994;**35**(1):73-76

[37] Santhanam S, Venkatraman A, Ramakrishna BS. Impairment of mitochondrial acetoacetyl CoA thiolase activity in the colonic mucosa of patients with ulcerative colitis. Gut. 2007;**56**(11):1543-1549

[38] Novak EA, Mollen KP. Mitochondrial dysfunction in inflammatory bowel disease. Frontiers in Cell and Development Biology. 2015;**3**:62

[39] d'Ettorre G, Rossi G, Scagnolari C, Andreotti M, Giustini N, Serafino S, et al. Probiotic supplementation promotes a reduction in T-cell activation, an increase in Th17 frequencies, and a recovery of intestinal epithelium integrity and mitochondrial morphology in ART-treated HIV-1-positive patients. Immunity, Inflammation and Disease. 2017;**5**(3):244-260

[40] Mitchell P, Moyle J. Chemiosmotic hypothesis of oxidative phosphorylation. Nature. 1967;**213**(5072):137-139

[41] Taanman JW. The mitochondrial genome: Structure, transcription, translation and replication. Biochimica et Biophysica Acta. 1999;**1410**(2):103-123

[42] Galluzzi L, Kepp O, Kroemer G. Mitochondria: Master regulators of danger signalling. Nature Reviews. Molecular Cell Biology. 2012;**13**(12):780-788

[43] Delpre G, Avidor I, Steinherz R, Kadish U, Ben-Bassat M. Ultrastructural abnormalities in endoscopically and histologically normal and involved colon in ulcerative colitis. American Journal of Gastroenterology (Springer Nature). Sep 1989;**84**(9):1038-1046. 9 Black and White Photographs, 4 Charts

[44] Nazli A, Yang P-C, Jury J, Howe K, Watson JL, Söderholm JD, et al. Epithelia under metabolic stress perceive commensal bacteria as a threat. The American Journal of Pathology. 2004;**164**(3):947-957

[45] Rodenburg W, Keijer J, Kramer E, Vink C, Van Der Meer R, Bovee-Oudenhoven IM. Impaired barrier function by dietary fructo-oligosaccharides (FOS) in rats is accompanied by increased colonic mitochondrial gene expression. BMC Genomics. 2008;**9**(1):1-15

[46] Rossi G, Gavazza A, Vincenzetti S, Mangiaterra S, Galosi L, Marchegiani A, et al. Clinicopathological and fecal proteome evaluations in 16 dogs presenting chronic diarrhea associated with lymphangiectasia. Veterinary Science. 2021;**8**(10):242

[47] Schürmann G, Brüwer M, Klotz A, Schmid KW, Senninger N, Zimmer K-P. Transepithelial transport processes at the intestinal mucosa in inflammatory bowel disease. International Journal of Colorectal Disease. 1999;**14**(1):41-46

[48] Halestrap AP, Dunlop JL. Intramitochondrial regulation of fatty acid β-oxidation occurs between flavoprotein and ubiquinone A role for changes in the matrix volume. The Biochemical Journal. 1986;**239**(3):559-565

[49] Söderholm JD, Olaison G, Peterson KH, Franzén LE, Lindmark T, Wirén M, et al. Augmented increase in tight junction permeability by luminal stimuli in the non-inflamed ileum of Crohn's disease. Gut. 2002;**50**(3):307-313

[50] Rath E, Berger E, Messlik A, Nunes T, Liu B, Kim SC, et al. Induction of dsRNA-activated protein kinase links mitochondrial unfolded protein response to the pathogenesis of intestinal inflammation. Gut. 2012;**61**(9):1269-1278

[51] Gersemann M, Stange EF, Wehkamp J. From intestinal stem cells to inflammatory bowel diseases. World Journal of Gastroenterology: WJG. 2011;**17**(27):3198

[52] Bär F, Bochmann W, Widok A, Von Medem K, Pagel R, Hirose M, et al. Mitochondrial gene polymorphisms that protect mice from colitis. Gastroenterology. 2013;**145**(5):1055-1063

[53] Fratila OC, Craciun C. Ultrastructural evidence of mucosal healing after infliximab in patients with ulcerative colitis. Journal of Gastrointestinal and Liver Diseases. Jun 2010;**19**(2):147-53. PMID: 20593047

[54] Dickman KG, Hempson SJ, Anderson J, Lippe S, Zhao L, Burakoff R, et al. Rotavirus alters paracellular permeability and energy metabolism in Caco-2 cells. American Journal of Physiology. Gastrointestinal and Liver Physiology. 2000;**279**(4):G757-G766

[55] He D, Hagen SJ, Pothoulakis C, Chen M, Medina ND, Warny M, et al. Clostridium difficile toxin A causes early damage to mitochondria in cultured cells. Gastroenterology. 2000;**119**(1):139-150

[56] Mottawea W, Chiang C-K, Mühlbauer M, Starr AE, Butcher J, Abujamel T, et al. Altered intestinal microbiota–host mitochondria crosstalk in new onset Crohn's disease. Nature Communications. 2016;**7**(1):1-14

[57] Saint-Georges-Chaumet Y, Edeas M. Microbiota–mitochondria inter-talk: Consequence for microbiota–host interaction. Pathogenesis Diseases. 2016;**74**(1)

[58] Somasundaram S, Rafi S, Hayllar J, Sigthorsson G, Jacob M, Price AB, et al. Mitochondrial damage: A possible mechanism of the "topical" phase of NSAID induced injury to the rat intestine. Gut. 1997;**41**(3):344-353

[59] Somasundaram S, Sigthorsson G, Simpson RJ, Watts J, Jacob M, Tavares IA, et al. Uncoupling of intestinal mitochondrial oxidative phosphorylation and inhibition of cyclooxygenase are required for the development of NSAID-enteropathy in the rat. Alimentary Pharmacology & Therapeutics. 2000;**14**(5):639

[60] Basivireddy J, Vasudevan A, Jacob M, Balasubramanian KA. Indomethacin-induced mitochondrial dysfunction and oxidative stress in villus enterocytes. Biochemical Pharmacology. 2002;**64**(2):339-349

[61] Duchmann R, Schmitt E, Knolle P, Zum Büschenfelde K-HM, Neurath M. Tolerance towards resident intestinal flora in mice is abrogated in experimental colitis and restored by treatment with interleukin-10 or antibodies to interleukin-12. European Journal of Immunology. 1996;**26**(4):934-938

[62] Lewis K, McKay DM. Metabolic stress evokes decreases in epithelial barrier function. Annals of the New York Academy of Sciences. 2009;**1165**:327-337

[63] Deitch EA, Xu D, Kaise VL. Role of the gut in the development of injury- and shock induced SIRS and MODS: The gut-lymph hypothesis, a review. Frontiers in Bioscience. Jan 2006;**1**(11):520-528

[64] Sánchez S, Martín MJ, Ortiz P, Motilva V, Alarcón de la Lastra C. Effects of dipyrone on inflammatory infiltration and oxidative metabolism in gastric mucosa: Comparison with acetaminophen and diclofenac. Digestive Diseases and Sciences. 2002;**47**(6):1389-1398

[65] Mazalli MR, Bragagnolo N. Increase of cholesterol oxidation and decrease of PUFA as a result of thermal processing and storage in eggs enriched with n-3 fatty acids. Journal of Agricultural and Food Chemistry. 2009;**57**(11):5028-5034

[66] Biswas K, Bandyopadhyay U, Chattopadhyay I, Varadaraj A, Ali E, Banerjee RK. A novel antioxidant and antiapoptotic role of omeprazole to block gastric ulcer through scavenging of hydroxyl radical. The Journal of Biological Chemistry. 2003;**278**(13):10993-11001

[67] Babbs CF, Cregor MD, Badylak SF. Histochemical demonstration of endothelial superoxide and hydrogen peroxide generation in ischaemic and reoxygenated rat tissues. Cardiovascular Research. 1992;**26**(6):593-602

[68] Rezaie A, Parker RD, Abdollahi M. Oxidative stress and pathogenesis of inflammatory bowel disease: An epiphenomenon or the cause? Digestive Diseases and Sciences. 2007;**52**(9):2015-2021

[69] Borrelli F, Fasolino I, Romano B, Capasso R, Maiello F, Coppola D, et al. Beneficial effect of the non-psychotropic plant cannabinoid cannabigerol on experimental inflammatory bowel disease. Biochemical Pharmacology. 2013;**85**(9):1306-1316

[70] Arab HH, Al-Shorbagy MY, Abdallah DM, Nassar NN. Telmisartan attenuates colon inflammation, oxidative perturbations and apoptosis in a rat model of experimental inflammatory bowel disease. PLoS One. 2014;**9**(5):e97193

[71] Suchodolski JS. Diagnosis and interpretation of intestinal dysbiosis

in dogs and cats. Veterinary Journal. 2016;**1997**(215):30-37

[72] Nicholson JK, Holmes E, Kinross J, Burcelin R, Gibson G, Jia W, et al. Host-gut microbiota metabolic interactions. Science. 2012;**336**(6086):1262-1267

[73] Hylemon PB, Zhou H, Pandak WM, Ren S, Gil G, Dent P. Bile acids as regulatory molecules. Journal of Lipid Research. 2009;**50**(8):1509-1520

[74] Joyce SA, Gahan CGM. Bile acid modifications at the microbe-host interface: Potential for nutraceutical and pharmaceutical interventions in host health. Annual Review of Food Science and Technology. 2016;**7**:313-333

[75] Vighi G, Marcucci F, Sensi L, Di Cara G, Frati F. Allergy and the gastrointestinal system. Clinical and Experimental Immunology. 2008;**153**(Suppl. 1):3-6

[76] Hooper LV, Gordon JI. Commensal host-bacterial relationships in the gut. Science. 2001;**292**(5519):1115-1118

[77] Xu J, Xu C, Chen X, Cai X, Yang S, Sheng Y, et al. Regulation of an antioxidant blend on intestinal redox status and major microbiota in early weaned piglets. Nutrition. 2014;**30**(5):584-589

[78] Tan J, McKenzie C, Potamitis M, Thorburn AN, Mackay CR, Macia L. The role of short-chain fatty acids in health and disease. Advances in Immunology. 2014;**121**:91-119

[79] Han B, Lin C-CJ, Hu G, Wang MC. 'Inside Out'—A dialogue between mitochondria and bacteria. The FEBS Journal. 2019;**286**(4):630-641

[80] Lin C-CJ, Wang MC. Microbial metabolites regulate host lipid metabolism through NR5A-Hedgehog signalling. Nature Cell Biology. 2017;**19**(5):550-557

[81] Han B, Sivaramakrishnan P, Lin C-CJ, Neve IAA, He J, Tay LWR, et al. Microbial genetic composition tunes host longevity. Cell. 2017;**169**(7):1249-1262.e13

[82] Staley C, Weingarden AR, Khoruts A, Sadowsky MJ. Interaction of gut microbiota with bile acid metabolism and its influence on disease states. Applied Microbiology and Biotechnology. 2017;**101**(1):47-64

[83] Suchodolski JS. Analysis of the gut microbiome in dogs and cats. Veterinary Clinical Pathology. 2022;**50**(Suppl. 1):6-17

[84] Nie Y, Hu J, Yan X. Cross-talk between bile acids and intestinal microbiota in host metabolism and health. Journal of Zhejiang University. Science. B. 2015;**16**(6):436-446

[85] Kuipers F, Bloks VW, Groen AK. Beyond intestinal soap--bile acids in metabolic control. Nature Reviews. Endocrinology. 2014;**10**(8):488-498

[86] Caron AZ, He X, Mottawea W, Seifert EL, Jardine K, Dewar-Darch D, et al. The SIRT1 deacetylase protects mice against the symptoms of metabolic syndrome. FASEB Journal. 2014;**28**(3):1306-1316

[87] Kazgan N, Metukuri MR, Purushotham A, Lu J, Rao A, Lee S, et al. Intestine-specific deletion of SIRT1 in mice impairs DCoH2-HNF-1α-FXR signaling and alters systemic bile acid homeostasis. Gastroenterology. 2014;**146**(4):1006-1016

[88] Alvarez C-S, Giménez R, Cañas M-A, Vera R, Díaz-Garrido N, Badia J,

et al. Extracellular vesicles and soluble factors secreted by *Escherichia coli* Nissle 1917 and ECOR63 protect against enteropathogenic *E. coli*-induced intestinal epithelial barrier dysfunction. BMC Microbiology. 2019;**19**(1):166

[89] Jacouton E, Mach N, Cadiou J, Lapaque N, Clément K, Doré J, et al. Lactobacillus rhamnosus CNCMI-4317 modulates Fiaf/Angptl4 in intestinal epithelial cells and circulating level in mice. PLoS One. 2015;**10**(10):e0138880

[90] Scott KP, Gratz SW, Sheridan PO, Flint HJ, Duncan SH. The influence of diet on the gut microbiota. Pharmacological Research. 2013;**69**(1):52-60

[91] Rogatzki MJ, Ferguson BS, Goodwin ML, Gladden LB. Lactate is always the end product of glycolysis. Frontiers in Neuroscience. 2015;**9**:22

[92] Rossi G, Pengo G, Caldin M, Steiner JM, Suchodolski JS. Effects of the probiotic Vsl#3 on microbiological, histological, and immunomodulatory parameters in dogs with inflammatory bowel disease. In: Proceedings of ACVIM Forum. New Orleans. 2012

[93] Rossi G, Magi G, Cortese S, Tarantino G, Pengo G, Renzoni G. New acquisitions about canine lymphangiectasia during IBD. In: Proceedings of 26th Meeting of European Society of Veterinary Pathology; Dubrovnik, Croatia. 2008. pp. 195-197

[94] Leschelle X, Goubern M, Andriamihaja M, Blottière HM, Couplan E, Gonzalez-Barroso M-D-M, et al. Adaptative metabolic response of human colonic epithelial cells to the adverse effects of the luminal compound sulfide. Biochimica et Biophysica Acta. 2005;**1725**(2):201-212

[95] Vermeiren J, Van de Wiele T, Van Nieuwenhuyse G, Boeckx P, Verstraete W, Boon N. Sulfide- and nitrite-dependent nitric oxide production in the intestinal tract. Microbial Biotechnology. 2012;**5**(3):379-387

[96] Beaumont M, Andriamihaja M, Lan A, Khodorova N, Audebert M, Blouin J-M, et al. Detrimental effects for colonocytes of an increased exposure to luminal hydrogen sulfide: The adaptive response. Free Radical Biology & Medicine. 2016;**93**:155-164

[97] Ijssennagger N, van der Meer R, van Mil SWC. Sulfide as a mucus barrier-breaker in inflammatory bowel disease? Trends in Molecular Medicine. 2016;**22**(3):190-199

[98] Chiba H, Osanai M, Murata M, Kojima T, Sawada N. Transmembrane proteins of tight junctions. Biochimica et Biophysica Acta. 2008;**1778**(3):588-600

[99] Paris L, Tonutti L, Vannini C, Bazzoni G. Structural organization of the tight junctions. Biochimica et Biophysica Acta. 2008;**1778**(3):646-659

[100] Schneeberger EE, Lynch RD. The tight junction: A multifunctional complex. American Journal of Physiology. Cell Physiology. 2004;**286**(6):C1213-C1228

[101] Saitou M, Furuse M, Sasaki H, Schulzke JD, Fromm M, Takano H, et al. Complex phenotype of mice lacking occludin, a component of tight junction strands. Molecular Biology of the Cell. 2000;**11**(12):4131-4142

[102] Van Itallie CM, Tietgens AJ, Anderson JM. Visualizing the dynamic coupling of claudin strands to the actin cytoskeleton through ZO-1. Molecular Biology of the Cell. 2017;**28**(4):524-534

[103] Campbell HK, Maiers JL, DeMali KA. Interplay between tight junctions & adherens junctions. Experimental Cell Research. 2017;**358**(1):39-44

[104] Turner JR. Intestinal mucosal barrier function in health and disease. Nature Reviews Immunology. 2009;**9**(11):799-809

[105] Keita AV, Söderholm JD. The intestinal barrier and its regulation by neuroimmune factors. Neurogastroenterology Motility. 2010;**22**(7):718-733

[106] Rossi G, Pengo G, Caldin M, Palumbo Piccionello A, Steiner JM, Cohen ND, et al. Comparison of microbiological, histological, and immunomodulatory parameters in response to treatment with either combination therapy with prednisone and metronidazole or probiotic VSL#3 strains in dogs with idiopathic inflammatory bowel disease. PLoS One. 2014;**9**(4):e94699

[107] Prasad S, Mingrino R, Kaukinen K, Hayes KL, Powell RM, MacDonald TT, et al. Inflammatory processes have differential effects on claudins 2, 3 and 4 in colonic epithelial cells. Laboratory Investigation. 2005;**85**(9):1139-1162

[108] Zeissig S, Bürgel N, Günzel D, Richter J, Mankertz J, Wahnschaffe U, et al. Changes in expression and distribution of claudin 2, 5 and 8 lead to discontinuous tight junctions and barrier dysfunction in active Crohn's disease. Gut. 2007;**56**(1):61-72

[109] MacDonald TT, Hutchings P, Choy MY, Murch S, Cooke A. Tumour necrosis factor-alpha and interferon-gamma production measured at the single cell level in normal and inflamed human intestine. Clinical and Experimental Immunology. 1990;**81**(2):301-305

[110] Gangwar R, Meena AS, Shukla PK, Nagaraja AS, Dorniak PL, Pallikuth S, et al. Calcium-mediated oxidative stress: A common mechanism in tight junction disruption by different types of cellular stress. The Biochemical Journal. 2017;**474**(5):731-749

[111] Rao R. Oxidative stress-induced disruption of epithelial and endothelial tight junctions. Frontier in Bioscience. 2008;**1**(13):7210-7226

[112] Matsuzawa-Ishimoto Y, Shono Y, Gomez LE, Hubbard-Lucey VM, Cammer M, Neil J, et al. Autophagy protein ATG16L1 prevents necroptosis in the intestinal epithelium. The Journal of Experimental Medicine. 2017;**214**(12):3687-3705

[113] Ohta H, Yamaguchi T, Rajapakshage BKW, Murakami M, Sasaki N, Nakamura K, et al. Expression and subcellular localization of apical junction proteins in canine duodenal and colonic mucosa. American Journal of Veterinary Research. 2011;**72**(8):1046-1051

[114] Ridyard AE, Brown JK, Rhind SM, Else RW, Simpson JW, Miller HRP. Apical junction complex protein expression in the canine colon: Differential expression of claudin-2 in the colonic mucosa in dogs with idiopathic colitis. Journal of Histochemistry. 2007;**55**(10):1049-1058

[115] Kaser A, Lee A-H, Franke A, Glickman JN, Zeissig S, Tilg H, et al. XBP1 links ER stress to intestinal inflammation and confers genetic risk for human inflammatory bowel disease. Cell. 2008;**134**(5):743-756

[116] Watanabe T, Asano N, Murray PJ, Ozato K, Tailor P, Fuss IJ, et al. Muramyl dipeptide activation of nucleotide-binding oligomerization domain 2 protects mice from experimental colitis.

The Journal of Clinical Investigation. 2008;**118**(2):545-559

[117] Noth R, Stüber E, Häsler R, Nikolaus S, Kühbacher T, Hampe J, et al. Anti-TNF-α antibodies improve intestinal barrier function in Crohn's disease. Journal of Crohn's & Colitis. 2012;**6**(4):464-469

[118] Brown GR, Lindberg G, Meddings J, Silva M, Beutler B, Thiele D. Tumor necrosis factor inhibitor ameliorates murine intestinal graft-versus-host disease. Gastroenterology. 1999;**116**(3):593-601

[119] Marafini I, Monteleone I, Di Fusco D, Cupi ML, Paoluzi OA, Colantoni A, et al. TNF-α producing innate lymphoid cells (ILCs) are increased in active celiac disease and contribute to promote intestinal atrophy in mice. PLoS One. 2015;**10**(5):e0126291

[120] Shen L, Black ED, Witkowski ED, Lencer WI, Guerriero V, Schneeberger EE, et al. Myosin light chain phosphorylation regulates barrier function by remodeling tight junction structure. Journal of Cell Science. 2006;**119**(Pt 10):2095-2106

[121] Su L, Shen L, Clayburgh DR, Nalle SC, Sullivan EA, Meddings JB, et al. Targeted epithelial tight junction dysfunction causes immune activation and contributes to development of experimental colitis. Gastroenterology. 2009;**136**(2):551-563

[122] Bruewer M, Utech M, Ivanov AI, Hopkins AM, Parkos CA, Nusrat A. Interferon-gamma induces internalization of epithelial tight junction proteins via a macropinocytosis-like process. FASEB Journal. 2005;**19**(8):923-933

[123] Cario E, Rosenberg IM, Brandwein SL, Beck PL, Reinecker HC, Podolsky DK. Lipopolysaccharide activates distinct signaling pathways in intestinal epithelial cell lines expressing Toll-like receptors. Journal of Immunology. 2000;**164**(2):966-972

[124] Kong W, McConalogue K, Khitin LM, Hollenberg MD, Payan DG, Böhm SK, et al. Luminal trypsin may regulate enterocytes through proteinase-activated receptor 2. Proceedings of the National Academy of Sciences of the United States of America. 1997;**94**(16):8884-8889

[125] Darmoul D, Marie JC, Devaud H, Gratio V, Laburthe M. Initiation of human colon cancer cell proliferation by trypsin acting at protease-activated receptor-2. British Journal of Cancer. 2001;**85**(5):772-779

[126] Chin AC, Vergnolle N, MacNaughton WK, Wallace JL, Hollenberg MD, Buret AG. Proteinase-activated receptor 1 activation induces epithelial apoptosis and increases intestinal permeability. Proceedings of the National Academy of Sciences of the United States of America. 2003;**100**(19):11104-11109

[127] Mielants H, De Vos M, Goemaere S, Schelstraete K, Cuvelier C, Goethals K, et al. Intestinal mucosal permeability in inflammatory rheumatic diseases. II. Role of disease. Journal of Rheumatology. 1991;**18**(3):394-400

[128] Camilleri M, Gorman H. Intestinal permeability and irritable bowel syndrome. Neurogastroenterology Motility. 2007;**19**(7):545-552

[129] Mooradian AD, Morley JE, Levine AS, Prigge WF, Gebhard RL. Abnormal intestinal permeability to sugars in diabetes mellitus. Diabetologia. 1986;**29**(4):221-224

[130] Fasano A. Zonulin and its regulation of intestinal barrier function:

The biological door to inflammation, autoimmunity, and cancer. Physiological Reviews. 2011;**91**(1):151-175

[131] Fasano A, Baudry B, Pumplin DW, Wasserman SS, Tall BD, Ketley JM, et al. Vibrio cholerae produces a second enterotoxin, which affects intestinal tight junctions. Proceedings of the National Academy of Sciences of the United States of America. 1991;**88**(12):5242-5246

[132] Fasano A, Fiorentini C, Donelli G, Uzzau S, Kaper JB, Margaretten K, et al. Zonula occludens toxin modulates tight junctions through protein kinase C-dependent actin reorganization, in vitro. The Journal of Clinical Investigation. 1995;**96**(2):710-720

[133] Fasano A, Uzzau S, Fiore C, Margaretten K. The enterotoxic effect of zonula occludens toxin on rabbit small intestine involves the paracellular pathway. Gastroenterology. 1997;**112**(3):839-846

[134] Fasano A, Not T, Wang W, Uzzau S, Berti I, Tommasini A, et al. Zonulin, a newly discovered modulator of intestinal permeability, and its expression in coeliac disease. Lancet. 2000;**355**(9214):1518-1519

[135] Wang W, Uzzau S, Goldblum SE, Fasano A. Human zonulin, a potential modulator of intestinal tight junctions. Journal of Cell Science. 2000;**113** (Pt 24):4435-4440

[136] Tripathi A, Lammers KM, Goldblum S, Shea-Donohue T, Netzel-Arnett S, Buzza MS, et al. Identification of human zonulin, a physiological modulator of tight junctions, as prehaptoglobin-2. Proceedings of the National Academy of Sciences of the United States of America. 2009;**106**(39):16799-16804

[137] van der Merwe JQ, Hollenberg MD, MacNaughton WK. EGF receptor transactivation and MAP kinase mediate proteinase-activated receptor-2-induced chloride secretion in intestinal epithelial cells. American Journal of Physiology. Gastrointestinal and Liver Physiology. 2008;**294**(2):G441-G451

[138] Goldblum SE, Rai U, Tripathi A, Thakar M, De Leo L, Di Toro N, et al. The active Zot domain (aa 288-293) increases ZO-1 and myosin 1C serine/threonine phosphorylation, alters interaction between ZO-1 and its binding partners, and induces tight junction disassembly through proteinase activated receptor 2 activation. FASEB Journal. 2011;**25**(1):144-158

[139] McNeil E, Capaldo CT, Macara IG. Zonula occludens-1 function in the assembly of tight junctions in Madin-Darby canine kidney epithelial cells. Molecular Biology of the Cell. 2006;**17**(4):1922-1932

[140] Rossi G, Gioacchini G, Pengo G, Suchodolski JS, Jergens AE, Allenspach K, et al. Enterocolic increase of cannabinoid receptor type 1 and type 2 and clinical improvement after probiotic administration in dogs with chronic signs of colonic dysmotility without mucosal inflammatory changes. Neurogastroenteroogyl Motility. 2020;**32**(1):e13717

[141] Lee H, Park JH, Park DI, Kim HJ, Cho YK, Sohn CI, et al. Mucosal mast cell count is associated with intestinal permeability in patients with diarrhea predominant irritable bowel syndrome. Journal of Neurogastroenterology Motility. 2013;**19**(2):244-250

[142] Jacob C, Yang P-C, Darmoul D, Amadesi S, Saito T, Cottrell GS, et al. Mast cell tryptase controls paracellular permeability of the intestine. Role of protease-activated receptor 2 and

beta-arrestins. The Journal of Biological Chemistry. 2005;**280**(36):31936-31948

[143] Wilcz-Villega EM, McClean S, O'Sullivan MA. Mast cell tryptase reduces junctional adhesion molecule-A (JAM-A) expression in intestinal epithelial cells: Implications for the mechanisms of barrier dysfunction in irritable bowel syndrome. The American Journal of Gastroenterology. 2013;**108**(7):1140-1151

[144] Fasano A. Intestinal permeability and its regulation by zonulin: Diagnostic and therapeutic implications. Clinical Gastroenterology. 2012;**10**:1096-1100

[145] Banfi C, Brioschi M, Barbieri SS, Eligini S, Barcella S, Tremoli E, et al. Mitochondrial reactive oxygen species: A common pathway for PAR1- and PAR2-mediated tissue factor induction in human endothelial cells. Journal of Thrombosis and Haemostasis JTH. 2009;**7**(1):206-216

[146] Clemente MG, Musu MP, Troncone R, Volta U, Congia M, Ciacci C, et al. Enterocyte actin autoantibody detection: A new diagnostic tool in celiac disease diagnosis: Results of a multicenter study. The American Journal of Gastroenterology. 2004;**99**(8):1551-1556

[147] Lammers KM, Lu R, Brownley J, Lu B, Gerard C, Thomas K, et al. Gliadin induces an increase in intestinal permeability and zonulin release by binding to the chemokine receptor CXCR3. Gastroenterology. 2008;**135**(1):194-204.e3

[148] Drago S, El Asmar R, Di Pierro M, Grazia Clemente M, Tripathi A, Sapone A, et al. Gliadin, zonulin and gut permeability: Effects on celiac and non-celiac intestinal mucosa and intestinal cell lines. Scandinavian Journal of Gastroenterology. 2006;**41**(4):408-419

[149] Pizzuti D, Bortolami M, Mazzon E, Buda A, Guariso G, D'Odorico A, et al. Transcriptional downregulation of tight junction protein ZO-1 in active coeliac disease is reversed after a gluten-free diet. Digestive Liver Diseases. 2004;**36**(5):337-341

[150] Sander GR, Cummins AG, Henshall T, Powell BC. Rapid disruption of intestinal barrier function by gliadin involves altered expression of apical junctional proteins. FEBS Letters. 2005;**579**(21):4851-4855

[151] Moore PF, Rodriguez-Bertos A, Kass PH. Feline gastrointestinal lymphoma: Mucosal architecture, immunophenotype, and molecular clonality. Veterinary Pathology. 2012;**49**(4):658-668

[152] Bertiaux-Vandaële N, Youmba SB, Belmonte L, Lecleire S, Antonietti M, Gourcerol G, et al. The expression and the cellular distribution of the tight junction proteins are altered in irritable bowel syndrome patients with differences according to the disease subtype. The American Journal of Gastroenterology. 2011;**106**(12):2165-2173

Chapter 3

Paneth Cells: The Gatekeepers of the Gut

Thifhelimbilu E. Luvhengo and Mwangala Nalisa

Abstract

Although its most well-written functions are digestion and absorption of nutrients, the gastrointestinal tract (GIT) is the most significant player in the human immune system. The GIT is home to more than 60% of the active immune cells in the entire body. Notwithstanding, the human gut is continuously exposed to antigens ingested with food and resident microorganisms. The density of microorganisms in the lumen of GIT increases aborad and is much higher in the colon. Despite a relatively low bacterial load in the small intestine, the environment is more precarious because it is nutritious and exposed to digestive enzymes. Its lining is made up of a single layer of epithelial cells covered by a thin and attenuated layer of mucus. Despite the continual exposure to the luminal antigens, the gut's immune system is kept in a state of relative immunosuppression. The pathogenesis of some of the common non-communicable diseases includes a systemic inflammatory state initiated by dysbiosis in the gut, increased permeability of the intestinal epithelium, translocation of microbiomes or their products, and then a persistent pro-inflammatory state. Paneth cells are the key players in the innate immunity of the gut and are responsible for maintaining its integrity.

Keywords: Paneth cells, microbiome, dysbiosis, antimicrobial peptides, defensins, innate immunity, systemic inflammation

1. Introduction

The gastrointestinal tract (GIT) is the largest organ in the body and is constantly exposed. At least 60% of active immune cells reside in the tissues of the GIT. An everyday healthy lifestyle requires a structurally intact and normal functioning GIT. The integrity of the GIT is maintained through continuous replacement of surface epithelial cells, which exfoliate and have to be replaced every 4–5 days [1]. Other key factors which are important for the physical integrity of the epithelium throughout the GIT are tight intercellular junctions, fine-tuning of the gut microbiome and active dampening down of the immune response. Dysbiosis followed by an ongoing systemic inflammatory state is involved in the pathogenesis of several gastrointestinal and extra-gastrointestinal conditions [2]. Medical conditions linked with a sustained pro-inflammatory state are inflammatory bowel disease (IBD), obesity, malignancies, arthritis, diabetes mellitus and acute or chronic organ-system dysfunction [1–3].

Dysbiosis is a prelude in the pathogenesis of inflammatory bowel disease (IBD), obesity, and other diseases associated with the prolonged systemic inflammatory state. It is defined as a change in the number or type of luminal microorganisms. Dysbiosis leads to the appearance of allochthonous organisms at various niches in the GIT, especially in the lumen of the small intestine and or the colon [3]. Dysbiosis leads to translocation of luminal microorganism and their products such as endotoxins, immune activation, and initiation of systemic inflammation. This chronic systemic inflammatory state is resistant to insulin and is obesogenic, diabetogenic, carcinogenic, and thrombogenic [1]. Dysbiosis and systemic inflammation also play a role in the pathophysiology of some of the complications associated with chronic human immunodeficiency virus (HIV) infection before or during treatment with antiretroviral (ARV) drugs [2]. Dysbiosis and chronic stimulation of gut immunity and subsequently systemic inflammation are purported to be among the factors which induce progressive deterioration of systemic immunity and depletion of CD4 T-lymphocytes count, and why HIV is currently not curable.

The integrity of the epithelium throughout the GIT, especially at the small intestine region, has to remain intact to ensure a healthy life. The gut immune system can defend the body from a state of perpetual systemic pro-inflammation because of a robust innate immunity that functions in liaison with elements of adaptive immunity. Every plant and animal species has a built-in mechanism to secrete antimicrobials (defensins) that prevent invasion by pathogenic organisms. Neutrophils are responsible for the secretion of defensins throughout the entire body in mammals, including the skin and parts of the GIT. Human alpha defensins are only produced by Paneth cells in the small intestine. Paneth cells are the key player in the innate immunity of the small intestine and are responsible for the robustness of gut immunity [1, 3–9].

2. Origin, structure and function of the Paneth cell

Paneth cells are found in the small intestine of humans and other vertebrates, including horses, sheep and rodents. Paneth cells are one of the four main derivatives of the intestinal stem cell [10, 11]. The other cells derived from the intestinal stem cells are enterocytes, goblet cells and neuroendocrine cells. These cells are found towards the luminal surface of the epithelium of the small intestine, whilst Paneth cells are located at the base of the intestinal crypt. They mingle with the crypt-based, Leucine-rich repeat-containing G-protein coupled receptor 5 (Lgr5+) intestinal stem cells in a 1:1 ratio [4, 12, 13]. Refer to **Figure 1** for more detail on the position of Paneth cells. Paneth cells are distinguishable from the Lgr5+ stem cells and other cells in the epithelium of the small intestine, which are found at the base of the intestinal crypt because they are secretory and have eosinophilic apical cytoplasmic granules and an extensive network of endoplasmic retinaculi [1, 12, 14, 15].

Paneth cells appear in utero from the 13th week of gestation [6, 16]. The number of Paneth cells progressively increase with a much more rapid expansion after the 29th week of pregnancy. A sufficient number of matured and functional Paneth cells is only attained at term [16]. Although Paneth cells are found in the entire colon of human embryos, they disappear soon after birth and are only found in the caecum and the ascending colon [17]. However, the location of Paneth cells in the epithelium of the small intestine is initially haphazard and more towards the luminal surface of the epithelium. Unlike the other derivatives of the intestinal stem cell, the Paneth cells later migrate downwards and end up at the base of the intestinal crypts. The stable number of Paneth cells in the crypt of a particular individual gets established in early adulthood. The overall number of Paneth cells in the intestinal crypt of the small

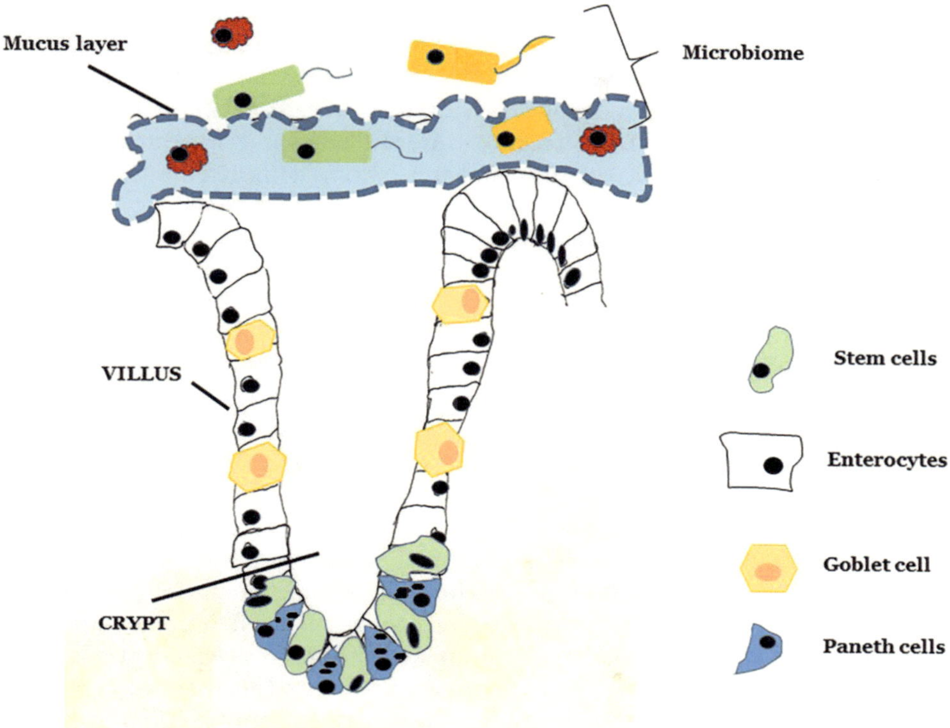

Figure 1.
A schematic showing the structure and position of Paneth cells in the villus.

intestine is influenced by factors such as the gestational age at birth, the mode of delivery, breastfeeding [6], weaning diet and period, dietary preference and disease states [17, 18]. The timing of colonization and the type of organisms involved play a role in both the development and number of Paneth cells [9, 19].

Once they are fully established, the Paneth cells are found at the base of the intestinal crypt, and in a healthy person, the number and distribution of Paneth cells remain relatively constant for up to 20 years [20]. Each intestinal crypt contains five to fifteen Paneth cells. The terminal ileum, which is the small intestine area with the highest load of microorganisms, contains the most Paneth cells. The overall number of Paneth cells in the small intestine may increase following a viral infection or a course of nutritional supplementation. Paneth cells may appear in ectopic sites due to metaplasia in certain diseased states, such as chronic inflammation [7, 21]. Similarly, the number of Paneth cells may be reduced due to malnutrition, the infestation of the GIT by parasites, chronic HIV disease, radiation enteritis, smoking, high fat diet, and aging [1–3].

There are three tiers of intestinal stem cells. The 1st tier of intestinal stem cells are found at the base of the crypt and are the so-called crypt base columnar cells (CBC cells) or Lgr5+ cells. The 1st tier stem cells are found in the most protected environment in the region where the most mature and furnished Paneth cells are. They are paired with Paneth cells on a 1:1 basis for intimate contact and direct communication. These Lgr5+ stem cells are the most vulnerable to radiation injury and require the most protection [22, 23]. It is the reason why they are closely associated with the Paneth cells. They rely on the Paneth cells for both protection and nourishment. The Lgr5+ intestinal stem cells also have the least capacity to repair any damage to their DNA. The second tier of the intestinal stem cells is found high up along the crypt-villous axis at around cell position 4. The 3rd tier of cells with stem cell capability is higher up and closer to the crypt-villous transition zone and are also called intermediate cells. The 2nd and 3rd tiers are normally quiescent and only become activated on-demand, such as following an injury to the intestinal epithelium.

Other cells found in the epithelium of the small intestine are enterocytes and goblet cells. The enterocytes are the most populous derivatives of the intestinal stem cells and are found on the luminal surface of the small intestine. The small intestine has microvilli to increase the surface area to absorb nutrients. The function of the enterocytes is not limited to digestion and absorption of nutrients, as they also participate in the innate immunity of the small intestine. Combined with junctional proteins, enterocytes provide an uninterrupted physical barrier [10, 16, 22]. The enterocytes also secrete cytokines and chemokines, which recruit immune response elements. The lifespan of enterocytes is around five days, and they are being replaced continuously. Apoptotic enterocytes remain structurally intact until they are replaced to prevent the creation of defects on the surface of the epithelium and thus increased permeability. The enterocytes can de-differentiate into the stem cells following the destruction of the intestinal stem cells and Paneth cells. The enterocytes and goblet cells can de-differentiate into stem cells if they have not yet undergone terminal division. However, the plasticity of Paneth cells is far greater as they can be de-differentiating and acquire stemness even after they have been terminally divided [22]. The responsibility of de-differentiating to stem cells is first reserved for the Paneth cells afferent task. The intermediate cells, i.e. intermediaries of the other derivatives, only get involved and regain stemness if the Paneth cells have been irreversibly damaged and cannot play the role [23, 24].

The goblet cells are the second most populous cell type in the epithelium of the small intestine. They are interspersed among the enterocytes on the luminal surface of the epithelium. The lifespan of goblet cells is around five days and thus similar to that of enterocytes. Goblet cells are also found in the colon, where they play a role similar to that performed by Paneth cells in the small intestine [14, 16, 17]. Goblet cells secrete various types of mucins. The mucins produced by the goblet cells are the main constituent of the mucus, which coats the mucosal surface of the epithelium of the small intestine and colon. Even though the layer of mucus which lines the luminal surface of the small intestine is thin and attenuated, it can assist in maintaining a high concentration of antimicrobial peptides and protein in the area adjacent to the surface epithelial cells—the other function of the goblet cells secretion of trophic factors such as the trefoil factor. The goblet cells can de-differentiate and acquire stemness if the stem cells and Paneth cells have been damaged [10, 17, 22].

Paneth cells are tall columnar cells that are pyramidal in shape because of a broader base, have supra-nuclear Golgi apparatus, and zinc-rich apical orientated cytoplasmic granules [6]. The granules contain more than 50 constituents, including antimicrobial peptides like human alpha defensin 5 and 6 (HD5 and HD6), lysozyme, secretory phospholipase A2, osteopontin, and associated pancreatitis peptide, trypsinogen, IgA, TNF-alpha and alpha 1-antitrypsin and catecholamines [6]. The predominant constituent which is contained in the granules of Paneth cell is HD5, which make up about 90% of the components. The HD5 is the main antimicrobial peptide that is secreted by the Paneth cells of the small intestine, which is responsible for the control of luminal microbiota [6]. Degranulation of Paneth cells is induced following stimulation, and the granules are replenished expeditiously, usually within 24 hours.

Paneth cells have autocrine, paracrine and endocrine functions. Similarly, Paneth cells respond to autocrine, paracrine or endocrine signals. Paneth cells also play a key role in innate and adaptive immunity by providing a direct line of communication between the two subsystems. They protect and regulate the functioning of the Lgr5+ intestinal stem cell and its derivatives, ensuring that exfoliating surface epithelial cells are regularly and timeously replaced [23]. Secretions from Paneth cells in the proximal parts of the small intestine influence the growth and function of distally situated small intestine stem cells and their derivatives [9]. The area at the base of the intestinal crypt (stem cell niche) is among the most active region in the body. The proliferation and differentiation of the Lgr5+ intestinal stem cells in the niche area is tightly regulated by secretions of Paneth cells and the cells situated in the connective tissue of the lamina propria around the base of the intestinal crypt [23]. The niche factors predominantly released by the Paneth cells influence the timing and type of division the intestinal stem cells should undergo while balancing the maintenance of the stemness and production of their progenies. The intestinal stem cell division may be symmetrical or asymmetrical [6, 10, 17].

A symmetrical division of the intestinal stem cells leads to the proliferation and production of daughter stem cells to expand the pool, whereas differentiation is prioritized during asymmetric division. The derivatives of the intestinal stem cell are generated following an asymmetric division. Both symmetrical and asymmetrical divisions are initiated and regulated by the niche factors which are produced by the Paneth cells and cells in the adjacent mesenchyme of the lamina propria in the peri-crypt space region. These niche factors, which include Wnt and Notch act as signals for the intestinal stem cells [23]. Even though they are in intimate physical contact, Paneth cells also have ligands for engagement with factors produced by the intestinal stem cells for cross-talk [23]. As indicated previously, When conditions are extremely hostile, and the stem cells have been destroyed, paneth cells can de-differentiate and acquire stem-ness.

Paneth cells have pattern recognition receptors that include nucleotide-binding oligomerization domain-like (NOD) and toll-like (TLR) receptors which they use to continuously sample the microbiome's composition in the lumen of the gut to prevent dysbiosis and or invasion by pathogenic organisms [24]. Paneth cells in the terminal ileum where Peyer's patches jointly sample the luminal contents and directly communicate with the M-cells. Paneth cells communicate remotely with the active immune cells and mesenchymal tissue in the lamina propria of the gut epithelium. Secretion from the Paneth cells is continuous with augmentation following a stimulus. The net effect of the Paneth cells' secretions depends on both their composition and volume. The secretions from the Paneth cells help to regulate and fine-tune the gut microbiome [9, 13, 24, 25]. The antimicrobial peptides secreted by the Paneth cells selectively kill pathogenic organisms while sparing the commensals. Ultimately, several microbial niches are created along the entire length of the small intestines where the microorganisms are naturalized and live in a symbiotic relationship with the host. Some of the organisms in the established niches assist during the digestion of food. Additionally, the composition of the flora freely floating in the lumen of the small intestine is different from the area close to the surface epithelial cells. The region closest to the epithelial cells' surface has the highest concentration of antimicrobial peptides and, therefore, the most repulsive to non-commensal organisms [26, 27].

Once established, the commensal organisms in the gut assist with digestion and absorption of essential nutrients in either the small intestine or colon, preventing the overgrowth of potentially pathogenic microorganisms and immune regulation [24]. Some of the ingested compounds in the food would not be digestible were it not for certain species of resident microorganisms. The overall number of nutrients available in a particular site is a secretion of antimicrobial molecules that creates areas of zonal dominance by some species of microorganisms. The task of the Paneth cells is to accept the dominant organisms in various niches and for the sustenance of the symbiotic relationship. A deviation from the established normality is detected by the Paneth cells through their pathogen recognition receptors leading to degranulation and release of antimicrobial peptides. The change in the microbiome may occur following the use of broadspectrum antibiotics, change in diet, change in the anatomy of the GIT, alteration of gut transient time or a state of suppressed systemic immunity [25].

The other role of the Paneth cells is to nourish the intestinal stem cells [11, 12, 22]. Paneth cells derive their energy from the glycolytic pathways, whereas the intestinal stem cells' Adenosine triphosphate (ATP) is derived from aerobic metabolism in their mitochondria. The lactic acid which is produced by a Paneth cell is shunted into adjacent intestinal stem cells for their metabolism. Paneth cells can sense the fed-state of the body and, after that, influence the intestinal stem cell activity accordingly. If the epithelium is damaged, they become more active. Paneth cells are found in the normal human small intestine from the duodenum to the terminal ileum [26, 27]. They are most abundant in the region of the terminal ileum. The hostility of the environment, which increases as one move distally along the small intestine due to increment in the number and species of microbes, is a plausible explanation of the need for more Paneth cells at the region of the terminal ileum [26]. Paneth cells are not found in a healthy stomach and colon, except for a few in the caecum and the ascending colon [26]. Paneth cells may develop following metaplasia associated with chronic gastritis and inflammatory bowel disease in the stomach and colon. The effects of paneth cell secretions are enumerable and continue to be added. The result of the secreted factor varies depending on the type, volume and concentration of secretions. These effects include antimicrobial activity, inflammation and regulation of intestinal stem cells'. The antimicrobial peptides secreted by Paneth cells help sterilize the intestinal crypt environment, the so-called "stem cell zone; and thus protect intestinal stem cells [7, 13, 22, 28–30].

The antimicrobial peptides which are produced by the Paneth cells are of three types: Type 1 is cationic, Type 2 is amphipathic, and Type 3 is composed of hydrophobic peptides [31, 32]. The micro-biocidal effects of Type 1 and Type 2 are based on induction of damage to the surface of the cell membrane and creation of large pores as it penetrates deeper into the hydrophobic cell membrane and its bi-layer, respectively. Type 3 peptides cause micellization of the cell membrane of pathogens. The antimicrobial peptides produced by various cells in humans include cathelicidins and alpha and beta-defensins. Beta-defensins are produced in almost every cell in the body, including neutrophils. The Paneth cell is the only source of cathelicidins, HD5 and HD6 in humans. The synthesis of cathelicidins ceases when the foetus reaches term and is then replaced with HD5 and HD6 [6, 24]. The antimicrobial peptides from human Paneth cells are secreted in a pro-active form and become activated in the lumen of the small intestine [32]. The Paneth cell secretions are secreted together with water and chloride ions and assisted by peristalsis to bathe the crypt environment to make it conducive to functioning the tiers of intestinal stem cells [6, 22, 33]. The solvent load and anionic composition in the secretion assist in the after-release potentiation of the antimicrobial peptides [31]. In addition, the bile salts in the small intestine influence the killing activity of paneth cell-derived antimicrobial peptides. The human alpha-defensins are active against bacteria, fungi, parasites, and viruses [31, 32].

The HD5 is microbicidal, and the HD6 binds the antigens to prevent invasion until it is eliminated. The action of HD6 is similar to that of the IgA antibodies. The binding of antigens by HD6 buys time for the other elements of innate immunity of the gut to arrive [31, 32].

Matured Paneth cells are found at the base of the intestinal crypt of the small intestine, where the most vulnerable but essential intestinal stem cells are found. The loss of senescent Paneth cells like absorptive enterocytes, goblet cells and neuroendocrine cells is programmed [6]. The senescent Paneth cells die and are removed through phagocytosis following apoptosis. The other mechanism involved in the death of Paneth cells is autophagy, which ensures that some essential constituents found in the cytoplasm are recycled. Apoptotic enterocytes located at the tip of villi remain structurally intact until they are replaced by a carpet of new cells arriving from the stem cell zone. A perfect balance between the rate of proliferation and loss through exfoliation is sustained to ensure that defects are not created, and the epithelium of the small intestine becomes permeable to microorganisms and their products [11, 16].

A healthy life without a normally functioning small intestine is not possible. The influence of the Paneth cells on the GIT starts soon after birth. Henceforth, the Paneth cell influences everything that happens in the gut, whether physical or biochemical barrier, absorption of nutrients, and linkage with the body's overall immune system. Paneth cells play a role in the development and maturation of the innate immunity of the gut and subsequently of the entire body. The immune system is a dominant player in systemic immunity, including adaptive immunity. Paneth cells maintain the integrity of the gut by controlling the microbiome, regulating proliferation and differentiation of the intestinal stem cells, influencing the quality of mucin in the mucous, and keeping the crypt environment relatively sterile for the protection of the intestinal stem cells. Paneth cells also release growth signals that influence the growth and function of the enterocytes, goblet cells, and neuroendocrine cells [7, 9, 11]. Chemokines and cytokines produced by Paneth cells can also recruit and influence components of adaptive immunity in the adjacent lamina propria [34]. Among the cytokines which are produced by the Paneth cells is TNF-alpha. There is also cross-talk between the Paneth cells and elements of adaptive immunity [34, 35].

A plethora of acute and chronic conditions are driven from the gut. These conditions may be initiated by a changing diet, starvation, trauma or sepsis. The normal development of the crypts and villi of the small intestine and control of the microbiome is dependent on the Paneth cells. Paneth cells continuously sample the luminal contents for the gut microbiota composition to prevent dysbiosis. Dysbiosis leads to increased gut permeability and translocation of bacteria and endotoxin [3]. Usually, when the body experiences significant physiological stress, the gut mucosa is strategically sacrificed; a typical example is the shunting of blood from the GIT in various shock states. If prolonged, what is meant to be a short-term survival strategy leads to dysbiosis, across the intestinal epithelium. Translocation of bacteria and endotoxins is the driver of systemic inflammation. The gut is the most trusted and potent site for eliminating invading pathogens [35, 36]. Shunting of invading pathogens to the GIT also applies to viruses, including HIV. Should the gut immune system fail to eliminate the pathogen, as it happens following HIV infection, the gut ultimately becomes a long-term reservoir and haven of mutated strains of the virus, which is currently impossible to eradicate despite the availability of potent antiviral drugs [36, 37].

3. Paneth cells and diseases

Paneth cells are involved in the fight against and pathogenesis of diseases. Paneth cells' effect on diseases is evidenced by changes in their granules, the total number, position, or distribution pattern. The health status of Paneth cells is assessed by checking their presence, position, number and intensity of staining of their granules [3, 10, 15, 27, 34, 38, 39]. The integrity of the intestinal epithelium is assessed directly by measuring the depth of the intestinal crypt, the height of the villi, mitotic count [27], Ki67 index, markers of apoptosis and epithelial integrity as evidenced by, for example; the presence of bacteria inside intestinal epithelial cells and translocation. Dysbiosis is one of the key steps in the pathogenesis of many medical and surgical diseases [19, 40]. Translating bacteria and endotoxins induces an immune response leading to a systemic inflammatory response. Immune activation and resulting systemic inflammation are deleterious to the body as it has been proven to be the underlying reason behind most metabolic diseases such as type 2 diabetes mellitus, cardiovascular disease, and obesity. Chronic immune activation and systemic inflammation are paradoxically the major drivers of complications associated with chronic HIV infection regardless of treatment with antiretroviral drugs [41–45]. A chronic inflammatory state is one of the main reasons why HIV currently can neither be cured nor eradicated.

Duodenal Paneth cells are reduced in individuals with idiopathic autoimmune enteropathy [21]. Cells with the characteristic of the Paneth cells and goblet cells, the so-called intermediate cells, may appear high up in the intestinal epithelium crypt above position eight in patients with inflammatory bowel disease [46]. The number and size of paneth cells are increased in individuals who have autism with gastrointestinal tract symptoms [47]. Among the conditions which are associated with dysbiosis and increased permeability of the intestinal epithelium are chronic HIV/AIDS [2, 42, 43, 48–53] and HIV-linked surgical conditions such as necrotizing enterocolitis [54], obstructive jaundice, inflammatory bowel disease [55–57], obesity [40, 58, 59] fasting and prolonged total parenteral nutrition [3, 38]. The above conditions have either been proven or are possibly linked to dysfunction of Paneth cells in the small intestine.

Diseases that have been conclusively linked with change in the appearance of paneth cells include necrotizing enterocolitis [1, 60], starvation [38] prolonged total parenteral nutrition [3], inflammatory bowel disease [7, 20, 36], mesenteric ischaemia, radiation enteritis [61], coeliac disease [27], colorectal carcinoma [62], autism [47] and HIV infection [63–65]. The other evidence of ill health in the gut include atrophy of small bowel mucosa, the appearance of defects in the mucosa, translocation of bacteria and toxins, systemic sepsis and development of inflammatory bowel diseases or the appearance of malignant neoplasms. In this chapter, we illustrate the direct role played by the Paneth cells in their pathogenesis using the diseases below.

3.1 Necrotizing enterocolitis

Current knowledge of the function of Paneth cells has enhanced the understanding of the crucial role in the pathogenesis of Necrotizing enterocolitis (NEC). NEC is a common condition that affects neonates and is manifested by acute inflammatory changes in the bowel wall. The risk factors of NEC include prematurity, low birth weight (<1.5Kg), formula feeding, antibiotic use and HIV status of the mother [2, 66]. Some studies attribute the pathogenesis of NEC to premature colonization of the gut in the neonate or an ischaemic event [60, 66–68]. Necrotizing enterocolitis is linked with untimely colonization of the gut when the gut immune system has not matured. A sufficient number of Paneth cells and their maturity only get established around the term. When they are well developed, Paneth cells can secrete an adequate amount of antimicrobial peptides, including HD5 and HD6, for innate immune defense. The switch from cathelicidins to defensins only occurs during the third trimester of pregnancy [60].

Consequently, preterm babies are prone to colonization by pathogenic organisms, followed by bowel invasion by pathogenic bacteria and inflammation. Early oral feeding of infants leads to dysbiosis, increased permeability, bacterial translocation, and an inflammatory state in the intestinal epithelium. The ongoing inflammation would then lead to ischaemia. The inflammatory process causes thrombosis of vessels in the submucosa and the lamina propria. In some cases, the initiating event for NEC is ischaemia and then inflammatory response. The most feared complication of NEC is bowel necrosis, leading to perforation and severe sepsis. In some cases of NEC, ischaemia is the initiating event. Both premature colonization and ischaemic events are associated with abnormality in the functioning of the Paneth cells [1, 60].

3.2 Colorectal carcinoma

Colorectal carcinoma is among the top five most common malignancies worldwide. The common sites involved in colon cancer are the rectosigmoid area, caecum and the transverse colon. The majority (>90%) of colorectal cancers are sporadic. The development of sporadic colorectal cancer is preceded by an adenomatous polyp, the so-called adenoma-carcinoma sequence. Colonocytes rely on nutrients derived from the fermentation of indigestible fibers by resident bacteria. Some sporadic cancers are associated with dysbiosis [69, 70]. Abnormalities of Paneth cells, including metaplasia, have been reported in colorectal cancer [62]. Furthermore, Paneth cells in the colon of patients diagnosed with adenomatous polyps or invasive cancer are associated with aggressive disease and poor prognosis [70].

3.3 Inflammatory bowel disease

Inflammatory bowel diseases (IBD) include Ulcerative colitis (UC) and Crohn's disease (CD). Although the main manifestations of ulcerative colitis are mainly limited to the colon, Crohn's disease is a systemic disease that can affect any organ in the body. The exact cause of either UC or CD remains unknown and is presumed to be environmental [46, 56, 57]. However, what is common in the pathogenesis of UC and CD is the failure of the innate immune system. In UC, there is dysbiosis with an overgrowth of pathogenic organisms. The pathogens are recognized as foreign by the Paneth cells, leading to an inflammatory response in the bowel wall [9]. Some cases of Crohn's disease are due to genetic defects, which diminishes the ability of Paneth cells to sense and regulate the gut microbiome. Mutation of NOD 2 receptor is linked with the development of CD as the Paneth cells of the affected individuals in CD cannot secrete sufficient antimicrobial peptides to prevent dysbiosis [57].

3.4 Paneth cells and the human immunodeficiency virus and other viral infections

The GIT is the trusted site for sequestration and subsequent elimination of pathogens following a viral infection. Immediately following the invasion by HIV or other viral infections, the virus is transported to active immune sites within the GIT. Some of the viruses are eliminated, whereas others remain dormant in tissues of the GIT. Paneth cells are involved during early and later phases following a viral infection. Regardless of the treatment status, diabetes mellitus, cardiovascular disease, dyslipidemia, and malignancies are more prevalent in HIV positive individuals than HIV negative individuals with the same conditions [51, 58]. Non-communicable diseases are responsible for a persistent reduction in the life expectancy of HIV positive individuals despite treatment with potent antiretroviral drugs [45]. A similar inflammatory process resulting from dysbiosis happens during chronic HIV infection and or treatment with ARVs, resulting in the same consequences of obesity and other non-communicable diseases [48, 51, 58, 71].

3.5 Paneth cells and obesity

The quality of Paneth cells is influenced by an individual's nutritional status and dietary intake. A diet that is high in fat is detrimental to the Paneth cell. Obesity affects a significant proportion of the citizens of most countries in the world regardless of their socioeconomic status [68, 72, 73]. Individuals who are obese are at an increased risk of diabetes mellitus, cardiovascular disease, liver dysfunction, and colorectal cancer, all characterized by a state of a persistent inflammatory response. The development of obesity is preceded by dysbiosis, bacterial translocation and persistent systemic inflammation [40]. The link between dysbiosis and Paneth cells has been established in the sections above.

3.6 Obstructive jaundice

The common causes of obstructive jaundice are biliary atresia in children and gall stones or malignancy in adults. Sepsis is the most dreaded complication of obstructive jaundice before or after surgical intervention. The sepsis in obstructive jaundice is preceded by dysbiosis, increased epithelial permeability and translocation of bacteria [74, 75]. A state of relative immunosuppression that generally exists in the gut is lost, and a pro-inflammatory state ensues as evidenced by increased baseline interleukin-6 and C-reactive protein [76, 77].

4. Conclusion

The large surface, a thin barrier, and a microbiome of approximately 100 trillion microorganisms make the GIT the area of the body most at risk of pathogenic invasion [78]. More than 60% of active immune cells are found in tissues of the GIT. The immune system of the GIT is kept dampened to prevent an immune response and persistent inflammation in the gut, which may spill over systemically. Hence the need to maintain gut integrity. The crucial role Paneth cells play in the provisioning and regulation of the innate immune system of the gut cannot be underestimated. The Paneth cell is a nurse, guardian and chaperone, fine-tuning the gut microbiome to prevent dysbiosis, controlling the physiological function, proliferation and differentiation of the intestinal stem cells, and acquiring stemness to replace damaged intestinal stem cells. These cells liaise with cells in the mesenchymal and recruitment of adaptive immunity to prevent pathogenesis, earning the term of gatekeeper of the gut.

Acknowledgements

The authors would like to acknowledge the Department of Surgery at the University of the Witwatersrand and Charlotte Maxeke Johannesburg Academic Hospital.

Conflict of interest

The authors have no conflict of interest.

Notes/thanks/other declarations

The authors would like to thank the University of the Witwatersrand for supporting this work and related academic projects.

Author details

Thifhelimbilu E. Luvhengo and Mwangala Nalisa*
Department of Surgery, The University of the Witwatersrand, Johannesburg, South Africa

*Address all correspondence to: mwangalan@gmail.com

References

[1] Gassler N. Paneth cells in intestinal physiology and pathophysiology. World Journal of Gastrointestinal Pathophysiology. 2017;**8**(4):150-160. DOI: 10.4291/wjgp.v8.i4.150

[2] Gootenberg DB, Paer JM, Luevano JM, Kwon DS. HIV-associated changes in the enteric microbial community: Potential role in loss of homeostasis and development of systemic inflammation. Current Opinion in Infectious Diseases. 2017;**30**:31-43. DOI: 10.1097/QCO.0000000000000341

[3] Heneghan AF, Pierre JF, Tandee K, et al. Parenteral nutrition decreases Paneth cell function and intestinal bactericidal activity while increasing susceptibility to bacterial Enteroinvasion. Journal of Parenteral and Enteral Nutrition. 2014;**38**(7):817-824. DOI: 10.1177/0148607113497514

[4] Jackson D, Theiss AL. Intestinal stem cell regulation via glycolytic activity of neighboring Paneth cells. Journal of Gastroenterology, Hepatology and Endoscopy. 2017;**2**(3):1019. DOI: not available

[5] Ouellette AJ, Selsted ME. Paneth cell defensins: Endogenous peptide components of intestinal host defense. The FASEB Journal. 1996;**10**(11):1280-1289. DOI: 10.1096/fasebj.10.11.8836041

[6] Porter EM, Bevins CL, Ghosh D, Ganz T. The multifaceted Paneth cell. Cellular and Molecular Life Sciences. 2002;**59**(1):156-170. DOI: 10.1007/s00018-002-8412-z

[7] Elphick DA, Mahida YR. Paneth cells: Their role in innate immunity and inflammatory disease. Gut. 2005;**54**(12):1802-1809. DOI: 10.1136/gut.2005.068601

[8] Ouellette AJ. Paneth cell α-defensins in enteric immunity. Cellular and Molecular Life Sciences. 2011. DOI: 10.1007/s00018011-0714-6

[9] Salzman NH. Paneth cell defensins and the regulation of the microbiome. Gut Microbes. 2010;**1**(6):401-406. DOI: 10.4161/gmic.1.6.14076

[10] Van der Flier LG, Clevers H. Stem cells, self-renewal, and differentiation in the intestinal epithelium. Annual Review of Physiology. 2009;**71**:241-260. DOI: 10.1146/annurev.physiol.010908.163145

[11] Barker N, van Oudenaarden A, Clevers H. Identifying the stem cell of the intestinal Crypt: Strategies and pitfalls. Cell Stem Cell. 2012;**11**(4):452-460. DOI: 10.1016/j.stem.2012.09.009

[12] The intestinal crypt, a prototype stem cell compartment. Cell. 2013;**154**(2): 274-284. DOI:10.1016/j.cell.2013.07.004

[13] Sankaran-Walters S, Hart R, Dills C. Guardians of the gut enteric Defensins. Frontiers in Microbiology. 2017;**8**:647. DOI: 10.3389/fmicb.2017.00647

[14] Mathan M, Hughes J, Whitehead R. The morphogenesis of the human Paneth cell: An immunocytochemical ultrastructural study. Histochemistry. 1987;**87**:91-96. DOI: 10.1007/BF00518730

[15] Ergun E, Ergun L, Asti RN, Kurum A. Light and electron microscopic morphology of Paneth cells in the sheep small intestine. Revue de Médecine Vétérinaire. 2003;**154**:351-355. DOI: not available

[16] Umar S. Intestinal stem cells. Different cell lines.

Compartmentalization of the epithelium into a proliferative zone (crypt). Current gastroenterology. Report. 2010;**12**:340-348. DOI: 10.1007/s11894-010-0130-3

[17] Cheng H, Leblond CP. Origin, differentiation and renewal of the four main epithelial cell types in the small intestine. V. Unitarian theory of the origin of the four epithelial cell types. American. Journal of Anatomy. 1974;**141**:537-562. DOI: 10.1002/aja.1001410407

[18] Crissey MAS, Guo RJ, Funakoshi S, Kong J, Liu J, Lynch JP. Cdx2 levels modulate intestinal epithelium maturity and Paneth cell development. Gastroenterology. 2011;**140**:517-528. DOI: 10.1053/j.gastro.2010.11.033

[19] Wang WL, Xu SY, Ren ZG, Tao L, Jiang JW, Zheng SS. Application of metagenomics in the human gut microbiome. World Journal of Gastroenterology. 2015;**21**(3):803-814. DOI: 10.10.3748/wjg.v21.i3.803

[20] Stappenbeck TS, McGovern DPB. Paneth cell alterations in the development and phenotype of Crohn's disease. Gastroenterology. 2017;**152**(2):322-326. DOI: 10.1053/j.gastro.2016.10.003

[21] Al Khalidi H, Streutker CJ. Enteropathy with loss of enteroendocrine and paneth cells in a patient with immune dysregulation: A case of adult autoimmune enteropathy. Human Pathology. 2006;**37**:373-376. DOI: 10.1016/j.humpath.2005.11.019

[22] Booth C, Potten CS. Gut instincts: Thoughts on intestinal epithelial stem cells. The Journal of Clinical Investigation. 2000;**105**(11):1493-1149. DOI: 10.1172/JCI10229

[23] Sato. Paneth cells constitute the niche for Lgr5 stem cells in intestinal crypts. Nature. 2011;**2011**(469):415-418. DOI: 10.1038/nature09637

[24] Lievin-Le Moal V, Servin AL. The frontline of enteric host defense against unwelcome intrusion of harmful microorganisms: Mucins, antimicrobial peptides, and microbiota. Clinical Microbiology Reviews. 2006;**19**(2):315-337. DOI: 10.1128/CMR.19.2.315-337.2006

[25] Salzman NH, Underwood MA, Bevins CL. Paneth cells, defensins, and the commensal microbiota: A hypothesis on intimate interplay at the intestinal mucosa. Seminars in Immunology. 2007;**19**:70-83. DOI: 10.1016/j.smim.2007.04.002

[26] Cunliffe RN, Rose FR, Keyte J, Abberley L, Chan WC, Mahida YR. Human defensin 5 is stored in precursor form in normal Paneth cells and is expressed by some villous epithelial cells and by metaplastic Paneth cells in the colon in inflammatory bowel disease. Gut. 2001;**48**(2):176-185. DOI: 10.1136/gut.48.2.176

[27] Di Sabatino A, Miceli E, Dhaliwal W, et al. Distribution, proliferation, and function of Paneth cells in uncomplicated and complicated adult celiac disease. American Journal of Clinical Pathology. 2008;**130**:34-42. DOI: 10.1309/5ADNAR4VN11TTKQ6

[28] Schroder JM. Epithelial antimicrobial peptides: Innate local host response elements. Cell and Molecular Life Science. 1999;**56**:32-46. DOI: 10.1007/s000180050004

[29] Bevins CL. Events at the host-microbial interface of the gastrointestinal tract. V. Paneth cell alpha-defensins in intestinal host defense. American Journal of Physiology. Gastrointestinal and Liver

Physiology. 2005;**289**(2):G173-G176. DOI: 10.1152/ajpgi.00079.2005

[30] Kim TH, Shivdasani RA. Stem cell niches: Famished Paneth cells, gluttonous stem cells. Current Biology. 2012;**22**(14):R580. DOI: 10.1016/j.cub.2012.06.017

[31] Henzler-Wildman KA, Lee D, Ramamoorthy A. Mechanism of lipid bilayer disruption by the human antimicrobial peptide, LL-37. Biochemistry. 2003;**42**:6545-6558

[32] Henzler-Wildman KA. Perturbation of the hydrophobic core of lipid bilayers by the human antimicrobial peptide LL-37. Biochemistry. 2004;**43**:8459-8469. DOI: 10.1021/bi036284s

[33] Bjerknes M, Cheng H. Gastrointestinal stem cells II. Intestinal stem cells. American journal of physiology gastrointestinal and liver. Physiology. 2005;**289**:G381-G387. DOI: 10.1152/ajpgi.00160.2005

[34] Mantani Y, Nishida M, Yuasa H, et al. Ultrastructural and histochemical study on the Paneth cells in the rat ascending colon. The Anatomical Record. 2014;**297**:1462-1471. DOI: 10.1002/ar.22937

[35] Mehandru S, Poles MA, Tenner-Racz K, Manuelli V, Jean-Pierre P, Lopez P. Mechanisms of gastrointestinal CD4+ T-cell depletion during acute and early human immunodeficiency virus type 1 infection. Journal of Virology. 2007;**81**(2):599-612. DOI: 10.1128/JVL.01739-06

[36] Liu TC, Gurram B, Baldridge MT, et al. Paneth cell defects in Crohn's disease patients promotes dysbiosis. JCI Insight. 2016;**1**(8):e86907. DOI: 10.1172/jci.insight.86907

[37] Guadalupe M, Reay E, Sankaran S, Prindiville T, Flamm J, McNeil A, et al. Severe CD4+ T-cell depletion in gut lymphoid tissue during primary human immunodeficiency virus type 1 infection and substantial delay in restoration following highly active antiretroviral therapy. Journal of Virology. 2003;**77**(21):11708-11717. DOI: 10.1128/JVI.77.21.11708-11717.2003

[38] Hodin CM, Lenaerts K, Grootjans J, et al. Starvation compromises Paneth cells. The American Journal of Pathology. 2011;**179**(6):2885. DOI: 10.1016/j.ajpath.2011.08.030

[39] Cazorla SI, Maldonado-Galdeano C, Weill R, De Paula J, Perdigon GDV. Oral Administration of Probiotics Increases Paneth Cells and Intestinal Antimicrobial Activity. Frontiers in Microbiology. 2018;**9**:736. DOI: 10.3389/fmicb.2018.00736

[40] Al-Assal K, Martinez AC, Torrinhas RS, Cardinelli C, Waitzberg D. Gut microbiota and obesity. Clinical Nutrition Experimental. 2018;**20**:60-64. DOI: 10.1016/j.yclnex.2018.03.001

[41] Zevin AS, McKinnon L, Burgener A, Klatt NR. Microbial translocation and microbiome dysbiosis in HIV-associated immune activation. Current Opinion. 2016;**11**(2):182-190. DOI: 10.1097/COH.0000000000000234

[42] Routy JP, Mehraj V. Potential contribution of gut microbiota and systemic inflammation on HIV vaccine effectiveness and vaccine design. AIDS Research and Therapy. 2017;**14**:48. DOI: 10.1186/s12981-017-0164-9

[43] Lu W, Feng Y, Jing F, Han Y, Lyu N, Liu F, et al. Association between gut microbiota and CD4 recovery in HIV-1 infected patients. Frontiers in

Microbiology. 2018;**9**:1461. DOI: 10.3389/fmicb.2018.01451

[44] Villaneuva-Millan MJ, Perez-Matute P, Recio-Fernandez E, Rosales JML, Oteo JA. Differential effects of antiretrovirals on microbial translocation and gut microbiota composition of HIV-infected patients. Journal of the International AIDS Society. 2017;**20**:21526. DOI: 10.7448/IAS.20.1.21526

[45] Pinto-Cardoso S, Klatt NR, Reyes-Teran G. Impact of antiretroviral drugs on the microbiome: Unknown answers to important questions. Current Opinion in HIV and AIDS. 2018;**13**:53-60. DOI: 10.1097/COH.0000000000000428

[46] Wehkamp J, Stange EF. An update review on the Paneth cell as key to Ileal Crohn's disease. Frontiers in Immunology. 2020;**11**(646):1-9. DOI: 10.3389/fimmu.2020.00646

[47] Horvath K, Papadimitriou JC, Rabsztyn A, Drachenberg C, Tildon JT. Gastrointestinal abnormalities in children with autistic disorder. The Journal of Pediatrics. 1999;**135**(5):559-563. DOI: 10.1016/s0022-3476(99)70052-1

[48] Rajasuriar R, Khoury G, Kamarulzaman A, French MA, Cameron PU, Lewin SR. Persistent immune activation in chronic HIV infection: Do any interventions work? AIDS. Wolters Kluwer; 2013;**27**(8):1199-1208. DOI: 10.1097/QAD.0b013e32835ecb8b

[49] Rajasuriar R, Wright E, Lewin SR. Impact of antiretroviral therapy (ART) timing on chronic activation/inflammation and end-organ damage. Current Opinion in HIV and AIDS. 2015;**10**(1):35-42. DOI: 10.1097/COH.0000000000000118

[50] Routy JP, Mehraj V, Cao W. HIV immunotherapy comes of age: Implications for prevention, treatment and cure. Expert Review of Clinical Immunology. 2015;**12**(2):91-94. DOI: 10.1586/1744666X.2016.1112269

[51] Ribeiro ABDTM, Heimesaat MM, Bereswill S. Changes of the intestinal microbiome-host homeostasis in HIV-infected individuals-a focus on bacterial gut microbiome. European Journal of Microbiology and Immunology. 2017;**7**(3):158-167. DOI: 10.1556/1886.2017.00016

[52] Yongjia J, Zhang F, Zhang R, et al. Changes in intestinal microbiota in hIV-1-infected subjects following cART initiation: Influence of CD4++ T cell count. Emerging Microbes & Infections. 2018;**7**:113. DOI: 10.1038/s41426-018-0117-y

[53] El-Far M, Tremblay CL. Gut microbial diversity in HIV infection post combined antiretroviral therapy: A key target for prevention of cardiovascular disease. Current Opinion in HIV and AIDS. 2018;**13**:38-44. DOI: 10.1097/COH.0000000000000426

[54] Sherman MP, Bennet SH, Hwang FFY, Sherman J, et al. Paneth cells and antibacterial host defense in neonatal small intestine. Infection and Immunity. 2005;**73**(9):6143-6146. DOI: 10.1128/IAI.73.9.6143-6146.2005

[55] Wehkamp J, Salzman NH, Porter E, Nuding S, Weichenthal M, Petras RE, et al. Reduced Paneth cell α-defensins in ileal Crohn's disease. PNAS. 2005;**102**(50):18129-18134. DOI: 10.1073/pnas.0505256102

[56] Sanders DSA. Mucosal integrity and barrier function in the pathogenesis of early lesions in Crohn's disease. Journal

of Clinical Pathology. 2005;**17**:568-572. DOI: 10.1136/jcp.2004.021840

[57] Simms LA, Doecke JD, Walsh MD, Huang N, et al. Reduced (alpha)-defensin expression is associated with inflammation and not NOD2 mutation in ileal Crohn's disease. Gut. 2008;**27:57**(7):903-910. DOI: 10.1136/gut.2007.142588

[58] Tseng CH, Wu CY. The gut microbiome in obesity. Journal of the Formosan Medical Association. 2018;**11**(1):S3-S9. DOI: 10.1016/j.jfma.2018.07.009

[59] Qin Y, Roberts JD, Grimm SA, et al. An obesity-associated gut microbiome reprograms the intestinal epigenome and leads to altered colonic gene expression. Genome Biology. 2018;**19**:7. DOI: 10.1186/s13059-018-1389-1

[60] Lueschow SR, Stumphy J, Gong H, Kern SL, Elgin TG, Underwood MA, et al. Loss of murine Paneth cell function alters the immature intestinal microbiome and mimics changes seen in neonatal necrotizing enterocolitis. PLoS One. 2018;**13**(10):e0204967. DOI: 10.1371/journal.pone.0204967

[61] Yu J. Intestinal stem cell injury and protection during cancer therapy. Translational Cancer Research. 2013;**2**(5):384-396 DOI: not available

[62] Szumilo J, Swatek J, Chroscicki A, Dudka J, Korobowicz E. Colonic adenocarcinoma with numerous Paneth cells and endocrine cells. Polish Journal of Pathology. 2005;**56**:89-92 DOI: not available

[63] Lima AA, Silva TMJ, Gifoni AMR. Mucosal injury and disruption of intestinal barrier in HIV-infected individuals with and without diarrhoea and cryptosporidiosis in Northeast Brazil. The American Journal of Gastroenterology. 1997;**92**:1861-1866 DOI: not available

[64] Hirao LA, Grishina I, Bourry O, et al. Early mucosal sensing of SIV infection by Paneth cells induces IL-1βProduction and initiates gut Epithehelial disruption. PLOS Pathogens. 2014;**10**(8):e1004311. DOI: 10.1371/journal.ppat.1004311

[65] Kelly P, Feakins R, Domizio P, et al. Paneth cell granule depletion in the human small intestine under infective and nutritional stress. Clinical and Experimental Immunology. 2004;**135**:303-309. DOI: 10.1111/j.1365-2249.2004.02374

[66] Berkhout DJC, Klaassen P, Niemarkt HJ, de Boode WP, Cossey V, van Goudoever JB, et al. Risk factors for necrotizing enterocolitis: A prospective multicenter Case_Control study. Neonatology. 2018;**114**:277-284. DOI: 10.1159/000489677

[67] Mukerjee S, Hooper LV. Antimicrobial defense of the intestine. Immunity. 2015;**42**:28. DOI: 10.1016/j.immuni.2014.12.028

[68] Nino DF, Sodhi CP, Hackam DJ. Necrotizing enterocolitis: New insights into pathogenesis and mechanisms. Nature Reviews. Gastroenterology & Hepatology. 2016;**13**(10):590-600. DOI: 10.1038/nrgastro.2016.119

[69] Kwaan MR, Jones-Webb R. Colorectal cancer screening in black men: Recommendations for best Practises. American Journal of Preventive Medicine. 2018;**55**(5S1):S95-S102. DOI: 10.1016/j.amepre.2018.05.008

[70] Zou S, Fang L, Lee MH. Dysbiosis of gut microbiota in promoting the development of colorectal cancer. Gastroenterology Report. 2018;**6**(1):1-12. DOI: 10.1093/gastro/gox031

[71] Sharma S, Tripathi P. Gut microbiome and type 2 diabetes: Where we are and where to go? Journal of Nutritional Biochemistry. 2018;**63**:101-108. DOI: 10.1016/j.jnutbio.2018.10.003

[72] Bhurosy T, Jeewon R. Overweight and obesity epidemic in developing countries: A problem with diet, physical activity, or socioeconomic status? The Scientific World Journal. 2014;**2014**:964236. DOI: 10.1155/2014/964236

[73] Ford ND, Patel SA, Narayan KMV. Obesity in low- and middle-income countries: Burden, drivers, and emerging challenges. Annual Review of Public Health. 2017;**38**:145-164. DOI: 10.1146/annurev-publhealth-031816-044604

[74] Clements WDB, McCaigue M, Erwin P, Halliday I, Rowlands BJ. Biliary decompression promotes Kupffer cell recovery in obstructive jaundice. Gut. 1996;**38**:925-931. DOI: 10.1136/gut.38.6.925

[75] Assimakopoulos SF, Scopa CD, Vagianos CE. Pathophysiology of increased intestinal permeability in obstructive jaundice. World Journal of Gastroenterology. 2007;**13**(48):6458-6464. DOI: 10.3748/wjg.v13.i48.6458

[76] Welsh FKS, Ramsden CW, MacLennan K, Sheridan MB, Barclay GR, Guillou PJ, et al. Increased intestinal permeability and altered mucosal immunity in Cholestatic jaundice. Annals of Surgery. 1998;**227**(2):205-212. DOI: 10.1097/00000658-199802000-00009

[77] Parks RW, Halliday MI, McCrory DC, Erwin P, Smye M, Diamond T, et al. Host immune responses and intestinal permeability in patients with jaundice. British Journal of Surgery. 2003;**90**(2):239-245. DOI: 10.1002/bjs.4029

[78] Vaishnava S, Behrendt CL, Ismail AS, Eckmann L, Hooper LV. Paneth cells directly sense gut commensals and maintain homeostasis at the intestinal host-microbial interface. PNAS. 2008;**105**(52):20858-20863. DOI: 10.1073/pnas.0808723105

Section 2

Dysbiosis, Helicobacter Pylori

Chapter 4

Dysbiosis, Tolerance, and Development of Autoimmune Diseases

Gratiela Gradisteanu Pircalabioru, Octavian Savu, Grigore Mihaescu, Corneliu Ovidiu Vrancianu and Mariana-Carmen Chifiriuc

Abstract

The pathogenesis of autoimmune diseases (AIDS) is not only attributed to genetic susceptibility, but also to environmental factors, among which, those disturbing gut microbiota have attracted increasing attention lately. Healthy gut microbiota has beneficial effects on the development and activity of the immune system, playing a central role in peripheric tolerance. Compositional and functional changes in gut microbiota were reported in various AIDS, and increasing evidence suggests that disturbed gut microbiota contributes to their immunopathogenesis. Thyroid and intestinal diseases prevalently coexist—for instance, Hashimoto's thyroiditis and Graves' disease are the most common autoimmune thyroid diseases and often co-occur with celiac disease. This association can be at least explained by increased intestinal permeability, allowing antigens to cross the barrier more easily and activate the immune system. The passage of microbial antigens into the internal environment may break the self-tolerance, generating the production of autoantibodies and/or autoreactive T cells. In this chapter, we briefly present the roles of intestinal microbiota in human physiology, with a focus on the role of microbiota in immune tolerance.

Keywords: microbiome, gut immunity, dysbiosis, immune tolerance, autoimmunity

1. Introduction

Immune tolerance is a physiological condition, characterized by the absence of an immune response to a specific antigen and the retention of the ability to develop an immune response to other different antigens. Tolerance to self-components develops both during embryonic development (i.e., central tolerance, which occurs in the primary lymphoid organs, along with the process of lymphocyte differentiation), and after birth (i.e., peripheral tolerance) [1].

Currently, the microbiota is considered an anatomically integrated meta-organ that performs functions through which it interferes with the host's physiology [2]. Thus, microbiota eubiosis is a major parameter of physiological homeostasis.

Human microbiota establishes three types of relationships with the host—symbiotic, commensal, and pathobiontic, respectively [3]. The terms "microbiota" and "microbiome" are equivalent, but not identical. The first refers to the population of microorganisms residing on the mucous membranes of the digestive, urogenital and respiratory tract, as well as on the skin, and the second designates the collective genome of the microbiota, called the metagenome [4]. The community microbiome was evaluated at 3.3 million redundant bacterial genes, about 150 times larger than the human gene complex [5]. The gut microbiota is influenced by various conditions, such as diet, health, mental stress, gender, or exercise, and conversely, it influences all body metabolism, immune reactivity, and behavior [6]. The microbiota contributes to the peripheral tolerance of the immune system toward autoantigens, with the retention of the immune reactivity against all antigens that do not cross-react with the tolerated antigen. Interruption of tolerance initiates an immune response to self-antigens characterized by the production of autoantibodies or autoreactive lymphocytes, which trigger an autoimmune conflict [7]. The purpose of this chapter is to highlight the role of the normal microbiota in the state of immune tolerance and to investigate the correlations of dysbiosis with endocrine AIDS.

2. The role of the intestinal microbiota in the physiology of the human body

The microbiota of the digestive tract consists of about 3×10^{13} to 40×10^{13} (3–40 trillion) bacterial cells, counting at least 10 times more than their host cells. The groups of *Bacteroidetes* and *Firmicutes* predominate in a numerical proportion of 90%, aside from lower density of *Proteobacteria, Actinobacteria, Fusobacteria*, and *Verrucomicrobia* and small populations of fungi, *Archaea*, and viruses, all exerting major functional effects on different organs. The bacterial microbiota belongs to 1000–1160 types of species [8]. The individual microbiota is evaluated in 150 to 160 species by the 16S RNA (rDNA analysis) ribotyping method [9]. The population composition of the intestinal microbiome stabilizes at the age of 3 years and is determined by various conditions, such as genetic factors, the maternal microbiota, the mode of birth (i.e., natural or by cesarean section), the antigenic exposure during early life, and is reconfigured mostly by diet [10, 11].

The microbiota is considered a virtual organ, whose functions must be integrated into general physiology. The host-microbiota interaction is primarily a symbiotic relationship, in which the host organism provides the ecological niche and nutrients for microbiota survival. The microbiota carries out fermentative and biosynthetic metabolic activities, thereby influencing systemic physiology [12]. The metabolism of the microbiota functions as a bridge between the diet with the human body. The intestinal microbiota increases the energy efficiency of the diet by fermenting the fibrous components, providing essential metabolites for organ systems, especially short-chain fatty acids (SCFA), such as acetic, propionic, and butyric acid. A proportion of 50% of the energy needs of epithelial cells is provided by SCFA [13, 14]. The modern diet is 7–10 times poorer in the fibrous component, compared to the traditional Mediterranean one. Microbiota synthesizes vitamin K and B, synthesizes amines through which it modifies endocrine function, stimulates the inflammatory process, has a protective role against the invasion of enteric pathogens (*Shigella flexneri*), metabolizes some drugs to their active form, ferments indigestible components of the diet (complex polysaccharides, amino acids, xenobiotics) [15], and modulates the lipid metabolism. The bile acids synthesized in the liver from cholesterol, facilitate the absorption of lipids and fat-soluble vitamins and maintain cholesterol balance. Also, the biliary acids have a signaling function through specific hepatocyte receptors [16–19]. All these functions are impaired in patients with endocrine AIDS.

3. Role of the microbiota in the development of the mucosa-associated immune system

From an immunological perspective, the mucous membranes which cover a total area of about 400 m^2, represent both an anatomical and functional entity, because they are populated by a large number of immune cells.

The intestinal microbiota, epithelium and digestive, respiratory, genital, urinary mucosa-associated immune system form a functional triad whose components influence each other close interactions, with a rapid dynamic of change, induced by population changes of the microbiota, due to diet variation and/or administration of antibiotics. The modification of the functional parameters of a component of the triad has major influences on the physiology of the whole organism. The microbiota interacts directly with the epithelium of the adjacent mucosa and influences its permeability, and both local and systemic inflammatory responses [10] The interaction of the microbiota with the mucosal immune system (gut-associated lymphoid tissue—GALT) induces the synthesis of a wide set of cytokines, with local regulatory action of intestinal physiology [20, 21].

The microbiota has an essential role in the functional modulation (education), first of all of the GALT structures. Germ-free and gnotobiotic animal studies have made a decisive contribution to understanding the functional relationships of the microbiota-epithelium-immune system triad and provided new evidence for the role of the intestinal microbiota as a whole, but also of different groups of bacteria in the functional development and maturation of the systemic immune system, especially GALT. Germ-free mice have structural and functional defects of the immune system—decreased TCD4 lymphocyte count and Th-2 predominance in the spleen, altered Th-17 and T-reg differentiation in the lamina propria, and restoration of deficiencies after colonization with *Bacteroides* and segmented filamentous bacteria (SFB). The balance of effector T lymphocytes is disturbed in intestinal dysbiosis and accelerates or suppresses the autoimmune reactions [22, 23]. The constant interaction of the microbiota with the cells of adaptive immunity prevents bacterial invasion and pathogenesis, but also the systemic immune response with detrimental effects against the microbial antigens [24]. The structural but especially functional peculiarities of GALT tend to delimit it more and more from the systemic immune system.

M cells that cover the subepithelial immune structures engulf the luminal antigens, through the mechanism of pinocytosis and transfer them unaffected to the immune structures in the underlying follicles (i.e., macrophages, dendritic cells, T and B lymphocytes). Macrophages and dendritic cells respond to microbiota antigens in a nonspecific manner by TLR recognition followed by cytokines release (i.e., IFNα, IL-18, and IL-22), which stimulate the epithelial cells to synthesize antimicrobial peptides.

The microbiota, through the composition of bacterial phyla, has a major influence on the development of T lymphocyte subpopulations and in maintaining the numerical balance of Th-2/Th-1 lymphocyte populations in lymphoid organs. The differentiation of T lymphocyte sets is influenced by the antigenic specificity of the dominant bacterial population and its metabolic properties—(i) some bacteria stimulate the predominant differentiation of proinflammatory TCD4 lymphocytes that synthesize IFNγ and IL-17A [25]; and (ii) others stimulate the differentiation of regulatory CD25+ and Foxp3+ TCD4 lymphocytes (T-reg), the essential mediator of immune tolerance by decreasing Th-17 lymphocytes [26, 27]. The direct relationship between the concentration of butyric acid and the number of T-reg lymphocytes is well known. SCFA, particularly butyric acid harbor important roles, that is, stimulate gene transcription for mucin synthesis, strengthen the intestinal barrier and render it impermeable to toxins and bacterial cell translocation, thus preventing chronic systemic inflammation, inhibiting the synthesis of pro-inflammatory interleukins (IL) (TNFα and IL-6) induced by LPS and regulate the innate and adaptive immunity [13, 14]. Th-17 lymphocytes play an essential role in anti-bacterial and anti-fungal defense, but at the same time have an important role in the initiation of inflammatory diseases, through the synthesis of pro-inflammatory IL-17 and IL-22 and the recruitment of neutrophils. In germ-free animals, the lamina propria is populated by a very small number of Th-17 lymphocytes [9]. Th-17 lymphocytes also decrease after antibiotic treatment [27]. The group of *Clostridium* SFB, following the colonization of the epithelium, induces an increase in the number of Th-17 lymphocytes, whose proinflammatory IL can promote the onset of rheumatoid arthritis and multiple sclerosis in gnotobiotic animals. In patients suffering from inflammatory bowel disease, which manifests clinically similar to Crohn's disease and ulcerative colitis, the number of T-reg lymphocytes with immunosuppressive function decreases in the lamina propria and the population of lymphocytes that have TCR for the bacterial microbiota antigens increases abnormally. The density of T-reg lymphocytes increases in gnotobiotic animals colonized by *Clostridium* SFB group, while the polysaccharide A of *Bacteroides fragilis* (which is attributed to symbiotic factor status) induces the differentiation of TCD4 lymphocytes to T-reg lymphocytes [9, 28]. TCD4 Foxp3 + lymphocytes secrete IL-10, the main anti-inflammatory cytokine, thus being involved in tolerance to microbial antigens. In germ-free animals, the dominance of Th-2 subpopulation in the spleen that favors allergic manifestations is restored by the polysaccharide A of *Bacteroides fragilis*.

The microbiota has also a profound influence on the development of B lymphocytes—it stimulates the synthesis of antibodies, especially of IgA type, targeted against thymus-dependent (Td), and thymus-independent (Ti) antigens. The *Clostridium* SFB and *Alcaligenes* group of bacteria are potential inducers of IgA synthesis specific for the intestinal microbiota antigens. In the absence of IgA, the *Clostridium* group is enriched, whereas the *Alcaligenes* group is diminished [9].

In germ-free animals, GALT structures play a key role in inducing immune tolerance against auto-antigens from the intestinal mucosa, are less developed and indicators of immune response activation are lacking. In these animals, the number of TCD4 lymphocytes and IgA-secreting plasma cells decreases in Peyer's patches, while in the spleen and lymph nodes, the number of B lymphocytes and germinal centers decreases.

In conclusion, the development, maturation, and function of the immune system are closely associated with the level of exposure to microbial antigens during early life, and as an opposite, insufficient exposure to various antigens increases the risk of autoimmune disorders occurrence [29].

4. Intestinal microbial antigens as inductors of central and peripheral tolerance

Despite its much-diversified antigen panel, the microbiota is tolerated by the immune system. Central tolerance is induced during fetal life, as immature lymphocytes are exposed to various antigenic peptides, and is essentially dependent on the specific process of antigenic peptide selection and presentation in association with the Human Leucocytes Antigen/ Major Histocompatibility Complex (HLA/MHC) molecules [30]. The occurrence of peripheral tolerance breaks results from a functional adaptation of the immune system to specific antigenic peptides that have not (sufficiently) been exposed to lymphocytes in the bone marrow or thymus during embryonic development. It is now considered that the T lymphocyte antigen receptor (TCR) is the major mediator of immune tolerance. That is why, from an evolutionary perspective, TCR recognizes both the genetic and microbial self [31].

The immune tolerance to commensal intestinal microbiota is peripheral and results from both an immediate neonate colonization of the digestive tract and a progressive co-evolution in which the interactions of gut-associated lymphoid tissue (i.e., GALT) with bacterial antigens have been modulating innate and adaptive primarily local immune reactivity. Commensal antigens, on contact with the intestinal mucosa, induce the state of tolerance, in which dendritic cells play an essential role, while the effectors are the epithelial cells with their covering molecular complex (i.e., antimicrobial peptides, mucin layer, surface immunoglobulin A—sIgA) [32]. Bacterial cells or their components (i.e., lipopolysaccharides, polysaccharides, peptidoglycans, teichoic acids, and DNA) that cross the intestinal barrier and reach the internal environment, activate the immune response [33].

4.1 Causes of losing immune tolerance to microbiota antigens

Interruption of immune tolerance to microbiota antigens is determined by several factors—genetic factors, the host's immune system, disturbance of the diversity, and physiology of the microbiota—as triggering events [34].

Mechanisms that modulate immune tolerance loss to the intestinal microbiota include: (i) abnormal translocation of bacteria in the internal environment due to permeability of the intestinal barrier, (ii) antigenic similarity of some bacterial peptides with epithelial molecules. Immune cells are activated by bacterial peptides and become autoreactive; and (iii) disorder of local and systemic immunity under the stimulating action of some bacterial derivatives (nucleic acids, polysaccharides, metabolites, and toxins). Aberrant activation of the immune system leads to the excessive synthesis of proinflammatory IL (IFN type I, IL-12, IL-23) and a decreased rate of synthesis of anti-inflammatory cytokines (IL-10, TGF-β—transforming growth factor) (**Figure 1**) [35].

4.2 Consequences of losing immune tolerance

Although the autoimmune conflict occurs most of the time without clinical manifestations, it can generate under certain conditions, such as AIDS, that are characterized by the appearance of tissue lesions or disruption of physiological processes. AIDS have a multifactorial etiology involving genetic, epigenetic, and environmental factors. It is estimated that 70% of AIDS are due to environmental factors [36]. Among the multiple cellular and molecular mechanisms, yet not well

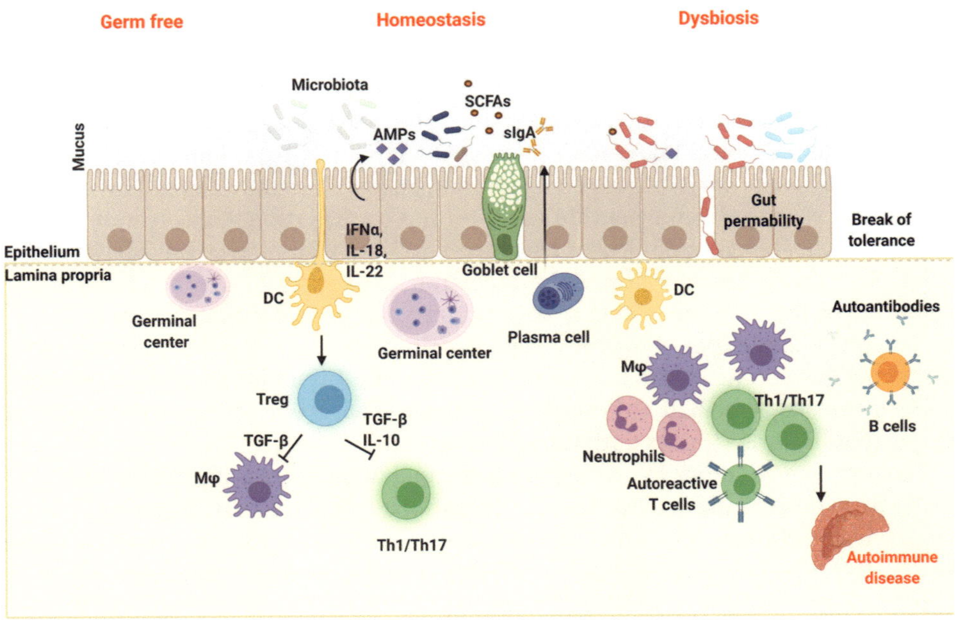

Figure 1.
The role of microbiota in mucosal homeostasis and immunological tolerance in healthy gut and activated inflammatory cascades in endocrine autoimmune disease. In germ-free animals, GALT structures are less developed and the microbiota has a major influence on the development of T lymphocyte subpopulations and in maintaining the numerical balance of Th-2/Th-1 lymphocyte populations in lymphoid organs. The healthy gut environment is characterized by high levels of antimicrobial peptides and metabolites (SCFAs), and the commensal-specific IgA is produced by plasma cells in the lamina propria, mediated by DCs in a T cell-independent mechanism. During homeostasis, gut microorganisms induce an immune tolerance phenotype in the host, whilst in inflammatory conditions, antigens from dysbiotic microorganisms activate Th1 and Th17 cells leading to decreased mucus layer, tissue injury, and microbial penetration and persistence in the intestinal tissues. This mucosal injury results in further uptake of microbial antigens that further perpetuate detrimental immune responses. Figure created with https://biorender.com/.

established, by which the state of immune self-tolerance is disturbed, we can mention—(i) the genetic predisposition that may explain the familial character of AIDS, which, in general, have a polygenic determinism. The risk of a certain autoimmune disease for monozygotic twins is about 12 to 60%, and for dizygotic twins is 5%. The most important are certain specific polymorphisms generated by the change of a nucleotide, that is, SNP (single nucleotide polymorphism) in MHC genes [9]. For example, over 90% of Caucasians with ankylosing spondylitis express an allele of the HLA-B27 family, differing from that of normal individuals by two amino acids located in the peptide binding groove [37]; (ii) release of sequestered antigens after trauma, surgery, infectious processes, etc., become accessible to lymphocytes, triggering the autoimmune conflict and tissue damage (e.g., basic myelin protein in the central nervous system becomes the antigenic target in multiple sclerosis; crystalline proteins induce autoimmune ophthalmopathy; sperm proteins, in cases of sperm stasis, induce the synthesis of immobilizing or binder autoantibodies of sperm, leading to autoimmune infertility) [38, 39]; (iii) modification of the chemical structure of autoantigens (so-called altered self-theory), which occurs under the influence of some physical factors (such as burns or radiation), biological (i.e., bacteria, viruses, fungi), or chemical (i.e., drugs, alcohol) factors, with the exposure of some new antigenic determinants

[40]; (iv) infectious agents, which may have an important role in triggering AIDS by various mechanisms, such as the antigenic resemblance of non-self to self-molecules and their cross-reactivity (e.g., protein M from *Streptococcus pyogenes* is antigenically similar to cardiomyocyte's membrane proteins); (v) stimulation of the proinflammatory cytokines production that cause nonspecific activation of self-reactive immune cells; superantigens of infectious agents (i.e., Epstein-Barr virus, mycoplasmas, *Staphylococcus aureus*, *Streptococcus pyogenes*) that induce polyclonal activation of lymphocytes [41]; (vi) loss of peripheral immune tolerance, due to either mutation that generates the appearance of immunocompetent, self-reactive T or B lymphoid cell clones, or T-reg cell deficiency, or Th cell activation; (vii) disruption of the equilibrium state of the idiotypic network by the synthesis of anti-idiotypic antibodies, which may be autoantibodies [42]; and (viii) hormonal imbalances that may be involved in triggering AIDS, therefore explaining their increased frequency in women (8: 1 ratio) except for ankylosing spondylitis, or in men with higher levels of estrogen hormones. Moreover, pregnancy is associated with an improvement in the severity of AIDS, especially in rheumatoid arthritis cases.

AIDS resemble some general features—the pathological process has an individual intensity, dynamics, and evolution, may overlap with the same patient, and are rare in childhood, except for type 1 diabetes mellitus.

Regardless of the triggering mechanism, AIDS is characterized by the synthesis of autoantibodies (that are antibodies specific to self-tissue components) or by the generation of autoreactive T lymphocytes. Tissue injuries following the action of immune effectors occur in one of the above-mentioned scenarios—(i) autoantibodies recognize the tissue antigens and form immune complexes, the complement is activated, and the result is the cell lysis, or (ii) indirect action, in which case, the antigen-antibody—complement immune complexes are deposited in small vessels (arterioles, capillaries) from various organs and produces inflammatory reactions, with the consequence of tissue destruction; the AIDS that are mediated by various antibodies have a common feature, that is the target tissue is damaged by a chronic inflammatory reaction without a known infectious cause; and (iii) the lesions in the target tissue occur under the action of infiltrated Tc lymphocytes [43, 44].

Some AIDS are characterized by strictly localized pathological processes, that is, effectors (especially antibodies) have specific action against antigens specific to the target tissue (such is the case for autoantibodies specific only to B cells from Langerhans islands in type 1 diabetes mellitus, or autoantibodies specific to thyroid epithelial cell in Hashimoto's thyroiditis) [45], sometimes the lesions are localized in a single organ, but autoantibodies do not have organ specificity (for instance, anti-mitochondrial antibodies in primary cirrhosis, or type IV anti-collagen autoantibodies in Goodpasture syndrome) [46, 47] while some AIDS are disseminated, characterized by the synthesis of autoantibodies to antigens with wide tissue distribution (e.g., antinuclear antibodies in systemic/disseminated lupus erythematosus) [48].

Often, in pathological cases, the body synthesizes auto-antibodies specific for components of the endocrine system, especially antibodies specific for a certain hormone receptor. The pathophysiological effects of these antibodies generated against hormone receptors are varied—they can stimulate the activity of the receptor, and the effect is to intensify the secretory activity of the gland (hormonal mimetic effect) or block the receptor, and the effect is to inhibit the secretory activity. Both antibodies can coexist in the same patient.

5. The role of microbiota in autoimmune-mediated endocrine diseases

The role of the microbiota in autoimmune pathology has been highlighted by experimental data collected from germ-free mice. The intestinal microbiota maintains the balance of protective reactions to pathogens and tolerance to commensals aimed at maintaining intestinal homeostasis [49, 50]. Alterations produced in the balance of the microbiota (that is dysbiosis) activate the proinflammatory immune response and favor the progression of autoimmune disorders, such as multiple sclerosis, inflammatory bowel disease, T1DM, rheumatoid arthritis, and other pathologies of the digestive tract and ancillary glands, including malignancies. However, the intimate mechanism of microbiota involvement in this pathogenesis remains unknown [51–53].

AIDS are caused primarily by predisposing genetic factors but also by other endogenous or environmental triggers. There is a permanent interaction of the local immune system with bacterial antigens in humans, and therefore dysbiosis of the microbiome is associated with autoimmune disorders and metabolic syndromes. Dysbiosis means, in fact, the numerical alteration, diversity, and physiology of the intestinal microbiota (the transcriptome, proteome, and metabolome change) [54].

Experimental results in germ-free or induced dysbiotic animals support either the microbiota's direct or indirect involvement in the pathogenesis of some AIDS. Hence, in patients with type 1 diabetes mellitus, rheumatoid arthritis, multiple sclerosis, or lupus, as in those suffering from inflammatory bowel disease (both Chron's disease and ulcerative colitis), Sjögren's syndrome, Behcet's disease, autoimmune skin diseases (such as vitiligo, psoriasis, atopic dermatitis), the digestive microbiota is altered in terms of diversity and numerical representation of some species [9, 18, 35]. Kriegel et al. consider that dysbiosis is an essential trigger of autoimmunity both at the mucosal and systemic levels [9]. The spread of autoimmune response seems to be generated either by disseminating bacterial antigens but mostly by cross-immune reactivity under homeostasis conditions [55]. Such a mechanism is supported by a rheumatic fever induced by M and SLO antigens of *Streptococcus pyogenes*, or Guillain-Barre syndrome induced by *Campylobacter jejuni* infection, both as transient autoimmune syndromes. Cross-reactivity of lipopolysaccharides, bacterial polysaccharides, or D amino acid polymers would be an important mechanism for initiating the autoimmune conflict. Patients with autoimmune disorders often have vitamin D deficiency; its administration in experimental settings to animals improves the course of the disease. Vitamin D deficiency is also associated with an increased risk of infectious diseases. Inflammatory cells convert vitamin D to its active form, which is calcitriol. Vitamin D is an essential factor for the activation and proliferation of inflammatory cells (macrophages, neutrophils) [56]. The probiotics could also reverse the chronic systemic inflammation associated with AIDS.

5.1 Type I diabetes mellitus

Type 1 diabetes mellitus (T1DM) has a well-defined autoimmune component, characterized by selective immune aggression against β-cells that secrete insulin [57]. The genetic predisposition for T1DM is unanimously accepted, but the interaction of genetic factors with environmental ones explains the sudden increase in incidence in Western countries [58]. More than 50% of monozygotic twins who have a sibling with T1DM remain healthy, showing that environmental factors (such as infectious agents, consumption of cow's milk in early childhood, or ingestion of contaminated food) play a major role in triggering the disease. Hence, out of 50 individuals suffering from congenital rubella virus infection [59, 60], nine developed diabetes at an average age of 28 years. However, some infections (i.e., *M. tuberculosis*, viruses, or parasites) exert a nonspecific inhibition on the onset of T1DM, probably by stimulating regulatory T cells [61–64].

The pathological mechanisms leading to the autoimmune destruction of pancreatic beta-cells in T1DM are very complex and incompletely elucidated. The pancreatic beta-cells express MHC II and co-stimulatory molecules, suggesting their role as antigen-presenting cells to TCD4 cells. Auto-antigens that stimulate the specific immune reactivity against pancreatic beta-cell are represented by insulin, glutamic acid decarboxylase—isoform 2 of 65 kD from beta-cell cytoplasm, a Zn transporter protein (ZnT8) involved in active secretion of insulin from islet granules, insulinoma-associated antigen 2 (alpha and beta), and a membrane protein acting as tyrosine phosphatase. The presence of humoral autoimmunity defines the risk of T1DM; antibodies against insulin were identified in 40% of children with the overt disease [65].

In patients with T1DM, it has been shown by immunohistochemical staining that the islets are infiltrated with macrophages, dendritic cells, TCD4, TCD8, NK, and fewer B lymphocytes, which can act as antigen-presenting cells for TCD4 cells. The immune response against islet antigens is associated with an inflammatory one in which IL-1, TNFα, and IFNγ are released [66]. The immune and inflammatory process destroys the beta cells. When about 80% of the beta-cell mass has been destroyed, the disease overt. This silent period may last for several years, sometimes decades. Along with the progressive destruction of β cells, the humoral antibody response and decreased glucose tolerance are documented until the clinical onset of the disease. Immune effectors selectively lyse insular β cells, leaving the other cell types intact. After the onset of hyperglycemia, the degree of mononuclear infiltration decreases [67].

The inflammatory diseases of the pancreas (such as chronic pancreatitis, neoplasia) are characterized by mast cells infiltrates into the acinar parenchyma, which releases various proteases (chymase, tryptase), acting as direct destroyers on islet's beta cells. The B4 type of leukotrienes, which derives from mast cells, exerts a chemoattractant effect on T lymphocytes [68].

Loss of pancreatic beta cells leads to insulin secretion deficiency, while the glucagon secretion becomes excessive and disrupts metabolism, resulting (in the absence of insulin) in diabetic ketoacidosis [69].

5.1.1 Experimental studies

The experimental results argue for the interference of the microbiota and T1DM pathological mechanisms—the incidence of diabetes is higher in mice raised in aseptic conditions, and the antibiotics administered to conventional animals accelerate the evolution of diabetic pathology. The NOD (non-obese diabetes) mice have a distinct microbiota from other resistant lines, and the incidence of type 1 diabetes mellitus is higher in specific pathogen-free animals [70].

5.1.2 Analytical results

Dysbiosis is shaped by host-related individual factors and early-life exposure to certain microorganisms, and its alterations undergo extensive changes with the change in diet. The permeability of the intestinal barrier plays an important role in the initiation and evolution of autoimmune conflict, aside from the background of genetic predisposition. The intercellular tight junctions control the permeability of the epithelium, allowing the absorption of nutrients, but preventing the passage of various environmental antigens (i.e., food, bacterial, viral, and fungal). Dysbiosis decreases intestinal permeability and facilitates the translocation of bacterial antigens [52].

Microbiota derangements have been implicated in the evolution of both T1DM and T2DM [71]. Dysbiosis occurs very early in subjects with a genetic predisposition for T1DM, probably since the neonatal period [51]. It is unknown whether the genetic predisposition to T1DM shapes the microbiota of high-risk individuals or whether the microbiota is the cause or effect of the disease [71].

As stated above, the human microbiota stabilizes during the first 3 years of life, while three parallel phenomena occur—(i) development of the immune system, (ii) maturation of the microbiota, and (iii) seroconversion to T1DM-associated autoantibodies. The possible conditioning of the two (i.e., seroconversion and T1DM occurrence) events is unknown. In a longitudinal study, Kostic et al. showed a decrease in the bacterial diversity of the microbiota that occurs before the development of the clinical disease in children positive for anti-insulin antibodies [70]. The *Clostridium*, *Veillonella Bacteroides* increase in abundance, while *Lactobacillus*, *Bifidobacterium*, *Prevotella* genera decrease compared to healthy subjects, suggesting the correlation between microbiota disturbance and T1DM [18, 19]. Different authors reported other changes associated with T1DM. Increases of *Bacteroidetes* (Gram-negative) and decreases of *Prevotella* and *Firmicutes* (Gram-positive producing SCFA) observed in children with T1DM when compared to healthy subjects suggest an increased intestinal barrier permeability and decreased SCFA production [70]. The microbiota of children with T1DM is unstable and has a smaller population of butyrate-producing bacteria, which correlates with an increased barrier permeability. Healthy children have higher levels of *Lactobacillus rhamnosus* and *Bifidobacterium dentium*, while the group of *Streptococcus mitis/oralis/pneumoniae* is abundant in subjects with T1DM.

Furthermore, the microbiota changes evolve with disease progression [65].

The fungal microbiome of the human population is evaluated in 267 species, with the most commonly represented by g. *Candida*, *Saccharomyces*, *Penicillium*, and *Aspergillus*. The individual mycobiome rarely contains more than one genus, but this panel is enough to influence the entire composition of the microbiota population, either directly by interactions with bacterial cells or indirectly by immune modulation. In patients with type 1 diabetes and those with inflammatory bowel disease, there was an overgrowth of *Candida* [70].

Despite the abundance of experimental and clinical results suggesting a bidirectional relationship between dysbiosis and T1DM onset and progress, there are questions that still need an answer—(i) is their relationship causal or simultaneous? and (ii) the condition of causality is that the change of one variable leads to the change of another repeatedly and generally? [65].

5.2 Autoimmune thyroid diseases

Thyroid AIDS are conditioned as other auto-immunities by a genetic predisposition, but other factors play an important role in triggering and evolving the autoimmune pathological process [72]. They occur with a frequency of about 4% in the human population and express by either hyper- or hypothyroidism. In both cases, the thyroid may increase in volume (goiter), while ophthalmopathy may develop in hyperthyroidism only [73]. Autoimmune thyroid disease affects especially women and from an immunological point of view, it is characterized by the presence of circulating autoantibodies, activated T cells against thyroid antigens, and by lymphocytic infiltration of the organ. Three specificities of anti-thyroid autoantibodies have been described—anti-thyroid peroxidase (microsomal antigen); anti-thyroglobulin; anti-TSH receptor of thyroid acinar cells [74, 75].

AIDS that cause thyroid failure, generically called thyroiditis, are characterized by lymphocytic infiltration. Depending on the clinical aspects there are two pathological conditions—Hashimoto's thyroiditis and atrophic thyroiditis (primary myxedema). In both cases, the thyroid tissue is lysed. Autoimmune thyroid disease is influenced by various factors, such as age, sex, race, and hormonal status [76, 77].

Autoimmune thyroid diseases (Graves and Hashimoto's thyroiditis) often coexist with intestinal diseases, especially celiac disease. The composition of the microbiota population is influenced by diet, affects the thyroid function, mostly by providing the micronutrients essential for the synthesis of thyroid hormones—iodine, iron, and copper. Selenium and zinc are essential for the conversion of T4 to T3, and vitamin D has an immune regulatory effect. Probiotic supplementation favorably influences the secretion of thyroid hormones [26].

Autoimmune thyroiditis is the most common thyroid disorder, with a prevalence of 10–12%. It is triggered by genetic and environmental factors (viral infections) and has an increased prevalence in patients with celiac disease. The commensal microbiota activates the proinflammatory response through innate immunity receptors from the toll-like receptor family and disrupts the intestinal permeability, which may be a triggering factor for Hashimoto's thyroiditis [78].

Hashimoto's thyroiditis is the most common endocrine AIDS (i.e., 10–12% of total autoimmune endocrinopathies), which is characterized by autoimmune destruction of thyroid follicles. The incidence increases with age and is 10 times higher in women. In the serum of patients with Hashimoto's thyroiditis are detected various specific autoantibodies, such as anti-thyroglobulin and/or anti-TPO (thyroid-peroxidase), anti-TSH receptor. Definitive for Hashimoto's disease is the replacement of thyroid tissue with lymphoid tissue. An impressive increase in thyroid volume may be observed, but no hormones are synthesized instead (dry goiter). The symptoms of Hashimoto's thyroiditis and celiac disease often overlap and share epidemiological, clinical, serological, pathological, hormonal, genetic, and immune similarities. Microbiome analysis performed on patients with this ailment revealed that abundance levels of *Blautia, Roseburia, Ruminococcus torques groups, Dorea, Fusicatenibacter, and Eubacterium hallii* group genera were significantly higher whereas *Faecalibacterium, Prevotella, and Bacteroides* genera were decreased [79–82].

Celiac disease (CD) is an autoimmune condition characterized by a specific serological and histological profile triggered by gluten ingestion in genetically predisposed individuals [83]. CD is the only AID known to be triggered by an exogenous antigen, that is, wheat gluten. Gluten is a mixture of proteins grouped in the fraction of gliadin and glutenin, which is the source of carbon and nitrogen for germinating seedlings. Gliadin triggers specific auto-antibody synthesis, the clinical feature being strictly dependent on dietary exposure to gluten and homologous proteins from other cereals. CD is one of the most common autoimmune disorders, with a reported prevalence of 0.5–1% of the general population, except in areas showing a low frequency of CD-predisposing genes and low gluten consumption [84]. Studies have shown that most CD cases remain undetected in the absence of serological screening due to heterogeneous symptoms and/or poor disease awareness. CD has a strong hereditary component confirmed by its high familial recurrence (~10–15%) and the high concordance of the disease among monozygotic twins (75–80%) [85]. Also common to other AIDS, the HLA class II heterodimers, specifically DQ2 and DQ8, have a relevant role, in the heritability of CD. HLA-DQ2 homozygosis confers a much higher risk (25–30%) of developing early-onset CD in infants with a first-degree family member affected by the disease [86].

Dysbiosis is considered an important factor in the interaction of intestinal and thyroid AIDS. The mechanisms that mediate the interaction of microbiota imbalance and thyroid auto-immunities include: (i) intestinal dysbiosis, which interrupts self-tolerance and tolerance to non-pathogenic bacteria, by post-translational modification of proteins. The bacterial enzymatic apparatus can transform the self or nonself peptide into initiators of the autoimmune reaction, (ii) lipopolysaccharides-induced TLR activation, which is associated with thyroiditis and synthesis of anti-thyroglobulin antibodies, (iii) induction of Th-2 lymphocyte differentiation, inhibition of Th-17 lymphocyte differentiation and induction of oral acid tolerance to retinoic acid, which can activate an immune response of tolerance at intestinal level, (iv) permeabilization of the intestinal barrier through injuries of the integrity of tight junctions, deficiency of butyric acid produced by the fermented components in the microbiota or excess of ingested proteins that are metabolized by the microbiota with an increase of putrefaction components; all these factors increase the permeability of the intestinal barrier, facilitating the passage of gliadin and activation of the immune response [26]; (v) changes in the transcriptome, proteome and metabolome of the microbiota [34].

Hashimoto's thyroiditis and CD share common antibodies, that is anti-tissular transglutaminase (anti-tTg). In patients with CD, tTg binds to the thyroid follicles and the extracellular matrix of the follicles, therefore amplifying the interactions of the microbiota with the thyroid tissue. There is a direct correlation between serum titers of anti-tTg anti-TPO antibodies. DR3-DQ2 and DR4-DQ8 alleles, involved in CD, are reported as common genes that predispose to endocrine AIDS [80].

6. The microbiota interference with other autoimmune-mediated diseases

Rheumatoid arthritis is characterized by a severe and chronic inflammatory condition of the joints. The clinical course of the disease underlines the potential role of dysbiosis in triggering an inflammatory process that involves autoimmune components [87]. Germ-free animals are protected from rheumatoid arthritis in experimental settings. However, the disease is induced in mice exposed to *Clostridium* SFB, which may act as pathobiont or symbiont, depending on conditions that are host-dependent or independent. Clostridial antigens stimulate Th17 (proinflammatory) lymphocytes that contribute to the progressive evolution of rheumatoid arthritis. Conversely, neutralization of Th17 lymphocytes halts the evolution of the disease. The microbiome of patients with rheumatoid arthritis is altered, with the abundance of *Prevotella copri*. Citrullinated peptides and specific anti-citrullinated proteins antibodies (ACPA) have been identified in patients suffering from rheumatoid arthritis. Citrullinated peptides result from the peptidyl-Arg-deiminase (PAD)-catalyzed deamination reaction. The enzyme is mainly released after the lysis of granulocytes, monocytes, and macrophages that accumulate in the inflammatory spreads. However, it is also produced by *Porphyromonas gingivalis* and *Aggregatibacter actinomycetemcomitans*, which citrulinate human fibrinogen, synovial fluid proteins [88]. Citrullination is a common physiological process, especially associated with inflammatory processes. Citrullinated proteins, identified in the inflamed synovial membrane of the arthritic joint, exhibit new epitopes and induce the synthesis of ACPA. Circulating ACPAs incorporate into immune complexes aside from citrullinated peptides originating in the joints.

Periodontitis that is caused by oral microbiota bacteria progresses similarly to rheumatoid arthritis—leukocyte infiltration and the progressive destruction of alveolar bone. Leukocytes release the set of proinflammatory interleukins (such as TNF, Il-1, IL-6, IL-12, IL-17, IL-18, and IL-33), growth factors (such as colony-stimulating factors—i.e., GM-CSF, monocyte-CSF), activator receptor of nuclear factor kappa-β ligand (RANKL), metalloproteases, nitric oxide, and PG E2 [89].

In 2013, Rinaldi identified auto-antibodies against the cellular wall of *Saccharomyces cerevisiae* in rheumatoid arthritis, lupus erythematosus, and antiphospholipid syndrome. These antibodies are also observed in the sera of 32% of patients with celiac disease before its clinical occurrence, and they are considered as a specific serological marker of the disease [90].

Behcet's disease is a chronic, multisystemic inflammation that is characterized by uveitis, which is a major cause of blindness, and recurrent ulcerative lesions involving the mouth and genital mucosa. There have been reported changes in Th-1, Th-17, and T-reg lymphocytes, whose activity is regulated by the microbiome [91], as well as the diversification of potentially pathogenic bacteria and the decrease of those that produce butyrate (*Clostridium*).

The pathological change in ulcerative colitis consists of diffuse inflammation, with limited ulcers in the chorion of the colonic mucosa. The pathological process is extended over the entire mucosa of the intestinal epithelium [92].

In Crohn's disease, the inflammatory infiltrates often generate extensive granulomas in the submucosa and even in the muscular layer of the colon and small intestine. The pathological process of Crohn's disease is localized, with the damaged areas of the intestine alternating with the healthy ones [93].

Crohn's disease and ulcerative colitis are not AIDS in the strict sense, because triggering antigens appear to be components of the intestinal microbiota translocated into the chorion, but are the consequence of a large immune response in non-pathogenic antigens, which occurs in people with a genetic predisposition. The inflammatory condition increases the permeability of the colonic epithelium, and the microbiome is modified—the method of 16S rDNA sequencing has shown a decrease in bacterial diversity, especially of the non-pathogenic population, in favor of potentially pathogenic ones [94].

Lupus erythematosus is the prototype of systemic autoimmune disease—an autoimmune response characterized by hyper-reactivity of B lymphocytes and the presence of a large spectrum of serum antibodies [95]. As its name, the disease involves many organs and systems and has various clinical manifestations. Lupus erythematosus affects especially women (female/male ratio = 9/1), with the highest risk during pregnancy [96]. The intestinal microbiota is altered—depletion of lactobacilli, increased *Lachnospiraceae* density and general diversity, compared to healthy individuals. A large proportion (over 65%) of patients have periodontitis [97], which is always associated with extensive changes in the oral microbiota, in which species with potential pathogens predominate—*Fusobacterium nucleatum*, *Actinomyces naeslundii*, *Ps. anaerobius*, *Bacteroides intermedius*, and *Porphyromonas gingivalis* [98].

Multiple sclerosis is a chronic demyelinating inflammatory disease of the central nervous system, characterized by destruction of the integrity of the haemato-encephalic barrier, T lymphocyte infiltrates, and autoimmune reaction against myelin proteins [99]. The immune response in experimental autoimmune encephalitis is mediated by Th-1 and Th-17 cells. The causative agent is not known, but the modification of the microbiota may be important in the onset and/or progression of autoimmune disease. The autoimmune encephalitis diminishes to extinction in *germ-free* mice, and colonization with *Clostridium* SFB restores the severity of the disease, as it stimulates the growth of the population of Th-17 (proinflammatory) lymphocytes [100]. Conversely, the administration of *Bacteroides* protects against demyelination and expansion of tissue-specific inflammation induced by Treg Foxp3 + [55].

The liver autoimmune disease appears to have a direct connection to the microbial load (cells, lipopolysaccharides, peptidoglycans, flagellin, DNA, RNA, toxins, and metabolites) that reaches the Kupffer cells and sinusoidal capillaries by passaging the portal vein. The immune response to these antigens can initiate liver damage and fibrosis [55, 101].

Vitiligo is a systemic autoimmune disease, which is characterized by areas of skin depigmentation, as a result of melanocyte lysis under the action of TCD8 lymphocytes. Melanocytes are located at the border between the epidermis and the dermis, but the disease is systemic because melanocytes are also found in other tissues. The number of melanocytes is the same in different individuals, but differences in pigmentation result from the number, distribution, and size of melanosomes in keratinocytes. The intestinal microbiota in patients with vitiligo is altered and is characterized by decreased taxonomic diversity [18, 102].

Atopic dermatitis is an inflammatory skin disease, clinically characterized by pruritus and xerosis (dry skin). The underlying cause is delayed hypersensitivity mediated by T lymphocytes. The local trigger is the colonization of the skin with *Staphylococcus aureus*. The toxins released by *S. aureus* exert a cytotoxic effect [103, 104].

7. Conclusions

Intestinal dysbiosis alters the permeability of the intestinal barrier. The passage of the microbiota antigens into the internal environment may induce the loss of self-tolerance with the generation of autoantibodies and/or autoreactive T cells, leading to the occurrence of cross-reactions. The microbiota alterations lead to an increase in enteric barrier permeability and the occurrence of lymphocyte infiltrates into the epithelial layer, augmenting the risk of cell-mediated auto-immune response. Many questions still need an answer about the role of the microbiota in triggering AIDS, such as—what are the roles of sex hormones and the role of X-linked genes expression in correlation with the microbiome in the polarization of gender-dependent AIDS. Do the changes in the microbiota, which are reported by many authors, contribute to the onset of AIDS by breaking the peripheric tolerance or they are the consequence of AIDS?

Acknowledgements

This research was funded by projects PD224/2021 (PD-2019-0499) and PFE-CDI.2021-587.

Conflict of interest

The authors declare no conflict of interest.

Author details

Gratiela Gradisteanu Pircalabioru[1*], Octavian Savu[2], Grigore Mihaescu[3], Corneliu Ovidiu Vrancianu[1] and Mariana-Carmen Chifiriuc[1,3]

1 Research Institute of University of Bucharest (ICUB), Bucharest, Romania

2 NIDNMD N.C. Paulescu, Bucharest, Romania

3 Faculty of Biology, Department of Microbiology and Immunology, University of Bucharest, Bucharest, Romania

*Address all correspondence to: gratiela.gradisteanu@icub.unibuc.ro

References

[1] Rayner F, Isaacs JD. Therapeutic tolerance in autoimmune disease. Seminars in Arthritis and Rheumatism. 2018;**48**:558-562. DOI: 10.1016/j.semarthrit.2018.09.008

[2] Liu H-H, Lin Y-C, Chung C-S, Liu K, Chang Y-H, Yang C-H, et al. Integrated counts of carbohydrate-active protein domains as metabolic readouts to distinguish probiotic biology and human fecal metagenomes. Scientific Reports. 2019;**9**:16836. DOI: 10.1038/s41598-019-53173-7

[3] Chen MX, Wang S-Y, Kuo C-H, Tsai I-L. Metabolome analysis for investigating host-gut microbiota interactions. Journal of the Formosan Medical Association. 2019;**118**(Suppl):S10-S22. DOI: 10.1016/j.jfma.2018.09.007

[4] Young VB. The role of the microbiome in human health and disease: An introduction for clinicians. BMJ. 2017;**356**:j831. DOI: 10.1136/bmj.j831

[5] Qin J, Li R, Raes J, Arumugam M, Burgdorf KS, Manichanh C, et al. A human gut microbial gene catalogue established by metagenomic sequencing. Nature. 2010;**464**:59-65. DOI: 10.1038/nature08821

[6] Zhu S, Jiang Y, Xu K, Cui M, Ye W, Zhao G, et al. The progress of gut microbiome research related to brain disorders. Journal of Neuroinflammation. 2020;**17**:25. DOI: 10.1186/s12974-020-1705-z

[7] Dehner C, Fine R, Kriegel MA. The microbiome in systemic autoimmune disease: Mechanistic insights from recent studies. Current Opinion in Rheumatology. 2019;**31**:201-207. DOI: 10.1097/BOR.0000000000000574

[8] Zhao Q, Elson CO. Adaptive immune education by gut microbiota antigens. Immunology. 2018;**154**:28-37. DOI: 10.1111/imm.12896

[9] Ruff WE, Kriegel MA. Autoimmune host-microbiota interactions at barrier sites and beyond. Trends in Molecular Medicine. 2015;**21**:233-244. DOI: 10.1016/j.molmed.2015.02.006

[10] Iebba V, Totino V, Gagliardi A, Santangelo F, Cacciotti F, Trancassini M, et al. Eubiosis and dysbiosis: The two sides of the microbiota. The New Microbiologica. 2016;**39**:1-12

[11] Cox AJ, West NP, Cripps AW. Obesity, inflammation, and the gut microbiota. The Lancet Diabetes and Endocrinology. 2015;**3**:207-215. DOI: 10.1016/S2213-8587(14)70134-2

[12] Adak A, Khan MR. An insight into gut microbiota and its functionalities. Cellular and Molecular Life Sciences. 2019;**76**:473-493. DOI: 10.1007/s00018-018-2943-4

[13] Dalile B, Van Oudenhove L, Vervliet B, Verbeke K. The role of short-chain fatty acids in microbiota-gut-brain communication. Nature Reviews. Gastroenterology & Hepatology. 2019;**16**:461-478. DOI: 10.1038/s41575-019-0157-3

[14] Sun M, Wu W, Chen L, Yang W, Huang X, Ma C, et al. Microbiota-derived short-chain fatty acids promote Th1 cell IL-10 production to maintain intestinal homeostasis. Nature Communications. 2018;**9**:3555. DOI: 10.1038/s41467-018-05901-2

[15] Zhao J, Zhang X, Liu H, Brown MA, Qiao S. Dietary protein

and gut microbiota composition and function. Current Protein & Peptide Science. 2019;**20**:145-154. DOI: 10.2174/1389203719666180514145437

[16] Greiner T, Bäckhed F. Effects of the gut microbiota on obesity and glucose homeostasis. Trends in Endocrinology and Metabolism. 2011;**22**:117-123. DOI: 10.1016/j.tem.2011.01.002

[17] Sandrini S, Aldriwesh M, Alruways M, Freestone P. Microbial endocrinology: Host-bacteria communication within the gut microbiome. The Journal of Endocrinology. 2015;**225**:R21-R34. DOI: 10.1530/JOE-14-0615

[18] Gianchecchi E, Fierabracci A. Recent advances on microbiota involvement in the pathogenesis of autoimmunity. International Journal of Molecular Sciences. 2019;**20**(2):E283. DOI: 10.3390/ijms20020283

[19] Rizzetto L, Fava F, Tuohy KM, Selmi C. Connecting the immune system, systemic chronic inflammation and the gut microbiome: The role of sex. Journal of Autoimmunity. 2018;**92**:12-34. DOI: 10.1016/j.jaut.2018.05.008

[20] Evans JM, Morris LS, Marchesi JR. The gut microbiome: The role of a virtual organ in the endocrinology of the host. The Journal of Endocrinology. 2013;**218**:R37-R47. DOI: 10.1530/JOE-13-0131

[21] Bäckhed F. Host responses to the human microbiome. Nutrition Reviews. 2012;**70**(Suppl 1):S14-S17. DOI: 10.1111/j.1753-4887.2012.00496.x

[22] Round JL, O'Connell RM, Mazmanian SK. Coordination of tolerogenic immune responses by the commensal microbiota. Journal of Autoimmunity. 2010;**34**:J220-J225. DOI: 10.1016/j.jaut.2009.11.007

[23] Round JL, Mazmanian SK. The gut microbiota shapes intestinal immune responses during health and disease. Nature Reviews. Immunology. 2009;**9**:313-323. DOI: 10.1038/nri2515

[24] Ost KS, Round JL. Communication between the microbiota and mammalian immunity. Annual Review of Microbiology. 2018;**72**:399-422. DOI: 10.1146/annurev-micro-090817-062307

[25] Fung TC, Olson CA, Hsiao EY. Interactions between the microbiota, immune and nervous systems in health and disease. Nature Neuroscience. 2017;**20**:145-155. DOI: 10.1038/nn.4476

[26] Knezevic J, Starchl C, Tmava Berisha A, Amrein K. Thyroid-Gut-Axis: How does the microbiota influence thyroid function? Nutrients. 2020;**12**(6):1769. DOI: 10.3390/nu12061769

[27] de Oliveira GLV, Leite AZ, Higuchi BS, Gonzaga MI, Mariano VS. Intestinal dysbiosis and probiotic applications in autoimmune diseases. Immunology. 2017;**152**:1-12. DOI: 10.1111/imm.12765

[28] Josefowicz SZ, Lu L-F, Rudensky AY. Regulatory T cells: Mechanisms of differentiation and function. Annual Review of Immunology. 2012;**30**:531-564. DOI: 10.1146/annurev.immunol.25.022106.141623

[29] Stiemsma LT, Reynolds LA, Turvey SE, Finlay BB. The hygiene hypothesis: Current perspectives and future therapies. ImmunoTargets and Therapy. 2015;**4**:143-157. DOI: 10.2147/ITT.S61528

[30] Truckenbrod EN, Jameson SC. The virtuous self-tolerance of virtual memory T cells. The EMBO Journal. 2018;**37**(14):e99883. DOI: 10.15252/embj.201899883

[31] Root-Bernstein R. Autoimmunity and the microbiome: T-cell receptor mimicry of "self" and microbial antigens mediates self tolerance in holobionts: The concepts of "holoimmunity" (TcR-mediated tolerance for the holobiont) and "holoautoimmunity" (loss of tolerance for the holobiont) are introduced. BioEssays. 2016;**38**:1068-1083. DOI: 10.1002/bies.201600083

[32] Waisman A, Lukas D, Clausen BE, Yogev N. Dendritic cells as gatekeepers of tolerance. Seminars in Immunopathology. 2017;**39**:153-163. DOI: 10.1007/s00281-016-0583-z

[33] Yamashiro Y. Gut microbiota in health and disease. Annals of Nutrition & Metabolism. 2017;**71**:242-246. DOI: 10.1159/000481627

[34] Takiishi T, Fenero CIM, Câmara NOS. Intestinal barrier and gut microbiota: Shaping our immune responses throughout life. Tissue Barriers. 2017;**5**:e1373208. DOI: 10.1080/21688370.2017.1373208

[35] Zhang X, Chen B, Zhao L-D, Li H. The gut microbiota: Emerging evidence in autoimmune diseases. Trends in Molecular Medicine. 2020;**26**:862-873. DOI: 10.1016/j.molmed.2020.04.001

[36] Khan MF, Wang H. Environmental exposures and autoimmune diseases: Contribution of gut microbiome. Frontiers in Immunology. 2019;**10**:3094. DOI: 10.3389/fimmu.2019.03094

[37] Reveille JD, Zhou X, Lee M, Weisman MH, Yi L, Gensler LS, et al. HLA class I and II alleles in susceptibility to ankylosing spondylitis. Annals of the Rheumatic Diseases. 2019;**78**:66-73. DOI: 10.1136/annrheumdis-2018-213779

[38] Tokuhara D, Kurashima Y, Kamioka M, Nakayama T, Ernst P, Kiyono H. A comprehensive understanding of the gut mucosal immune system in allergic inflammation. Allergology International. 2019;**68**:17-25. DOI: 10.1016/j.alit.2018.09.004

[39] Zininga T, Ramatsui L, Shonhai A. Heat shock proteins as Immunomodulants. Molecules. 2018;**23**(11):2846. DOI: 10.3390/molecules23112846

[40] Wolchok JD. Altered self: The not-so-neo-antigens. Nature Reviews. Immunology. 2018;**18**:152. DOI: 10.1038/nri.2018.7

[41] Lin DY-W, Tanaka Y, Iwasaki M, Gittis AG, Su H-P, Mikami B, et al. The PD-1/PD-L1 complex resembles the antigen-binding Fv domains of antibodies and T cell receptors. Proceedings of the National Academy of Sciences of the United States of America. 2008;**105**:3011-3016. DOI: 10.1073/pnas.0712278105

[42] Hampe CS. Protective role of anti-idiotypic antibodies in autoimmunity—Lessons for type 1 diabetes. Autoimmunity. 2012;**45**:320-331. DOI: 10.3109/08916934.2012.659299

[43] Daha NA, Banda NK, Roos A, Beurskens FJ, Bakker JM, Daha MR, et al. Complement activation by (auto-) antibodies. Molecular Immunology. 2011;**48**:1656-1665. DOI: 10.1016/j.molimm.2011.04.024

[44] Bovin N, Obukhova P, Shilova N, Rapoport E, Popova I, Navakouski M, et al. Repertoire of human natural anti-glycan immunoglobulins. Do we have auto-antibodies? Biochimica et Biophysica Acta. 2012;**1820**:1373-1382. DOI: 10.1016/j.bbagen.2012.02.005

[45] Sørgjerd EP. Type 1 diabetes-related autoantibodies in different forms of

diabetes. Current Diabetes Reviews. 2019;**15**:199-204. DOI: 10.2174/15733998 14666180730105351

[46] Greco A, Rizzo MI, De Virgilio A, Gallo A, Fusconi M, Pagliuca G, et al. Goodpasture's syndrome: A clinical update. Autoimmunity Reviews. 2015;**14**:246-253. DOI: 10.1016/j. autrev.2014.11.006

[47] Nguyen HH, Shaheen AA, Baeza N, Lytvyak E, Urbanski SJ, Mason AL, et al. Evaluation of classical and novel autoantibodies for the diagnosis of primary biliary cholangitis-autoimmune hepatitis overlap syndrome (PBC-AIH OS). PLoS One. 2018;**13**:e0193960. DOI: 10.1371/journal.pone.0193960

[48] Li QY, Yu F. Advances in autoantibodies against C-reactive protein in systemic lupus erythematosus. Zhongguo Yi Xue Ke Xue Yuan Xue Bao. 2019;**41**:678-684. DOI: 10.3881/j. issn.1000-503X.10870

[49] Wen L, Ley RE, Volchkov PY, Stranges PB, Avanesyan L, Stonebraker AC, et al. Innate immunity and intestinal microbiota in the development of type 1 diabetes. Nature. 2008;**455**:1109-1113. DOI: 10.1038/ nature07336

[50] Wu H-J, Ivanov II, Darce J, Hattori K, Shima T, Umesaki Y, et al. Gut-residing segmented filamentous bacteria drive autoimmune arthritis via T helper 17 cells. Immunity. 2010;**32**:815-827. DOI: 10.1016/j.immuni.2010.06.001

[51] McLean MH, Dieguez DJ, Miller LM, Young HA. Does the microbiota play a role in the pathogenesis of autoimmune diseases? Gut. 2015;**64**:332-341. DOI: 10.1136/gutjnl-2014-308514

[52] Durazzo M, Ferro A, Gruden G. Gastrointestinal microbiota and type 1 diabetes mellitus: The state of art. Journal of Clinical Medicine. 2019;**8**. DOI: 10.3390/jcm8111843

[53] Mori K, Nakagawa Y, Ozaki H. Does the gut microbiota trigger Hashimoto's thyroiditis? Discovery Medicine. 2012;**14**:321-326

[54] Kriss M, Hazleton KZ, Nusbacher NM, Martin CG, Lozupone CA. Low diversity gut microbiota dysbiosis: Drivers, functional implications and recovery. Current Opinion in Microbiology. 2018;**44**:34-40. DOI: 10.1016/j.mib. 2018.07.003

[55] Li B, Selmi C, Tang R, Gershwin ME, Ma X. The microbiome and autoimmunity: A paradigm from the gut-liver axis. Cellular & Molecular Immunology. 2018;**15**:595-609. DOI: 10.1038/cmi.2018.7

[56] Colotta F, Jansson B, Bonelli F. Modulation of inflammatory and immune responses by vitamin D. Journal of Autoimmunity. 2017;**85**:78-97. DOI: 10.1016/j.jaut.2017.07.007

[57] DiMeglio LA, Evans-Molina C, Oram RA. Type 1 diabetes. Lancet (London, England). 2018;**391**:2449-2462. DOI: 10.1016/S0140-6736(18)31320-5

[58] Siljander H, Honkanen J, Knip M. Microbiome and type 1 diabetes. eBioMedicine. 2019;**46**:512-521. DOI: 10.1016/j.ebiom.2019.06.031

[59] Gale EAM. Congenital rubella: Citation virus or viral cause of type 1 diabetes? Diabetologia. 2008;**51**:1559-1566. DOI: 10.1007/s00125-008-1099-4

[60] Korkmaz HA, Ermiş Ç. A case of immune-mediated type 1 diabetes mellitus due to congenital rubella ınfection. Annals of Pediatric Endocrinology &

Metabolism. 2019;**24**:68-70. DOI: 10.6065/apem.2019.24.1.68

[61] El-Senousy WM, Abdel-Moneim A, Abdel-Latif M, El-Hefnawy MH, Khalil RG. Coxsackievirus B4 as a causative agent of diabetes mellitus type 1: Is there a role of inefficiently treated drinking water and sewage in virus spreading? Food and Environmental Virology. 2018;**10**:89-98. DOI: 10.1007/s12560-017-9322-4

[62] Hayakawa T, Nakano Y, Hayakawa K, Yoshimatu H, Hori Y, Yamanishi K, et al. Fulminant type 1 diabetes mellitus associated with Coxsackievirus type B1 infection during pregnancy: A case report. Journal of Medical Case Reports. 2019;**13**:186. DOI: 10.1186/s13256-019-2130-8

[63] Levet S, Charvet B, Bertin A, Deschaumes A, Perron H, Hober D. Human endogenous retroviruses and type 1 diabetes. Current Diabetes Reports. 2019;**19**:141. DOI: 10.1007/s11892-019-1256-9

[64] Dworzański J, Drop B, Kliszczewska E, Strycharz-Dudziak M, Polz-Dacewicz M. Prevalence of Epstein-Barr virus, human papillomavirus, cytomegalovirus and herpes simplex virus type 1 in patients with diabetes mellitus type 2 in South-Eastern Poland. PLoS One. 2019;**14**:e0222607. DOI: 10.1371/journal.pone.0222607

[65] Regnell SE, Lernmark Å. Early prediction of autoimmune (type 1) diabetes. Diabetologia. 2017;**60**:1370-1381. DOI: 10.1007/s00125-017-4308-1

[66] Katsarou A, Gudbjörnsdottir S, Rawshani A, Dabelea D, Bonifacio E, Anderson BJ, et al. Type 1 diabetes mellitus. Nature Reviews Disease Primers. 2017;**3**:17016. DOI: 10.1038/nrdp.2017.16

[67] Andersen MK, Hansen T. Genetic aspects of latent autoimmune diabetes in adults: A Mini-review. Current Diabetes Reviews. 2019;**15**:194-198. DOI: 10.2174/1573399814666180730123226

[68] Sayed BA, Christy A, Quirion MR, Brown MA. The master switch: The role of mast cells in autoimmunity and tolerance. Annual Review of Immunology. 2008;**26**:705-739. DOI: 10.1146/annurev.immunol.26.021607.090320

[69] Alam U, Asghar O, Azmi S, Malik RA. General aspects of diabetes mellitus. Handbook of Clinical Neurology. 2014;**126**:211-222. DOI: 10.1016/B978-0-444-53480-4.00015-1

[70] Kostic AD, Gevers D, Siljander H, Vatanen T, Hyötyläinen T, Hämäläinen A-M, et al. The dynamics of the human infant gut microbiome in development and in progression toward type 1 diabetes. Cell Host & Microbe. 2015;**17**:260-273. DOI: 10.1016/j.chom.2015.01.001

[71] Mullaney JA, Stephens JE, Costello M-E, Fong C, Geeling BE, Gavin PG, et al. Type 1 diabetes susceptibility alleles are associated with distinct alterations in the gut microbiota. Microbiome. 2018;**6**:35. DOI: 10.1186/s40168-018-0417-4

[72] Bäckhed F. Changes in intestinal microflora in obesity: Cause or consequence? Journal of Pediatric Gastroenterology and Nutrition. 2009;**48**(Suppl 2):S56-S57. DOI: 10.1097/MPG.0b013e3181a11851

[73] Rayman MP. Multiple nutritional factors and thyroid disease, with particular reference to autoimmune thyroid disease. The Proceedings of the Nutrition Society. 2019;**78**:34-44. DOI: 10.1017/S0029665118001192

[74] Antonelli A, Ferrari SM, Corrado A, Di Domenicantonio A, Fallahi P. Autoimmune thyroid disorders. Autoimmunity Reviews. 2015;**14**:174-180. DOI: 10.1016/j.autrev.2014.10.016

[75] Effraimidis G, Wiersinga WM. Mechanisms in endocrinology: Autoimmune thyroid disease: Old and new players. European Journal of Endocrinology. 2014;**170**:R241-R252. DOI: 10.1530/EJE-14-0047

[76] Kunc M, Gabrych A, Witkowski JM. Microbiome impact on metabolism and function of sex, thyroid, growth and parathyroid hormones. Acta Biochimica Polonica. 2016;**63**:189-201. DOI: 10.18388/abp.2015_1093

[77] Fröhlich E, Wahl R. Microbiota and thyroid interaction in health and disease. Trends in Endocrinology and Metabolism. 2019;**30**:479-490. DOI: 10.1016/j.tem.2019.05.008

[78] Peng S, Li C, Wang X, Liu X, Han C, Jin T, et al. Increased toll-like receptors activity and TLR ligands in patients with autoimmune thyroid diseases. Frontiers in Immunology. 2016;**7**:578. DOI: 10.3389/fimmu.2016.00578

[79] Zaletel K, Gaberšček S. Hashimoto's thyroiditis: From genes to the disease. Current Genomics. 2011;**12**:576-588. DOI: 10.2174/138920211798120763

[80] Lerner A, Jeremias P, Matthias T. Gut-thyroid axis and celiac disease. Endocrine Connections. 2017;**6**:R52-R58. DOI: 10.1530/EC-17-0021

[81] Mori K, Yoshida K. Viral infection in induction of Hashimoto's thyroiditis: A key player or just a bystander? Current Opinion in Endocrinology, Diabetes, and Obesity. 2010;**17**:A

[82] Kyritsi EM, Kanaka-Gantenbein C. Autoimmune thyroid disease in specific genetic syndromes in childhood and adolescence. Frontiers in Endocrinology (Lausanne). 2020;**11**:543. DOI: 10.3389/fendo.2020.00543

[83] Caio G, Volta U, Sapone A, Leffler DA, De Giorgio R, Catassi C, et al. Celiac disease: A comprehensive current review. BMC Medicine. 2019;**17**:142. DOI: 10.1186/s12916-019-1380-z

[84] Singh P, Arora S, Singh A, Strand TA, Makharia GK. Prevalence of celiac disease in Asia: A systematic review and meta-analysis. Journal of Gastroenterology and Hepatology. 2016;**31**:1095-1101. DOI: 10.1111/jgh.13270

[85] Lundin KEA, Wijmenga C. Coeliac disease and autoimmune disease-genetic overlap and screening. Nature Reviews. Gastroenterology & Hepatology. 2015;**12**:507-515. DOI: 10.1038/nrgastro.2015.136

[86] Vriezinga SL, Auricchio R, Bravi E, Castillejo G, Chmielewska A, Crespo Escobar P, et al. Randomized feeding intervention in infants at high risk for celiac disease. The New England Journal of Medicine. 2014;**371**:1304-1315. DOI: 10.1056/NEJMoa1404172

[87] De Luca F, Shoenfeld Y. The microbiome in autoimmune diseases. Clinical and Experimental Immunology. 2019;**195**:74-85. DOI: 10.1111/cei.13158

[88] Klareskog L, Rönnelid J, Lundberg K, Padyukov L, Alfredsson L. Immunity to citrullinated proteins in rheumatoid arthritis. Annual Review of Immunology. 2008;**26**:651-675. DOI: 10.1146/annurev.immunol.26.021607.090244

[89] Araújo VMA, Melo IM, Lima V. Relationship between periodontitis and rheumatoid arthritis: Review of the literature. Mediators of

Inflammation. 2015;**2015**:259074. DOI: 10.1155/2015/259074

[90] Rinaldi M, Perricone R, Blank M, Perricone C, Shoenfeld Y. Anti-Saccharomyces cerevisiae autoantibodies in autoimmune diseases: From bread baking to autoimmunity. Clinical Reviews in Allergy and Immunology. 2013;**45**:152-161. DOI: 10.1007/s12016-012-8344-9

[91] Tanabe S. The effect of probiotics and gut microbiota on Th17 cells. International Reviews of Immunology. 2013;**32**:511-525. DOI: 10.3109/08830185.2013.839665

[92] Feuerstein JD, Moss AC, Farraye FA. Ulcerative colitis. Mayo Clinic Proceedings. 2019;**94**:1357-1373. DOI: 10.1016/j.mayocp.2019.01.018

[93] Feld L, Glick LR, Cifu AS. Diagnosis and Management of Crohn Disease. JAMA. 2019;**321**:1822-1823. DOI: 10.1001/jama.2019.3684

[94] Ott SJ, Musfeldt M, Wenderoth DF, Hampe J, Brant O, Fölsch UR, et al. Reduction in diversity of the colonic mucosa associated bacterial microflora in patients with active inflammatory bowel disease. Gut. 2004;**53**:685-693. DOI: 10.1136/gut.2003.025403

[95] Narváez J. Review: Systemic lupus erythematosus 2020. Medicina Clínica (Barcelona). 2020;**155**(11):494-501. DOI: 10.1016/j.medcli.2020.05.009

[96] Zhang H, Liao X, Sparks JB, Luo XM. Dynamics of gut microbiota in autoimmune lupus. Applied and Environmental Microbiology. 2014;**80**:7551-7560. DOI: 10.1128/AEM.02676-14

[97] Mutlu S, Richards A, Maddison P, Scully C. Gingival and periodontal health in systemic lupus erythematosus. Community Dentistry and Oral Epidemiology. 1993;**21**:158-161. DOI: 10.1111/j.1600-0528.1993.tb00742.x

[98] Loyola-Rodriguez JP, Martinez-Martinez RE, Abud-Mendoza C, Patiño-Marin N, Seymour GJ. Rheumatoid arthritis and the role of oral bacteria. Journal of Oral Microbiology. 2010;**2**. DOI: 10.3402/jom.v2i0.5784

[99] Thompson AJ, Baranzini SE, Geurts J, Hemmer B, Ciccarelli O. Multiple sclerosis. Lancet (London, England). 2018;**391**:1622-1636. DOI: 10.1016/S0140-6736(18)30481-1

[100] Lebrun C, Vukusic S. Immunization and multiple sclerosis: Recommendations from the French multiple sclerosis society. Multiple Sclerosis and Related Disorders. 2019;**31**:173-188. DOI: 10.1016/j.msard.2019.04.004

[101] Abe K, Fujita M, Hayashi M, Okai K, Takahashi A, Ohira H. Gut and oral microbiota in autoimmune liver disease. Fukushima Journal of Medical Science. 2020;**65**:71-75. DOI: 10.5387/fms.2019-21

[102] Boniface K, Seneschal J, Picardo M, Taïeb A. Vitiligo: Focus on clinical aspects, Immunopathogenesis, and therapy. Clinical Reviews in Allergy and Immunology. 2018;**54**:52-67. DOI: 10.1007/s12016-017-8622-7

[103] Torres T, Ferreira EO, Gonçalo M, Mendes-Bastos P, Selores M, Filipe P. Update on atopic dermatitis. Acta Médica Portuguesa. 2019;**32**:606-613. DOI: 10.20344/amp.11963

[104] Iwamoto K, Moriwaki M, Miyake R, Hide M. Staphylococcus aureus in atopic dermatitis: Strain-specific cell wall proteins and skin immunity. Allergology International. 2019;**68**:309-315. DOI: 10.1016/j.alit.2019.02.006

Chapter 5

Immune System, Gut Microbiota and Diet: An Interesting and Emerging Trialogue

Marianna Roselli, Raffaella Canali, Alberto Finamore, Andrea Ghiselli and Chiara Devirgiliis

Abstract

The present chapter provides a comprehensive overview of the multifaceted links connecting the immune system, the intestinal microbiota, and the diet, covering also some recent, less explored, and emerging topics such as the "trained immunity" and the immune cell metabolic activity. The main characteristics of the innate and adaptive immune system are described, as well as the gut-associated lymphoid tissue (GALT). Gut microbiota structure and function are also presented. Particular emphasis is given to the diet as a modulator of the microbiota-immune system crosstalk, focusing on the impact of the three main dietary components (carbohydrates, proteins, and fats) and the different dietary profiles on the gut microbiota, by shaping its composition and the deriving microbial metabolites that influence host health, also through interaction with the immune system. Western and Mediterranean diets are described and chosen as representative models of detrimental and beneficial dietary patterns, respectively.

Keywords: innate immunity, GALT, intestinal microbiome, Western diet, Mediterranean diet, metabolic inflammation

1. Introduction

The immune system is fundamental to protect the organism from pathogens and toxic exogenous agents, by discriminating between "self" and "nonself" antigens, and in normal physiological conditions it is programmed to react against "nonself." At the intestinal level, the gut-associated lymphoid tissue (GALT) is a key component of immune defense, protecting the body from foreign antigens and pathogens, while allowing tolerance to commensal bacteria and dietary antigens [1]. The intestinal microbiota, defined as the complex microbial community residing in the host's digestive tract, is recognized as an effective integral component of the host immune system, capable of finely tuning both the innate and adaptive immune responses during the entire lifespan. Indeed, the intimate relationship set up between microbiota and immune cells in the intestine is crucial for the maintenance of immunological homeostasis and, mostly, for the "education" of the immune system during the early stages of life [2]. Diet has a strong influence on the gut microbiota, acting both as a modulator able to select specific microbial groups and providing substrates that can be metabolized by the microbiota producing metabolites that impact host health also through interaction with the immune system [3, 4]. Therefore, there is a close connection between diet, gut microbiota, and immune system, orchestrated by a fine-tuning of the complex mechanisms underlying this cross-talk.

2. The immune system

The immune system is a complex network designed to react against harmful foreign agents as well as pathogens. The immune system is immature during fetal and neonatal life. The fetus receives passive protection from the mother *in utero*, thanks to maternal immunoglobulins (Ig)G, which can cross the placental barrier. This protection continues until the first months of the life of the newborn, as maternal IgG are transferred also through breastfeeding. The development of the immune system begins early in life and is greatly influenced by the type of feeding and environmental exposure (including factors such as the presence of domestic animals, antibiotic use, and timing of introduction of different foods from weaning). In the elderly, a progressive decline of the immune function is observed, a process known as immunosenescence [5]. Two types of immune response exist: the nonspecific or innate response, which is the first line of defense and operates in a nonselective way against foreign antigens, and the specific or adaptive one, which is triggered after exposure to a particular antigen. Both responses have a cellular and a humoral component. The two responses are interconnected for several reasons: 1–The cytokines (soluble mediator molecules) secreted in the early stages of the innate response influence the type of adaptive immune response that will develop; 2–Macrophages and dendritic cells, activated during the innate response, act as antigen-presenting cells (APC) for naive (i.e., having not yet encountered the antigen) T lymphocytes, inducing their differentiation into effector T lymphocytes; 3 - In some cases, the phagocytosis performed by innate immune cells is more efficient, if the microorganism to be cleared has been previously bound and surrounded by antibodies (opsonization). The nature of the antigen determines which of the two responses is preferentially activated; however, a complete immune response requires the coordinated participation of both types, and it ends once the trigger is resolved (self-limiting capacity) [6].

2.1 Innate immune response

The innate immune response is less specialized and generally less effective than the adaptive one. The cellular mechanisms of innate immunity are characterized by phagocytic and cytotoxic activities, while the humoral component is based on the complement system.

2.1.1 Cellular component

Neutrophil granulocytes are normally found in the bloodstream. During the acute phase of inflammation, they are among the first inflammatory cells migrating from blood vessels to the inflamed site, recruited by chemical signals such as interleukin-8 (IL-8), through a process called chemotaxis. Similarly, monocytes migrate from the bloodstream to tissues in response to chemokine release at infection sites, become activated, and differentiate into macrophages. Macrophages constitute, together with neutrophils, the largest group of cells endowed with phagocytic activity, they carry out their defense action by surrounding foreign microorganisms with pseudopodia, i.e., extroversions of the plasma membrane, forming the phagosome. The phagosome then merges with the numerous granules present in the cytoplasm, containing various compounds toxic for microorganisms, such as defensins, cathelicidins, lysozyme, and lactoferrin, forming the phagolysosome. Alternatively, the so-called respiratory or oxidative burst is activated, resulting in the formation of reactive oxygen species finally producing hypochlorite, hypobromite, and hypoiodite that kill microorganisms. In addition to the phagocytic function, macrophages are also responsible for the processing and presentation of antigens to T cells, as mentioned above [7]. Natural killer (NK) cells eliminate virus-infected and tumor cells through a cytotoxic activity, mediated by perforin-containing granules and granzymes. The former form pores in the plasma membrane and the latter, entering through these pores, induce the caspase cascade, leading to apoptosis of target cells. NK cells can also kill target cells through another mechanism, referred to as antibody-dependent cellular cytotoxicity (ADCC), in which NKs recognize target cells to which IgG has been previously bound [7, 8]. Many cells of the innate immune system are activated through receptors expressed on their membrane, namely pattern recognition receptors (PRR), with a long evolutionary history, which are able to recognize conserved structural patterns expressed by microorganisms, such as the microbe- and pathogen- associated molecular patterns (MAMP and PAMP, respectively). In particular, the type of PRR recognizing MAMP and PAMP is represented by the toll-like receptors (TLRs), a family of transmembrane proteins primarily expressed on the surface of immunocompetent cells, i.e., monocytes, macrophages, and dendritic cells, but also on intestinal epithelial cells. When a TLR recognizes a MAMP, a complex protein signal transduction cascade is triggered generating the appropriate immune response for that microorganism [9]. In most cases, the inflammatory response activated by TLRs leads to the activation of the nuclear factor-kappaB (NF-κB), which induces the transcription of numerous pro-inflammatory genes, including IL-8 [10]. TLRs can also be activated by endogenous danger signals such as the damage-associated molecular patterns (DAMP), molecules that are released in the intracellular or extracellular space following tissue injury, cellular stress, or apoptosis. Some innate responses can activate the inflammasome, a multiprotein complex resident in the cytosol as an inactive form, particularly in macrophages. A 2-hit-theory has been postulated, stating that for inflammasome activation two distinct signals are required. The first signal, triggered by PAMPs or DAMPs, activates the TLR signaling cascade, leading to the expression of some pro-inflammatory cytokines in an inactive form, such as proIL-1β and proIL-18. The second stimulus activates the inflammasome and generates caspase-1. Only thereafter proIL-1β and pro-IL-18 are cleaved by caspase-1 to mature IL-1β and IL-18, which can be secreted by macrophages and promote the inflammatory response [11]. Dendritic cells (DCs) are specialized to "sample" the entry sites of potential infectious agents, so they are found as immature cells in nonlymphoid tissues where antigens can be encountered, such as skin and other mucosal sites. The antigen recognition, through TLRs' activation, initiates the maturation process of DCs, which are induced to secrete various pro-inflammatory cytokines. After the encountering, the antigen is internalized through phagocytosis or pinocytosis and processed by the DCs, which migrate to secondary lymphoid organs (lymph nodes, spleen), where the exposed antigens are presented to populations of T lymphocytes, both naive and memory cells [6]. Mast cells and basophil granulocytes, similarly to monocytes, circulate in the blood as immature progenitor cells, differentiating into mature cells in different tissues in response to cytokine secretion. Mast cells and basophils are particularly found in association with blood vessels and nerves, in close proximity to mucosal surfaces that interface with the external environment, where they are able to detect infectious agents through TLRs. Upon activation, mast cells and basophils immediately extrude histamine from granules and, within a few minutes, release lipid mediators (such as prostaglandins, leukotrienes, and thromboxane), promoting vascular permeability, vasodilation, and rapid recruitment of eosinophils, neutrophils, and other immune cells [7]. Eosinophils are another type of circulating granulocytes that can be recruited to sites of inflammatory reactions, where their numbers can be 100-fold higher than in the blood. When activated, eosinophils release the contents of their granules (numerous enzymes, major basic protein, eosinophilic cationic protein), which act primarily on extracellular helminthic parasites. Eosinophils also actively participate in allergic diseases [8]. Innate lymphoid cells (ILCs), although lacking antigen-specific receptors, play an important role in the inflammatory response and the maintenance of immune homeostasis, particularly in mucosal tissues. Based on their phenotypic and functional features, ILCs have been grouped into three major subsets. Among them, group 3 ILC (ILC3) are implicated in intestinal homeostasis as they produce IL-22, a key regulator of the intestinal barrier [12]. Recent studies have shown that some myeloid cells of the innate immune system, essentially macrophages and NK cells, can develop a nonspecific immunological memory, i.e., these cells, after a first stimulus, acquire the ability to respond effectively to a subsequent stimulus, different from the first. Effector stimuli for such "innate memory" are represented by various components of bacteria or fungi, such as lipopolysaccharide (LPS) or β-glucans, as well as viruses, and such innate memory is called "trained immunity" [13]. Following activation, the cells involved in this phenomenon undergo processes of chromatin unfolding, which thus becomes more accessible for gene transcription. These processes, globally referred to as "epigenetic reprogramming," include methylations, acetylations, and phosphorylations at specific chromatin sites. The activation of gene transcription following the first stimulus is therefore accompanied by the acquisition of specific "epigenetic profiles," which are only partially lost after the elimination of the stimulus. In this way, a kind of nonspecific "memory" is developed, which makes some innate immune cells more easily and rapidly activated, following a subsequent heterologous stimulus. Trained immunity has gained increasing scientific relevance in recent years, for the hypothesis that previous infections can induce a metabolic and epigenetic reprogramming of some cells of innate immunity, leading to an improved defense response during subsequent infections of various types, at the same time trained immunity could also be negatively involved in hyperactivation of the immune system leading to chronic inflammation, as in atherosclerosis [14].

2.1.2 Humoral component

The complement system represents a set of plasma and membrane proteins endowed with enzymatic activity that can result in direct lysis of the foreign agent. These proteins circulate in the blood as functionally inactive molecules, called components. Complement activation occurs by cascade mechanism events, leading to sequential activation of the various inactive components. There are distinct pathways of complement activation: classical (activated by antigen–antibody binding), alternative, and lectinic, which are triggered by different mechanisms, but then converge in a common pathway leading to the formation of the membrane attack complex, which, by binding to the microorganism membranes, determines their osmotic lysis, through the formation of pores on the membrane itself [7].

2.2 Adaptive immune response

The adaptive immune response has the ability to recognize specific antigens and to remember those antigens in case of a subsequent exposure: this immunological memory allows a very rapid response, as the particular antigen has been already previously encountered and recognized. T and B lymphocytes, mediators of this response, undergo clonal expansion when they encounter the specific antigen they are programmed to recognize. At that moment, lymphocytes experience a real metabolic switch, increasing their metabolic needs for glucose and aminoacids, and passing from the normal oxidative phosphorylation typical of naive cells to aerobic glycolysis, in which pyruvate produced by glycolysis is reduced to lactic acid, with the simultaneous generation of NAD+ molecules, which promote the continuous production of 2 ATP molecules for each metabolized glucose molecule. This process, which occurs in the presence of oxygen, is less efficient than oxidative phosphorylation, but much faster, and thus able to meet the high ATP demand required to rapidly increase the biosynthesis of lipids, proteins, and nucleic acids of activated lymphocytes [15]. T lymphocytes exclusively recognize antigenic peptides exposed on the membrane of APCs via the major histocompatibility complex (MHC), whereas B lymphocytes recognize soluble, circulating antigens. Lymphocytes are present in an immature form in the primary lymphoid organs (bone marrow and thymus), where they differentiate into mature lymphocytes through a particular process of nonhomologous genetic recombination in the genes coding for antigen receptors (antibodies for B lymphocytes and T-cell receptors (TCR) for T lymphocytes). This somatic gene rearrangement accounts for the vast heterogeneity of lymphocytes, allows each individual to have a large and unique immunological repertoire, able to recognize a very large number of molecular configurations present in foreign agents, and thus counteract the majority of infections encountered during life [6].

2.2.1 T lymphocytes

T lymphocytes, effectors of the cell-mediated adaptive response, are divided into two major populations: T helper (Th) lymphocytes, bearing the CD4 receptor, which recognizes antigens presented by the MHC type II molecules, expressed on so-called "professional" APCs (dendritic cells, macrophages, and B lymphocytes), and cytotoxic T lymphocytes (Tc), with the CD8 receptor, which recognizes antigens presented by the MHC type I molecule, expressed on all nucleated cells [7]. CD4 and CD8 are co-stimulatory molecules, which bind to the MHC complex together with the TCR, contributing to T lymphocyte activation, which triggers different signaling cascades acting via various molecules and second messengers [16]. Th cells are critical in coordinating the immune response of other T cells and assist B cells in antibody secretion, whereas Tc cells are involved in the direct removal of damaged, pathogen-infected, or tumor cells. Th cells can develop into T helper 1 (Th1) or T helper 2 (Th2) cells, depending on the context in which antigen presentation occurs. Indeed, the cytokine secretion profile by APCs determines the "fate" of Th lymphocyte differentiation (**Figure 1**). The Th1 response is established in microenvironments where APCs produce essentially IL-12. This cytokine induces T lymphocytes to secrete IL-2 and interferon (IFN)-γ, through which cell-mediated responses, such as the antiviral

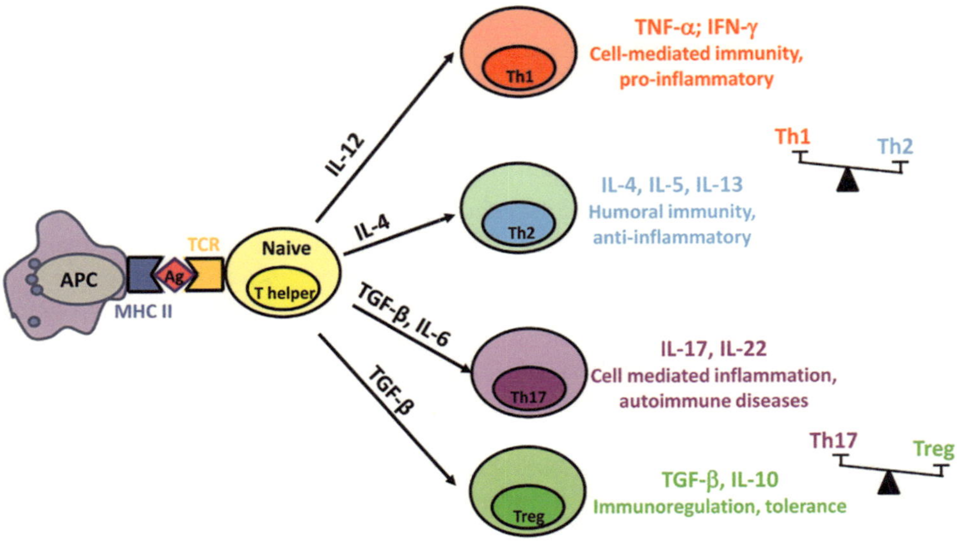

Figure 1.
Different differentiation fates of Thelper cells. Based on the context of antigen presentation and cytokine secretion, Thelper cells can differentiate in different subpopulations with pro- or anti-inflammatory features. The maintenance of the correct balance between them plays an important role for immune homeostasis. Ag, antigen; APC, antigen presentig cell; IFN, interferon; MHC, major histocompatibility complex; TCR, T cell receptor; TGF, transforming growth factor; Th, T helper; TNF, tumor necrosis factor.

response, are triggered. In particular, IFN-γ activates macrophages, inducing respiratory burst. Other important cytokines activated in this cascade, classically defined as pro-inflammatory, are IL-1, tumor necrosis factor (TNF)-α, and IL-8. In contrast, the Th2 response is associated with the production of IL-4, IL-5, and IL-13 cytokines, which mainly attract eosinophils and mast cells. The Th2 response, particularly involved in parasitic infections and allergies, is also associated with the expression of cytokines, such as IL-10, defined as anti-inflammatory, as they inhibit or reduce inflammatory-type responses. The maintenance of the correct balance between Th1 and Th2 subpopulations plays an important role in inflammation resolution [6]. Other relevant Th lymphocyte subtypes are the Th17, producing IL-17 and IL-22 cytokines, that are important in the response to extracellular pathogens such as fungi and bacteria, but also in some intestinal inflammatory responses; and the regulatory T (Treg) lymphocytes, involved in immune homeostasis especially in the intestine, as they contribute to the maintenance of oral tolerance to nonself harmless antigens, derived from food, but also environmental, such as pollen (**Figure 1**). The correct Th17/Treg balance is now recognized as fundamental for the maintenance of health status, in fact, this equilibrium results altered in many diseases with an autoimmune component, such as inflammatory bowel disease (IBD) [17].

2.2.2 B lymphocytes

B lymphocytes, expressing the surface receptor CD19, are responsible for the production of antibodies (or immunoglobulins), mediators of the adaptive humoral response. Upon encountering antigen, B lymphocytes can differentiate into short-lived antibody-producing plasma cells or long-lived memory cells. Plasma cells produce one of the five classes of immunoglobulins: IgG, IgM, IgA, IgD, and IgE, each with a specific role. IgM, representing about 10% total Ig in serum, are the first Ig produced in response to a foreign antigen. After 5–6 days from infection normally IgM reach their peak concentration and, thanks to the specific profiles of cytokines released, the so-called "isotypic switch" to IgG occurs. IgG are the predominant and most important class of immunoglobulins, in fact, they represent about 70–75% total Ig in serum. IgG reach the peak of secretion about 14 days after infection and persist for long periods. Normally IgG are the most effective in foreign antigen removal, through the above-mentioned opsonization process, as well as through activation of the complement system [6, 8]. IgD are found in serum at low concentrations (representing less than 1% of all plasma Ig) and their biological functions, related to the regulation of peripheral tolerance to self-antigens and in the maintenance of mucosal homeostasis, involving also host-microbiota interactions, have been elucidated only recently [18]. IgA represent 15–20% serum Ig, but their concentration is higher in secretions (saliva, breast milk, tears, sweat, respiratory, and intestinal secretions) and in the mucosa, i.e., those tissues covering the hollow organs and therefore in contact with the external environment (digestive, respiratory, and genital apparatus), where IgA contribute to preventing microorganisms from adhering to and penetrating inside the body through epithelial cells. In intestinal mucosa, IgA are found in dimeric form (secretory IgA, sIgA), particularly important for the protection against bacteria and viruses from the lumen, but also for the maintenance of oral tolerance to harmless food antigens, as detailed below. IgE, present in serum only in trace amounts, play a role in the removal of extracellular parasites (such as helminths) by opsonization, but are also important mediator in allergic responses. Indeed, IgE bind to a receptor expressed on the membrane of basophils and mast cells, stimulating the degranulation and release of histamine and lipid mediators into the intercellular space, triggering the allergic reaction, as described above [8].

2.3 Soluble mediators of the immune response

Soluble mediators, called cytokines, low molecular weight "messenger proteins" secreted by many cell types, both immune and nonimmune, are involved in both innate and adaptive immune responses. Cytokines send intracellular signals by binding to specific membrane receptors present on the same cells that produced them, or on other target cells, that can be in proximity or not, thus acting in an autocrine, paracrine, or endocrine manner. In general, it is possible to distinguish four different types of cytokines, on the basis of their biological effect: 1 - cytokines produced by leukocytes, having effects on the leukocytes themselves: interleukins; 2 - cytokines with chemoattractant properties, i.e., with a positive effect on cell motility: chemokines; 3 - cytokines that induce differentiation and proliferation of stem cells: colony-stimulating factors; 4 - cytokines that interfere with viral replication: interferons. Cytokines can exert a pro- or anti-inflammatory action, but often the outcome depends on the context of the microenvironment where they are secreted, and on the cells involved [7].

3. The intestinal immune system

The intestinal immune system is the most extensive lymphoid tissue, given the enormous surface area of the intestinal mucosa with which it is associated. It is called gut-associated lymphoid tissue (GALT) and is mainly composed of:

organized lymphatic follicles, called Peyer's patches (PPs); mesenteric lymph nodes (MLN); *lamina propria* lymphocytes (LPLs); intraepithelial lymphocytes (IELs) (**Figure 2**) [1].

3.1 Inductive sites

PPs represent the main sites where antigenic presentation occurs, called inductive sites, where the intestinal immune response is triggered. The PPs are covered by an epithelial layer, containing specialized membranous cells, the M cells, responsible for the transport of antigens, bacteria, and macromolecules from the intestinal lumen into the patches. These specific characteristics on the one hand make M cells designated for the transepithelial transport of antigens, and on the other hand make them more easily accessible by pathogens. In fact, many pathogens use M cells as a "gateway" to cross the intestinal barrier. M cells do not have brush border nor glycocalyx, but they have an extensive system of endocytic vesicles and a large intraepithelial pocket, where vesicles, containing antigens from the lumen, are released. In the pocket are APCs, which acquire the material carried by M cells and present the antigens to naive lymphocytes, present in the underlying subepithelial layer, organized in lymphatic follicles. In such follicles, B lymphocytes are located in the germinal centers, whereas T lymphocytes preferentially occupy the periphery and interfollicular spaces. DCs are also able to expose luminal antigens through various mechanisms: in the *lamina propria*, they can take antigens directly from the lumen, as they are able

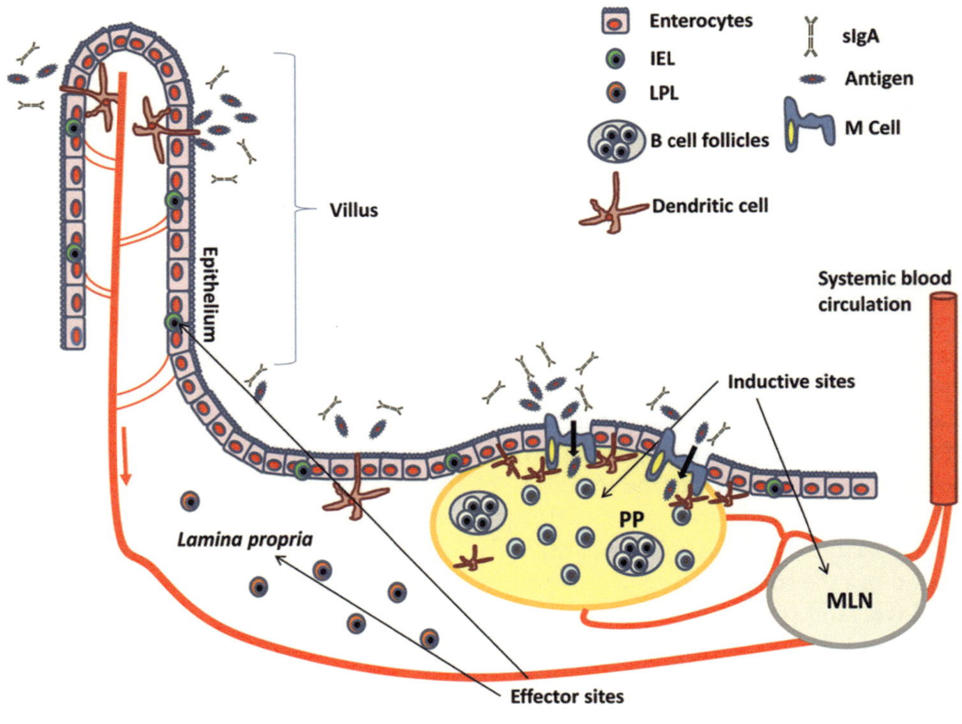

Figure 2.
Schematic representation of gut-associated lymphoid tissue (GALT) in small intestine. GALT is composed of organized lymphoid tissues of the Peyer's patches (PPs) and mesenteric lymph nodes (MLNs), the principal sites for induction of immune responses, while the lamina propria and epithelial layer are the effector sites. IEL, intraepithelial lymphocyte; LPL, lamina propria lymphocyte; MLN, mesenteric lymph nodes; PP, Peyer's patches.

to interdigitate between epithelial cells; or they can take luminal antigens that cross the intestinal barrier through transient "openings" [19]. After being primed, naive T and B cells become memory/effector cells and migrate from PP to MLN via efferent lymph and then via the thoracic duct to peripheral blood for subsequent extravasation at mucosal effector sites, both intestinal and extraintestinal, where the immune response will take place [20]. Other than a large number of lymphocytes, MLN also contain macrophages and APCs, which can themselves initiate immune responses against incoming antigens.

3.2 Effector sites

In the intestine, activated B and T lymphocytes essentially target two different lymphoid compartments: the *lamina propria* and the mucosal epithelium. The B lymphocytes of the *lamina propria* essentially produce IgA, the main class of antibodies secreted in the intestinal mucosa in large quantities, as mentioned above. In fact, it is estimated that 80% of plasma cells secreting antibodies reside in the intestinal *lamina propria*. The main function of IgA is to contribute to the intestinal barrier as the first line of defense, binding to antigens, neutralizing them, and removing them from the mucosa. IgA, unlike IgG, do not trigger an inflammatory response, as they do not bind to the complement system. As previously mentioned, IgA are found in the mucosal secretion as dimers, associating with a polypeptide present on the basement membrane of enterocytes, the secretory component (SC). Through the SC, IgA are transported through enterocytes and released into the intestinal lumen, becoming sIgA. This component gives sIgA resistance to proteases in the lumen, rendering them well-designated to perform their function in the intestine. Antigens able to bypass this first line of defense reach the *lamina propria*, where they encounter IgG, and the resulting immune complexes activate the complement system and trigger the inflammatory response. T lymphocytes in the *lamina propria* are effector cells, essentially CD4+ (helper/inducer phenotype) [21]. In the spaces between enterocytes, above the basement membrane (subepithelial space), there are populations of resident IELs, essentially CD8+ (suppressor/cytotoxic phenotype), acting as "sentinels", being the first components of the intestinal immune system exposed to food and microbial antigens. Indeed, IELs are among the most abundant lymphocyte populations in the body and play a key role in host defense against pathogens, wound repair, and intestinal homeostasis maintenance. IELs are composed of various cell subtypes bearing different TCRs, that can recognize antigenic peptides presented by conventional MHC molecules or by nonclassical MHC molecules, meaning that these cells are able to respond to some bacterial antigens in the absence of antigenic presentation by APCs [22].

4. The gut microbiota

4.1 Structure and function

Body surfaces facing the external environment, namely the skin and all mucosal surfaces (nasal, oral, gastrointestinal, etc.) are colonized by a huge number of microorganisms, collectively called microbiota. Most of them reside in the gut, in a *continuum* of extremely dynamic microbial communities. In terms of microbial density, it is estimated that approximately 10^{12} microorganisms per gram of content are present in colon and feces. These microorganisms belong to all three domains of life: Bacteria, which predominate, Archaea (methanogens, essentially belonging to *Methanobrevibacter* and *Methanosphaera* genera), and Eukarya (fungi and protists) [23]. The evolution of the intestinal microbiota starts at birth and is completed during the first years of life until it stabilizes in the adult phase. Immediately after birth, the gastrointestinal tract is rapidly colonized by a microbial consortium whose composition varies depending on several factors, such as the mode of delivery (vaginal or caesarian), the diet during infancy (breast or formula milk), and during adulthood (for example, vegetable or meat-based), the use of antibiotics. In particular, breastfeeding stimulates the maturation of the intestinal microbiota, as breast milk contains bifidogenic oligosaccharides (HMO, human milk oligosaccharides), which have a prebiotic action [24]. The maturation is then completed within the first years of life and occurs in parallel and synergistically with the development of the immune system. Perturbations of gut microbiota composition are associated with aging, and these changes favor the growth of pathogens and increase the susceptibility to gut-related diseases [25]. In this complex ecosystem, the collective genomes of bacteria and other microorganisms have been the focus of increasing interest over the past two decades, facilitated by the rapid development of culture-independent genomic approaches and advanced computational technologies. The gut microbiota is characterized by an enormous phylogenetic diversity, with more than 1000 bacterial species found in the entire human population, among which about 150 are present in a single individual. At higher phylogenetic levels this biodiversity is reduced, in fact, the human gut microbiota is composed of two main populations belonging to the Firmicutes and Bacteroidetes phyla, which collectively constitute over 90% of the known phylogenetic taxa. Other less abundant, but not less important phyla, such as Actinobacteria, Proteobacteria, and Verrucomicrobia, whose relative abundances are often below 1%, are also present. The advent of culture-independent methods, although detecting a high inter-individual variability in the composition of the intestinal microbiota, has allowed to identify a common "microbial core", with shared metabolic activities, characterizing healthy individuals [26]. Indeed, the relative proportions of the various phyla are maintained in balance under physiologic conditions (eubiosis), whereas changes in microbial composition and function, termed dysbiosis, associated to a lower overall microbial diversity, often occur in immune-mediated and metabolic disorders, thus proving the important role of the gut microbiota in maintaining host health status, which goes far beyond the initial experimental observations about relevance in regulating body fat tissue accumulation and energy balance [27]. The microbiome, defined as the collective genome of the gut microbiota, contains approximately 3.3 million genes, a number about 150-fold higher than that of the genes of the human genome, most of which are involved in both the metabolism of carbohydrates, amino acids, cofactors, and vitamins, and the biosynthesis of secondary metabolites. Thanks to this enormous genetic heritage, intestinal microorganisms exert a profound influence on the nutritional, metabolic, and immune responses of the host, so that the intestinal microbiota is considered an "accessory organ" and the higher organisms, with their associated microbial communities, are defined as "holobionts" [28]. As mentioned, the main function of the gut microbiota concerns metabolic activity. Intestinal bacteria are, in fact, able to produce essential nutrients such as vitamins and, mostly, to extract energy from complex polysaccharides, which are not digestible by the human enzymes present in the gastrointestinal tract. Indeed, the microbiota possesses the metabolic capacity to degrade a wide range of substrates that reach the colon. In particular, the fermentation of complex polysaccharides produces, among other substances, the short-chain fatty acids (SCFA), essentially acetate, propionate, and butyrate, which play a key and multifactorial role in the physiology of the host. Microbiota also contributes to the barrier effect, counteracting colonization by enteropathogens and opportunistic pathogens. The main mechanisms involved are both direct, such as competition for nutrient resources and adhesion sites to the intestinal mucosa, the inhibition of bacterial growth through the creation of microenvironments at acidic pH, and the production of bacteriocins (such as colicins, microcins, and nisin), and indirect, through stimulation of the host immune system and of maturation and growth of enterocytes [29]. Moreover, it is now universally recognized the existence of a gut-brain axis that envisages an active contribution of the intestinal microbiota in the regulation of anxiety, pain, and behavior by acting on the synthesis of neurotransmitters, and a possible contribution to the pathophysiology of disorders of the central nervous system. Finally, the gut microbiota is also able to interact and modulate the endocrine system, strongly influencing the levels of stress-related hormones and insulin, as well as appetite [30].

4.2 Influence of gut microbiota on the immune system

The intestinal microbiota is recognized as an effective integral component of the host immune system, capable of finely tuning the immune responses, innate and adaptive, in the different phases of life. Indeed, the close relationship established between bacteria and immune cells in the gut is crucial for the maintenance of immunological homeostasis and, mostly, for the "education" of the immune system during the early stages of life [2]. In fact, according to the most recent theories, the interaction between microbiota and the immune system is necessary to "train," first, and "keep trained," then, the various functions of the latter. Thanks to the continuous contact with the gut microorganisms, with the molecules they synthesize, with those they produce from undigested food components, the immune system satisfies two apparently conflicting needs: to defend the organism from real threats, and to tolerate microbes and molecules not harmful to the organism. Indeed, the large variety of microorganisms constituting the microbiota can be functionally distinguished into symbionts and pathobionts, also referred to as opportunistic pathogens, both fundamental, as the former educate the immune system to tolerance, while the latter train it to pathogen recognition and attack [31]. In the physiological condition of eubiosis, symbionts and pathobionts are present in equilibrium. If this balance is altered, for example, due to an excessive antibiotic treatment, one of the two groups becomes predominant, leading to the onset of one of two possible extreme conditions: hyperstimulation of the immune system (inflammation) or hypostimulation (immunosuppression) [32] (**Figure 3**). It is worth noting that pathobionts, that are not harmful and are even necessary to educate the immune system in physiological conditions, become dangerous when the equilibrium is altered, as in dysbiosis. The immunological surveillance of the intestinal microorganisms involves the abovementioned TLRs, which recognize MAMPs and PAMPs [9, 33]. These receptors differently act in distinct cellular compartments. Indeed, recognition of these receptors on the apical surface of the epithelium (i.e., the one in contact with the intestinal lumen) generally promotes tolerance towards commensal bacteria and foodborne antigens, and low (basal) inflammatory tone; conversely, activation of these same receptors on the basolateral side, in contact with the underlying mucosa, promotes strong inflammatory responses. Numerous microbial stimuli activate inflammatory cascades through signal transduction pathways that essentially involve the nuclear transcription factor NF-κB, with consequent production of pro-inflammatory cytokines, such as IL-1 and IL-6, or anti-inflammatory factors more directly related to the extinguishing of inflammation/immune response, such as IL-10, thus playing a crucial role in maintaining intestinal homeostasis [10]. The different communities of the intestinal microbiota, characterized by metabolic specialization, complementarity, and cooperation, constitute a very complex network of microbe-microbe and microbe-host interaction, in the form of a symbiotic or mutualistic relationship, resulting in a continuous cross-talk. The host derives substantial immunological and metabolic benefits from the physical proximity of microbial populations in the gut and underlying tissues, but at the same time, this proximity poses an ongoing threat to health. In fact, although the immune system is designed to establish the proper balance between tolerance to the gut microbiota, maintaining a low level of basal inflammation and surveillance against infectious agents and opportunistic pathogens, the disruption of this balance, for example, due to inflammatory diseases or following the excessive use of antibiotics, induces a malfunction of the intestinal barrier with consequent opening of the junctions between enterocytes. The assembly and maintenance of tight junctions are regulated by several signaling pathways, that can be altered by pro-inflammatory cytokines, in particular TNF-α, IFN-γ, and IL-1β. Thus, an increase of these cytokines due to an inflammatory status can induce a decrease in the expression of tight junction proteins, or alter their phosphorylation status, causing a "loosening" of tight junctions [34]. This condition, referred to as a leaky gut syndrome, facilitates the translocation of pathogenic bacteria or harmful antigens from the intestinal lumen to the underlying mucosa (**Figure 3**). This process determines the establishment of endotoxemia, i.e., the presence of LPS in the circulation. The LPS, present on the cell wall of Gram-negative bacteria, is one of the microbial components able to act as an immune activator, therefore, representing a MAMP that

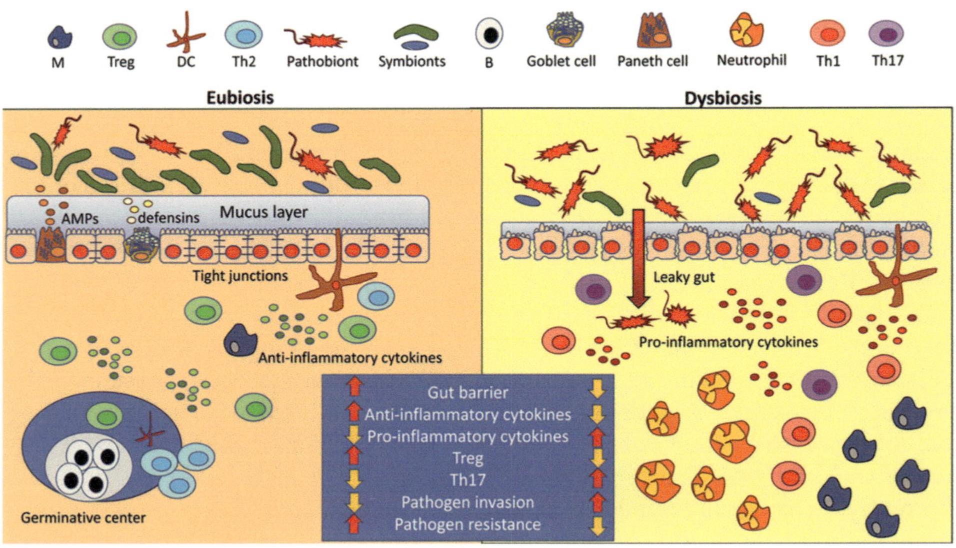

Figure 3.
Schematic representation of eubiosis and dysbiosis conditions. Gut immunological homeostasis is the result of a continuous cross-talk between microbiota and immune system. In eubiosis, the commensals predominate over pathobionts, maintaining the integrity of the intestinal barrier and an anti-inflammatory milieu. In dysbiosis, pathobionts take over and cross epithelial barrier inducing inflammation. AMPs, antimicrobial peptides; B, B lymphocyte; DC, dendritic cell; M, macrophage; Treg, regulatory T cell.

binds to TLR4 and triggers an inflammatory response, which from local becomes systemic. The polysaccharide A of *Bacteroides fragilis*, on the other hand, triggers an anti-inflammatory response, by stimulating IL-10 production and Treg proliferation [35]. Other bacteria, such as segmented filamentous bacteria (SFB), is pathobionts present exclusively during the first years of life, and play an important role in the immune system training, by inducing IL-17 secretion in the intestine and stimulating the production of IgA in the mucosal membranes of the oral and respiratory cavity [36]. Moreover, some evidence shows that SBF can promote IL-22 production by ILCs [37]. Fundamental to the maintenance of intestinal homeostasis is the proper balance of the different T lymphocyte subpopulations, mentioned in the 2.2.1 paragraph of the present chapter. In particular, the Th17/Treg balance appears crucial, and this balance is also modulated by the microbiota. It is worth noting that the interactions between microbiota and the immune system can have different outcomes, depending on the context of eubiosis or dysbiosis. Recently, the advent of high throughput molecular sequencing techniques has allowed the isolation from the human intestine of some bacteria with anti-inflammatory activity, which are of particular interest, especially for their possible applications in counteracting obesity and inflammatory bowel diseases, such as Crohn's disease and ulcerative colitis. Among the most important are *Akkermansia muciniphila* and *Faecalibacterium prausnitzii*, which have been defined as "next generation probiotics," as they are not yet commercially available, but candidates to be used as "biotherapeutics" [38].

5. Diet as a modulator of the microbiota-immune system cross-talk

An adequate and appropriate nutritional status and composition of the diet, in terms of foods, nutrients, and bioactive substances, are critical for the proper functioning of the immune system, which in turn is a fine sensor of the nutritional status of the individual [39]. When the immune system is activated to respond to nonself antigens, the demand for energy and nutrients increases and cells undergo the metabolic switch [40], as previously mentioned in the 2.2 paragraph. The dependence of the immune response on energy, and therefore the onset of immune deficits as a result of undernutrition, is known for a long time, but recently it has been observed that also the excessive consumption of food and excessive intake of calories alter the immune system. In fact, if on the one hand a serious caloric restriction impairs immune system functionality and increases the risk of infections (as observed in childhood malnutrition, still widely spread in developing countries), on the other hand, an unbalanced diet rich in high-calorie foods leads to negative consequences, inducing an inflammatory state and metabolic disorders. Many metabolic diseases are in fact characterized by a chronic low-grade systemic inflammation, called metaflammation (metabolic inflammation). Obesity and overnutrition are both associated with this inflammatory state leading to an increased risk of cardiovascular disease, heart attack, type 2 diabetes, and nonalcoholic steatohepatitis [41].

5.1 Impact of diet on gut microbiota

Diet has a profound influence also on the gut microbiota, acting both as a modulator able to select specific microbial groups, and as a provider of substrates that can be metabolized by the microbiota producing metabolites that impact on host health status, also through interaction with the immune system. Therefore, there is a close connection between diet, gut microbiota, and immune system, orchestrated by a fine tuning of the complex mechanisms underlying this cross-talk. The influence of diet in modulating gut microbiota composition is related to the concept of "enterotype." Indeed, although a wide inter-individual variability is observed among the bacterial groups present in the gut, the microbiota of most individuals can be classified into one of three variants or enterotypes, based on the dominant genera (*Bacteroides*, *Prevotella*, or *Ruminococcus*), which constitutes a relatively stable "core" [3, 26]. These enterotypes are associated with long-term dietary regimens [42]. In particular, enterotype 1, characterized by a predominance of the genus *Bacteroides*, able to extract the maximum energy from the fermentation of carbohydrates and proteins and to produce high amounts of vitamins B2 (riboflavin), B7 (biotin), and ascorbic acid (vitamin C), is associated with a diet rich in animal proteins and fats and low in fiber and vegetables, typical of the "Western Diet" profile. This enterotype may be related to increased intestinal inflammation and consequently to an increased state of general inflammation. Enterotype 2, dominated by the *Prevotella* genus, able to degrade complex polysaccharides and to produce high levels of vitamin B1 (thiamine) and vitamin B9 (folic acid), is instead correlated with a diet profile rich in fiber and carbohydrates. Finally, enterotype 3 is characterized by a predominance of bacteria of the genus *Ruminococcus* and is associated with a dietary profile rich in simple sugars [43]. Although most published papers demonstrate how long-term dietary regimen affects the structure and activity of the gut microbiota, there is still evidence suggesting the ability of the microbiota to respond to short-term dietary change in terms of macronutrients. For example, short-term consumption of diets composed entirely of animal or plant products has been shown to alter the structure of the gut bacterial community, minimizing inter-individual differences. Specifically, the most pronounced effect has been found for diets based on animal products, resulting in increased levels of bile-tolerant microorganisms (*Alistipes*, *Bilophila*, and *Bacteroides*), and decreased levels of Firmicutes capable of metabolizing plant polysaccharides (*Roseburia*, *Eubacterium rectale,* and *Ruminococcus bromii*) [44].

5.1.1 Effect of macronutrients

Among macronutrients, the effect of carbohydrates on the microbiota is the most described, while for proteins and lipids the mechanisms are less defined. Micronutrient intake is also critical for gut well-being; in fact, vitamin deficiencies have been associated with alterations in barrier function and GALT immune response. However, it is important to emphasize that modifications to the immune system and microbiota are primarily associated with the composition of the diet as a whole, and not with specific foods or nutrients [4]. Many complex carbohydrates are known to act as prebiotics, selectively stimulating, in the intestine, the growth of microorganisms beneficial to human health, such as bifidobacteria. Dietary fiber is a heterogeneous and complex mixture of different combinations of monosaccharides, with a minimum of 10 monomeric units or oligosaccharides containing from 3 to 9 monomeric units. A further classification of dietary fiber is related to its water solubility, viscosity, and fermentability. Polysaccharides are further categorized in non-starch polysaccharides and resistant starch, while oligosaccharides include resistant oligosaccharides. Soluble fiber is typically fermented to SCFA by the intestinal microbiota. A growing body of literature shows that dietary fiber has the potential to change the gut microbiota and alter metabolic regulation in humans. Most findings supporting the fiber hypothesis are based on short-term dietary interventions, while only sparse data evaluating the impact of long-term dietary fiber on the gut microbiome exist. Specific sources of dietary fiber were differentially associated with the gut microbiome. Fiber from fruit and vegetable intake was related to the gut microbiome composition, characterized by an increased abundance of Clostridia, an important class of dietary fiber fermenters producing SCFA. Other evidence showed an association between legume fiber intake and Actinobacteria abundance, particularly Bifidobacteriales [45]. A recent systematic review demonstrated that the most consistent results can be related to an increased abundance of SCFA-producers, alterations in microbiota diversity, and in the *Prevotella*/*Bacteroides* ratio. However, to what extent a dietary intervention with fiber may affect the human gut microbiota and hence metabolic regulation is currently not well described, due to differences in methodologies and lack of standardization that hamper the interpretation of the results [46]. It is known that also proteins can shape gut microbiome, and that different protein sources differently impact its profile. As an example, a diet rich in pea protein has been shown to increase *Bifidobacterium* and *Lactobacillus* levels [47]. Approximately 12–18 g of dietary protein reaches the human colon daily. Several gut microbiota species such as *Clostridum* spp., *Bacteroides* spp., and *Lactobacillus* spp. can metabolize proteins through different proteases [48]. Microbial metabolites deriving from dietary protein fermentation by gut microbiota include short branched chain fatty acids, sulfur-containing products, aromatic compounds, polyamines, and ammonia. Interestingly, several neuroactive compounds including neurotransmitters such as GABA, norepinephrine, dopamine, serotonin, and histamine are produced from amino acids by gut microbiota, and this is one of the most attractive topics to understand the role of microbiota in gut-brain axis [48]. On the other hand, the pro-atherogenic metabolite trimethylamine-N-oxide (TMAO) is produced by the combined activity of microbial and host enzymes after consumption of animal proteins, with a negative impact on health. Most of the ingested fatty acids are absorbed in the human small intestine, but a small fraction (about 7%) reaches the colon. With respect to carbohydrates and protein, the impact of dietary fats on gut microbiota profile is less reported. The most characterized effect of a high-fat diet is related to a decreased Bacteroidetes/Firmicutes ratio [49].

5.2 Effect of microbial metabolites on the host immune system

Most of the physiological effects of the microbiota are mediated by metabolites produced by the bacteria themselves or derived from the microbial transformation of host molecules. In fact, the gut microbiota has a high potential to synthesize bioactive compounds by acting on molecules of endogenous origin or derived from the diet. As previously mentioned, SCFAs are the principal metabolites derived from the microbial fermentation of complex polysaccharides. Acetate and propionate are mostly produced by Bacteroidetes, while Firmicutes are the principal butyrate-producing microorganisms [50]. While propionate and acetate reach the liver through the portal vein, where they contribute to gluconeogenesis and lipogenesis, respectively, butyrate, mainly produced by Firmicutes, plays a fundamental role in the intestine and represents the major fuel for enterocytes. SCFAs, especially butyrate, are molecules fully capable of transducing signals, as they are ligands of G-Protein Coupled Receptors (GPCRs). This interaction activates various molecular signaling pathways in the different intestinal cells, resulting in strengthening the intestinal barrier and exerting an anti-inflammatory action. In particular, Paneth cells are stimulated to release antimicrobial substances; intestinal endocrine L cells release satiety peptides, glucagon-like-1 (GLP-1) and peptide YY (PYY); goblet cells are stimulated to produce mucin, while in epithelial cells butyrate exerts a trophic effect, promoting the expression of junction proteins and cell regeneration. SCFAs also have important actions on both innate and adaptive immune cells present in the intestine, increasing IL-10 expression levels and promoting Treg cell differentiation. SCFAs are also epigenetic modulators, as they act as inhibitors of histone deacetylase enzymes, resulting in transcriptional activation of several genes, including a Treg cell-specific transcription factor, Foxp3, that leads to an anti-inflammatory phenotype, through inhibition of NF-κB. In other contexts, however, it has been observed that SCFAs may have opposite, pro-inflammatory effects, especially in the presence of LPS or TNF-α. This observation demonstrates how the same molecule can have beneficial or detrimental effects, depending on the concurrent conditions of eubiosis or dysbiosis [4, 51]. The microbiota plays an essential role also in the metabolism of bile acids, influencing their profile with over 20 different secondary bile acids produced. Such diversity of bile acids composition differently affects the physiology and metabolism of the entire body. Cholesterol-derived primary bile acids, essentially cholic and chenodeoxycholic acid, are first conjugated with taurine and glycine in the liver to form the corresponding conjugated bile salts which are stored in gallbladder. Released into the duodenum after an abundant meal, most bile salts (95%) are reabsorbed from the terminal ileum and colon and delivered back to the liver via the portal vein in a process known as enterohepatic circulation [52]. A small percentage of bile salts, estimated at around 5%, reaches the colon, where they are deconjugated in a reaction catalyzed by bile salt hydrolase (BSH), and mediated by a broad spectrum of aerobic and anaerobic bacteria (Gram-positive *Bifidobacterium*, *Lactobacillus*, *Clostridium*, and *Enterococcus*, and Gram-negative *Bacteroides*). Then bacterial dehydrogenase enzymes convert primary bile acids into the secondary bile acids deoxycholic and lithocholic acids. This reaction is mediated by a limited number of bacteria belonging to *Bacteroides, Clostridium, Eubacterium, Lactobacillus*, and *Escherichia* genera [53]. Thus, gut microbiota composition determines the profile of secondary bile acids that are produced. The secondary bile acids are absorbed into the colon, return to the liver and after being conjugated enter the enterohepatic circulation. Secondary bile acids can undergo epimerization, sulfation, glucuronidation, and conjugation with N-acetylglucosamine in the liver, kidneys, and gut to form tertiary bile acids [52]. Bile acids exert multiple physiological functions, which are: 1 - intestinal detergent activity that solubilizes dietary lipids and fat-soluble vitamins promoting their absorption; 2 - hormone- like properties by acting as signaling molecules via two independent pathways, farnesoid X receptor (FXR) and G protein-coupled bile acid receptor (TGR5) signaling. Binding FXR, bile acids can regulate their homeostasis, as well as lipogenesis, gluconeogenesis, tumor suppression, and intestinal barrier function; while through TGR5, they regulate glucose homeostasis, energy expenditure, and anti-inflammatory response. Different bile acids have different affinities towards these receptors, with secondary bile acids preferentially activating TGR5; 3- antibacterial properties providing protection against invasive microorganisms, and acting as mediators of gut innate defense. However, it is important to note that bile acids can become cytotoxic at high concentrations, and excessive accumulation can lead to oxidative stress, apoptosis, and liver damage [54]. In this context, any dietary component, which could influence gut microbiota composition, may also modulate bile acid homeostasis and the ability to impact host health. High dietary fat intake is known to increase primary bile acids release into the small intestine and stimulate secondary bile acid synthesis mediated by various bacteria, including *Lactobacillus*, *Bifidobacterium,* and *Bacteroides* [55]. Milk fat has been shown to induce shifts in hepatic conjugation of bile acids in mice, from glycocholic to taurocholic acid, compromising barrier integrity and resulting in increased abundance of *Bilophila wadsworthia*, a bile-tolerant pathobiont able to trigger a Th-1 immune response [56]. Carbohydrate intake affects bile acids metabolism, in particular, the role of soluble and insoluble dietary fiber in binding bile acids is well-documented in several studies. In a recent randomized cross-over clinical study, the consumption of a diet rich in whole grains, legumes, vegetables, and fruits was compared with a refined grain diet (high glycemic load) for the effect on circulating bile acids. The results showed a significant increase in the concentrations of specific bile acid ligands of FXR and TGR5 associated to a reduction of insulin resistance [57]. However, the role of the diet on bile acids composition and health is still partially known and needs to be confirmed and expanded in order to translate these findings into clinical settings. Lastly, tryptophan metabolites represent another important class of bacterial metabolites. Tryptophan is an essential amino acid and an important precursor of both microbial and host metabolites. Tryptophan can follow three different metabolic pathways, leading to the formation of serotonin, quinurenine, or indole and its derivatives, which represent the ligands of the aryl hydrocarbon receptor (AhR). In particular, the main microbial metabolite is indole, although the metabolic processes and pathways are complex and multiple. Indole derivatives are considered key mediators of intestinal homeostasis, as they act on epithelial renewal and barrier integrity through the activation of AhRs, which are expressed on many immune cell types, such as IELs, Th17 cells, ILCs, macrophages, dendritic cells, and neutrophils. The main effect is the production of IL-22 by ILC3, which in turn regulates metabolism by improving insulin sensitivity, modulating lipid metabolism in adipose tissue and liver, and promoting intestinal barrier integrity. Indole metabolites may also promote Th17/Treg reprogramming [4, 51].

5.3 Western diet and Mediterranean diet: examples of detrimental and beneficial dietary profiles

The scientific literature describes the diet as the most characterized factor capable of shaping gut microbiota and immune system. Indeed, the nutritional status of an individual and the composition of the diet, in terms of foods, nutrients, and bioactive substances, influence immunity. A recent analysis by Rinninella and colleagues [58] highlighted the effects of different dietary habits on gut microbiota composition, by comparing vegetarian/vegan, gluten-free, ketogenic, low FODMAP (i.e., low in highly fermentable but poorly absorbable carbohydrates and polyols), Western and Mediterranean diets. Overall, restrictive diets (gluten-free, ketogenic, low FODMAP) have been shown to exert negative effects on the intestinal microbiota, in terms of reduction of biodiversity and alteration of eubiosis, impacting also on the integrity of the intestinal epithelium (especially in the case of ketogenic diet), and on inflammatory status. Among the different dietary profiles, the most consistent evidence concerns the Western Diet and the Mediterranean Diet, indeed Western Diet was shown to negatively impact gut microbiota composition and diversity, and to reduce the intestinal mucus layer, thus favoring bacterial translocation and endotoxemia, while Mediterranean Diet was associated to increased bacterial diversity and improved gut barrier function [58]. The Western Diet is typically described as a diet high in calories and rich in ultra-processed foods with high levels of sugars, saturated and trans fats, salt and food additives, while complex carbohydrates, fiber, vitamins and minerals, and other bioactive molecules (such as polyphenols and omega-3 fatty acids) are scarcely present. The main effects of this diet concern the elevation of plasma glucose and insulin levels, with a consequent increase in the accumulation of lipids in adipose tissue, which induces a rapid weight gain compared to more balanced diets. Furthermore, recent rodent and human studies have established that the Western dietary pattern is associated with elevated levels of inflammatory biomarkers, suggesting a direct or indirect action on the immune system [59]. It is noteworthy that macronutrients in food are part of a complex microstructure from which the physical, sensorial and nutritional properties, and health implications derive. "The complex assembly of nutrients and non-nutrients interacting physically and chemically, that influences the release, mass transfer, accessibility, digestibility, and stability of many food compounds" has been described as food matrix [60]. Therefore, diverse food matrices can differently affect the digestion and absorption processes of food compounds and play a role in the microbial fermentation of unabsorbed components. Ultra-processed foods and beverages are considered an important hallmark of the Western Diet, and high consumption of these foods appears to be correlated with an increased risk of morbidity. Food processing involves applying controlled procedures in order to preserve, destroy, transform, and create edible structures, whose aim is to prolong the shelf -life of foods. Ultra-processed foods are microbiologically safe, highly palatable, ready-to-eat, and highly profitable products composed primarily of ingredients not routinely found in "real foods" (e.g., hydrogenated/de-esterified oils or additives designed to provide the previously mentioned characteristics). The poor and uncomplex matrix of these foods, together with their low fiber content, generates an unfavorable environment in the gut and microbiota, thus leading to dysbiosis and immune alterations. Therefore, the Western Diet, intended also as an incorrect lifestyle, would induce low-grade inflammatory processes, which are a risk factor for the development of various chronic inflammatory diseases, predisposing the individual to metabolic inflammation, through various mechanisms, acting at both levels of microbiota and intestinal permeability [61]. The action on the microbiota leads to the onset of dysbiosis, intended both as taxonomic (shifts in microbial groups composition), but especially as metabolic (changes in microbial function). Moreover, the decreased bacterial diversity makes the microbial ecosystem less resilient and more susceptible to external stressors. The increase in pathogenic bacteria also causes an increase in pro-inflammatory metabolites, which can influence the response at the level of GALT, with which they are intimately linked. When abnormalities occur in these interactions, intestinal permeability can increase and the leaky gut phenomenon occurs, leading to metabolic endotoxemia, as described previously. But metabolic inflammation arises primarily at the level of white adipose tissue, where adipocytes, cells almost entirely formed by a single large lipid droplet, release numerous adipokines, cytokine-like molecules, in response to changes in lipid accumulation and in local and systemic inflammation. Adipokines can be either pro- or anti-inflammatory and play a key role in linking metabolism with immune function [62]. In individuals with normal metabolic status, pro- and anti-inflammatory adipokines are correctly balanced, and Th2 lymphocytes, Treg cells, and macrophages with an anti-inflammatory phenotype predominate in adipose tissue. Treg cells secrete IL-10 and also stimulate IL-10 secretion by macrophages. Eosinophils secrete IL-4 and IL-13, further contributing to the anti-inflammatory and insulin-sensitive phenotype. A long-term hypercaloric diet causes an increase in the number and size of adipocytes, which become hypertrophic and dysfunctional, starting to secrete pro-inflammatory adipokines, especially TNF-α. In addition, circulating immune cells, mainly monocytes, are recruited from the bloodstream in response to chemotactic signals (particularly monocyte chemoattractant protein 1, MCP-1) produced in adipose tissue, transmigrate there, and differentiate into macrophages secreting high amounts of pro-inflammatory cytokines, such as TNF-α, IL-1β, IL-6. These cytokines act in a paracrine manner, inducing changes in T lymphocyte populations, with a decrease in Treg and an increase in Th1 cells, which in turn secrete pro-inflammatory cytokines, thus generating a vicious circle, where the inflammatory state becomes systemic. Indeed, cytokines and chemokines from adipose tissue can act in an endocrine way and promote inflammation in other tissues, also causing the onset of insulin resistance and other metabolic disorders associated with obesity [63]. Adipocytes in visceral adipose tissue are metabolically very active and very sensitive to lipolysis, so following a prolonged positive caloric balance, very high amounts of free fatty acids (FFAs) are generated and released into the portal system. Insulin resistance results from an excess of circulating FFAs and excess TNF-α in adipose tissue, as both molecules result in functional blockade of the insulin receptor and its associated signal transduction. In particular, FFAs and TNF-α block insulin receptor signaling by activating phosphorylation of the insulin receptor substrate (IRS)-1 at a serine residue. Serine phosphorylation of IRS-1 causes it to detach from the insulin receptor, resulting in functional blockade of the receptor and of insulin signal transmission itself. In addition, TNF-α, secreted by adipocytes and adipose tissue macrophages, also acts by another mechanism, namely by inducing dephosphorylation of IRS-1 at tyrosine residues. Tyrosine-dephosphorylation has the same effect as serine-phosphorylation, thus IRS-1 is inactivated and detached from the insulin receptor [64]. It is known that several components characterizing the Western Diet determine an inflammatory state through the activation of the innate immune response, for example, excess cholesterol is considered the main cause of inflammation in the atherosclerotic process. In addition, an excess of free cholesterol crystals causes damage to lysosomes, with subsequent release of the pro-inflammatory cytokines IL-1β and IL-18 through activation of the NLR family pyrin domain containing 3 (NLRP3) inflammasome, resulting in a systemic response characterized by a chronic low-grade inflammatory state, associated to insulin resistance and the onset of some related diseases, including colorectal cancer [65]. Saturated fatty acids carried through excessive consumption of animal-derived foods also have cytotoxic effects and can activate endoplasmic reticulum stress as well as the NLRP3 inflammasome. More recent theories suggest that saturated fatty acids induce dysbiosis and subsequent release of metabolites that alter intestinal permeability, inducing metabolic endotoxemia [59]. From a taxonomic point of view, an excess of fat causes an increase in the Firmicutes/Bacteroidetes ratio, while some unrefined oils, such as palm oil, may cause a decrease in *A. muciniphila* and *Clostridium leptum*. Through the consumption of red meat, eggs, and high-fat dairy products, dietary introduced L-carnitine and phosphatidylcholine are converted to the pro-atherogenic metabolite TMAO through a two-stage process, including first a fermentation process by the intestinal microbiota in an anaerobic environment, and then an oxidation catalyzed by the hepatic enzyme Monooxygenase containing Flavin 3. TMAO is a metabolite involved in the activation of inflammatory macrophages and the formation of atherosclerotic plaque foam cells, thus contributing to increased cardiovascular risk [59]. Consumption of excessive amounts of red meat also leads to elevated amounts of iron-eme, which has been associated with alteration of certain bacterial groups including *Escherichia coli* and *B. fragilis*. In general, levels of bacterial genera capable of metabolizing plant polysaccharides such as *Roseburia*, *Eubacterium*, *Ruminococcus,* and *Prevotella* are underrepresented in individuals on the Western Diet, whereas the relative abundance of bile-tolerant microorganisms increases [44]. The concept of Mediterranean Diet has been developed to describe the eating habits followed by the populations of the Mediterranean basin, mainly Greece and Southern Italy. The Mediterranean Diet is based on the consumption of fruits, vegetables, legumes, dried fruits, fish, olive oil, and whole grains which together provide a combination of complex carbohydrates, polyunsaturated fatty acids, and bioactive molecules such as polyphenols and other antioxidants. It is also characterized by a low consumption of proteins of animal origin. A large number of epidemiological studies have shown that the Mediterranean Diet is associated to increased life expectancy, improved quality of life, and lower prevalence of diseases related to chronic inflammation, such as coronary heart disease, type 2 diabetes, and some forms of cancer. These beneficial properties are mediated by different mechanisms, including lipid-lowering, anti-inflammatory, and anti-oxidant effects. Accumulating evidence suggests that such activities are not ascribable to single foods or nutrients, but to interactions and synergistic activities of different nutrients and bioactive compounds from a variety of diverse foods with intact matrices [66]. Among the many molecules found in these foods, omega-3 polyunsaturated fatty acids, polyphenols, as well as fiber, can be mentioned. In particular, omega-3 contributes to balancing the Firmicutes/Bacteroidetes ratio and to increase bifidobacteria and Lachnospiraceae [67], while some polyphenols, e.g., resveratrol or hydroxytyrosol, have been described as modulators of bacterial groups beneficial for human health [68]. Some nutritional intervention trials based on Mediterranean Diet have been proposed as a therapeutic approach to improve the composition of the microbiota and the state of the immune system, opening the perspective of a possible use of this dietary habit to modulate the microbiota, directing it towards a healthy profile. In fact, it has been demonstrated that adherence to the Mediterranean Diet correlates with a state of eubiosis, in which members of the phylum Bacteroidetes and beneficial bacteria belonging to the clostridia group increase, while Proteobacteria and Bacillaceae decrease. In addition, increased levels of lactic acid bacteria (mainly lactobacilli and bifidobacteria) have been observed, together with a more general increase in biodiversity and stability of the intestinal microbiota, suggesting a greater resilience to possible perturbations. A study focused on obese subjects also showed that an intervention with Mediterranean Diet increased the abundance of SCFA-producing gut bacteria *Roseburia* and *Oscillospira* [69]. In conclusion, the Mediterranean Diet, rich in foods of plant origin, provides polyphenols, high-quality fats (monounsaturated such as oleic acid and polyunsaturated with high content of omega 3), micronutrients, such as vitamins and trace elements, and dietary fiber that, carried by an adequate and complete dietary matrix, exert their beneficial properties in maintaining the eubiosis of the intestinal microbiota and its metabolites, together with the integrity of the intestinal barrier and immune tolerance. In contrast, the Western Diet and ultra-processed foods, characterized by low levels of dietary fiber or micronutrients, have a plethora of nutritional components, including refined carbohydrates, poor quality fats (trans fatty acids and an excessive omega 6/omega 3 ratio due to refined oils), unhealthy salt and additives (mainly sweeteners), and finally excessive consumption of red and processed meats. In addition, they comprise a poor food matrix that causes detrimental effects on the intestinal barrier, leading to increased permeability, dysbiosis, and altered metabolite profiles, resulting in local inflammation, systemic endotoxemia, and chronic inflammation. **Figure 4** summarizes these observations.

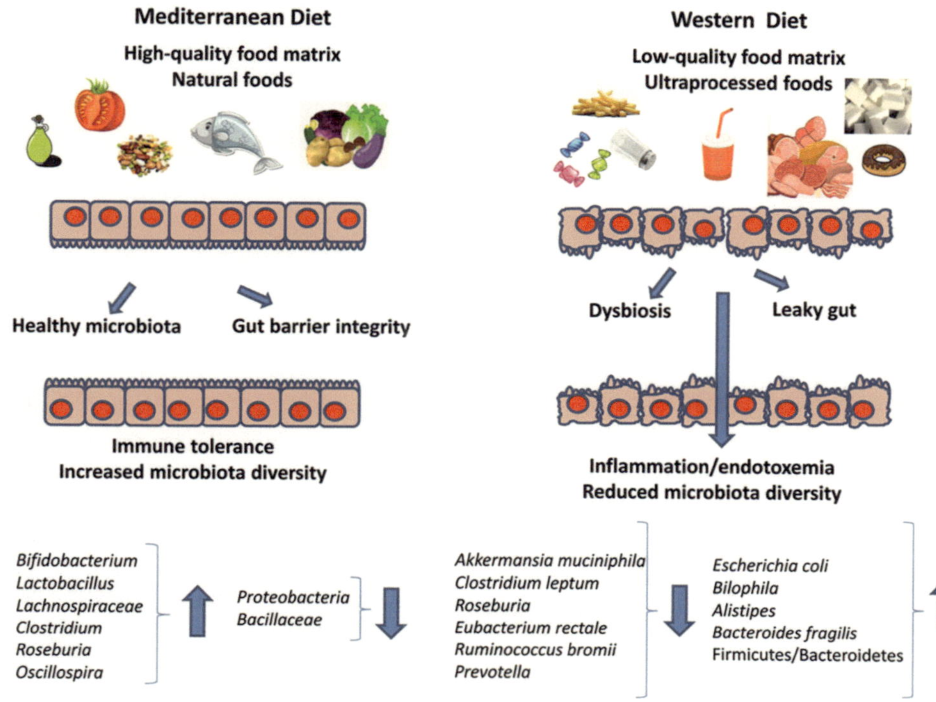

Figure 4.
Comparison of "Mediterranean" and "Western" dietary profiles. The main effects on gut integrity, immune status and microbiota composition are schematically represented.

Acknowledgements

This work was funded by the Italian Ministry of Agriculture, Food, and Forestry Policies (MiPAAF), within the trans-national project INTIMIC–Knowledge Platform on food, diet, intestinal microbiomics, and human health of the Joint Programming Initiative Healthy Diet for a Healthy Life (JPI-HDHL), Expression of Interest n. 795 (MICROFLUX Project, MiPAAF DM 36954/7303/18).

Conflict of interest

The authors declare no conflict of interest.

Author details

Marianna Roselli, Raffaella Canali*, Alberto Finamore, Andrea Ghiselli
and Chiara Devirgiliis
Research Centre for Food and Nutrition, Consiglio per la ricerca in agricoltura e
l'analisi dell'economia agraria (CREA), Rome, Italy

*Address all correspondence to: raffaella.canali@crea.gov.it

References

[1] Ahluwalia B, Magnusson MK, Ohman L. Mucosal immune system of the gastrointestinal tract: Maintaining balance between the good and the bad. Scandinavian Journal of Gastroenterology. 2017;**52**(11):1185-1193

[2] Salzman NH. The role of the microbiome in immune cell development. Annals of Allergy, Asthma & Immunology. 2014;**113**(6):593-598

[3] Arumugam M, Raes J, Pelletier E, Le Paslier D, Yamada T, Mende DR, et al. Enterotypes of the human gut microbiome. Nature. 2011;**473**(7346):174-180

[4] Garcia-Montero C, Fraile-Martinez O, Gomez-Lahoz AM, Pekarek L, Castellanos AJ, Noguerales-Fraguas F, et al. Nutritional components in Western diet versus Mediterranean diet at the gut microbiota-immune system interplay. Implications for health and disease. Nutrients. 2021;**13**(2)

[5] Rodrigues LP, Teixeira VR, Alencar-Silva T, Simonassi-Paiva B, Pereira RW, Pogue R, et al. Hallmarks of aging and immunosenescence: Connecting the dots. Cytokine & Growth Factor Reviews. 2021;**59**:9-21

[6] Abbas A, Lichtman A, Pillai S. Cellular and Molecular Immunology. 9th ed. Elsevier; 2016

[7] Parkin J, Cohen B. An overview of the immune system. Lancet. 2001;**357**(9270):1777-1789

[8] Stiehm ER, Joseph A. Bellanti (ed) immunology IV: Clinical applications in health and disease. Journal of Clinical Immunology. 2012;**32**(3):647

[9] Cario E. Bacterial interactions with cells of the intestinal mucosa: Toll-like receptors and NOD2. Gut. 2005;**54**(8):1182-1193

[10] Aluri J, Cooper MA, Schuettpelz LG. Toll-like receptor signaling in the establishment and function of the immune system. Cell. 2021;**10**(6):1374

[11] Shao BZ, Xu ZQ, Han BZ, Su DF, Liu C. NLRP3 inflammasome and its inhibitors: A review. Frontiers in Pharmacology. 2015;**6**:262

[12] Mjosberg J, Spits H. Human innate lymphoid cells. The Journal of Allergy and Clinical Immunology. 2016;**138**(5):1265-1276

[13] Netea MG. Training innate immunity: The changing concept of immunological memory in innate host defence. European Journal of Clinical Investigation. 2013;**43**(8):881-884

[14] Netea MG, Dominguez-Andres J, Barreiro LB, Chavakis T, Divangahi M, Fuchs E, et al. Defining trained immunity and its role in health and disease. Nature Reviews. Immunology. 2020;**20**(6):375-388

[15] Alwarawrah Y, Kiernan K, MacIver NJ. Changes in nutritional status impact immune cell metabolism and function. Frontiers in Immunology. 2018;**9**:1055

[16] Hwang JR, Byeon Y, Kim D, Park SG. Recent insights of T cell receptor-mediated signaling pathways for T cell activation and development. Experimental & Molecular Medicine. 2020;**52**(5):750-761

[17] Lee GR. The balance of Th17 versus Treg cells in autoimmunity. International Journal of Molecular Sciences. 2018;**19**(3):730

[18] Gutzeit C, Chen K, Cerutti A. The enigmatic function of IgD: Some answers at last. European Journal of Immunology. 2018;**48**(7):1101-1113

[19] Rescigno M, Urbano M, Valzasina B, Francolini M, Rotta G, Bonasio R, et al. Dendritic cells express tight junction proteins and penetrate gut epithelial monolayers to sample bacteria. Nature Immunology. 2001;**2**(4):361-367

[20] Brandtzaeg P. Mucosal immunity: Induction, dissemination, and effector functions. Scandinavian Journal of Immunology. 2009;**70**(6):505-515

[21] Silva-Sanchez A, Randall TD. Anatomical uniqueness of the mucosal immune system (GALT, NALT, iBALT) for the induction and regulation of mucosal immunity and tolerance. Mucosal Vaccines. 2020:21-54

[22] Van Kaer L, Olivares-Villagomez D. Development, homeostasis, and functions of intestinal intraepithelial lymphocytes. Journal of Immunology. 2018;**200**(7):2235-2244

[23] Matijasic M, Mestrovic T, Paljetak HC, Peric M, Baresic A, Verbanac D. Gut microbiota beyond bacteria-Mycobiome, Virome, Archaeome, and eukaryotic parasites in IBD. International Journal of Molecular Sciences. 2020;**21**(8):2668

[24] Ayechu-Muruzabal V, van Stigt AH, Mank M, Willemsen LEM, Stahl B, Garssen J, et al. Diversity of human Milk oligosaccharides and effects on early life immune development. Frontiers in Pediatrics. 2018;**6**:239

[25] Adak A, Khan MR. An insight into gut microbiota and its functionalities. Cellular and Molecular Life Sciences. 2019;**76**(3):473-493

[26] Clemente JC, Ursell LK, Parfrey LW, Knight R. The impact of the gut microbiota on human health: An integrative view. Cell. 2012;**148**(6):1258-1270

[27] Fan Y, Pedersen O. Gut microbiota in human metabolic health and disease. Nature Reviews. Microbiology. 2021;**19**(1):55-71

[28] Simon JC, Marchesi JR, Mougel C, Selosse MA. Host-microbiota interactions: From holobiont theory to analysis. Microbiome. 2019;**7**(1):5

[29] Ducarmon QR, Zwittink RD, Hornung BVH, van Schaik W, Young VB, Kuijper EJ. Gut microbiota and colonization resistance against bacterial enteric infection. Microbiology and Molecular Biology Reviews. 2019;**83**(3):e-00007-19

[30] Cryan JF, O'Riordan KJ, Cowan CSM, Sandhu KV, Bastiaanssen TFS, Boehme M, et al. The microbiota-gut-brain Axis. Physiological Reviews. 2019;**99**(4):1877-2013

[31] Hornef M. Pathogens, commensal symbionts, and Pathobionts: Discovery and functional effects on the host. ILAR Journal. 2015;**56**(2):159-162

[32] Zechner EL. Inflammatory disease caused by intestinal pathobionts. Current Opinion in Microbiology. 2017;**35**:64-69

[33] Behzadi P, Garcia-Perdomo HA, Karpinski TM. Toll-like receptors: General molecular and structural biology. Journal of Immunology Research. 2021;**2021**:9914854

[34] Al-Sadi R, Boivin M, Ma T. Mechanism of cytokine modulation of epithelial tight junction barrier. Front Bioscience (Landmark Ed). 2009;**14**:2765-2778

[35] Erturk-Hasdemir D, Kasper DL. Finding a needle in a haystack: Bacteroides fragilis polysaccharide a as the archetypical symbiosis factor. Annals of the New York Academy of Sciences. 2018;**1417**(1):116-129

[36] Hedblom GA, Reiland HA, Sylte MJ, Johnson TJ, Baumler DJ. Segmented filamentous bacteria - metabolism meets immunity. Frontiers in Microbiology. 2018;**9**:1991

[37] Flannigan KL, Denning TL. Segmented filamentous bacteria-induced immune responses: A balancing act between host protection and autoimmunity. Immunology. 2018;**154**:537

[38] O'Toole PW, Marchesi JR, Hill C. Next-generation probiotics: The spectrum from probiotics to live biotherapeutics. Nature Microbiology. 2017;**2**:17057

[39] Venter C, Eyerich S, Sarin T, Klatt KC. Nutrition and the immune system: A complicated tango. Nutrients. 2020;**12**(3):818

[40] Childs CE, Calder PC, Miles EA. Diet and Immune Function. Nutrients. 2019;**11**:8

[41] Zmora N, Bashiardes S, Levy M, Elinav E. The role of the immune system in metabolic health and disease. Cell Metabolism. 2017;**25**(3):506-521

[42] Gophna U. Microbiology. The Guts of Dietary Habits Science. 2011;**334**(6052):45-46

[43] Costea PI, Hildebrand F, Arumugam M, Backhed F, Blaser MJ, Bushman FD, et al. Enterotypes in the landscape of gut microbial community composition. Nature Microbiology. 2018;**3**(1):8-16

[44] David LA, Maurice CF, Carmody RN, Gootenberg DB, Button JE, Wolfe BE, et al. Diet rapidly and reproducibly alters the human gut microbiome. Nature. 2014;**505**(7484):559-563

[45] Dominianni C, Sinha R, Goedert JJ, Pei Z, Yang L, Hayes RB, et al. Sex, body mass index, and dietary fiber intake influence the human gut microbiome. PLoS One. 2015;**10**(4):e0124599

[46] Myhrstad MCW, Tunsjo H, Charnock C, Telle-Hansen VH. Dietary fiber, gut microbiota, and metabolic regulation-current status in human randomized trials. Nutrients. 2020;**12**(3):859

[47] Swiatecka D, Narbad A, Ridgway KP, Kostyra H. The study on the impact of glycated pea proteins on human intestinal bacteria. International Journal of Food Microbiology. 2011;**145**(1):267-272

[48] Portune KJ, Beaumont M, Davila A-M, Tomé D, Blachier F, Sanz Y. Gut microbiota role in dietary protein metabolism and health-related outcomes: The two sides of the coin. Trends in Food Science & Technology. 2016;**57**:213-232

[49] Murphy EA, Velazquez KT, Herbert KM. Influence of high-fat diet on gut microbiota: A driving force for chronic disease risk. Current Opinion in Clinical Nutrition and Metabolic Care. 2015;**18**(5):515-520

[50] Feng W, Ao H, Peng C. Gut microbiota, short-chain fatty acids, and herbal medicines. Frontiers in Pharmacology. 2018;**9**:1354

[51] Gasaly N, de Vos P, Hermoso MA. Impact of bacterial metabolites on gut barrier function and host immunity: A focus on bacterial metabolism and its relevance for intestinal inflammation. Frontiers in Immunology. 2021;**12**:658354

[52] Pushpass RG, Alzoufairi S, Jackson KG, Lovegrove JA. Circulating bile acids as a link between the gut microbiota and cardiovascular health: Impact of prebiotics, probiotics and polyphenol-rich foods. Nutrition Research Reviews. 2021;**1-20**:1

[53] Rodriguez-Morato J, Matthan NR. Nutrition and gastrointestinal microbiota, microbial-derived secondary bile acids, and cardiovascular disease. Current Atherosclerosis Reports. 2020;**22**(9):47

[54] Singh J, Metrani R, Shivanagoudra SR, Jayaprakasha GK, Patil BS. Review on bile acids: Effects of the gut microbiome, interactions with dietary fiber, and alterations in the bioaccessibility of bioactive compounds. Journal of Agricultural and Food Chemistry. 2019;**67**(33):9124-9138

[55] Yokota A, Fukiya S, Islam KB, Ooka T, Ogura Y, Hayashi T, et al. Is bile acid a determinant of the gut microbiota on a high-fat diet? Gut Microbes. 2012;**3**(5):455-459

[56] Devkota S, Chang EB. Interactions between diet, bile acid metabolism, gut microbiota, and inflammatory bowel diseases. Digestive Diseases. 2015;**33**(3):351-356

[57] Ginos BNR, Navarro SL, Schwarz Y, Gu H, Wang D, Randolph TW, et al. Circulating bile acids in healthy adults respond differently to a dietary pattern characterized by whole grains, legumes and fruits and vegetables compared to a diet high in refined grains and added sugars: A randomized, controlled, crossover feeding study. Metabolism. 2018;**83**:197-204

[58] Rinninella E, Cintoni M, Raoul P, Lopetuso LR, Scaldaferri F, Pulcini G, et al. Food components and dietary habits: Keys for a healthy gut microbiota composition. Nutrients. 2019;**11**(10):2393

[59] Christ A, Lauterbach M, Latz E. Western diet and the immune system: An inflammatory connection. Immunity. 2019;**51**(5):794-811

[60] Aguilera JM. The food matrix: Implications in processing, nutrition and health. Critical Reviews in Food Science and Nutrition. 2019;**59**(22):3612-3629

[61] Christ A, Latz E. The Western lifestyle has lasting effects on metaflammation. Nature Reviews. Immunology. 2019;**19**(5):267-268

[62] Cao H. Adipocytokines in obesity and metabolic disease. The Journal of Endocrinology. 2014;**220**(2):T47-T59

[63] Osborn O, Olefsky JM. The cellular and signaling networks linking the immune system and metabolism in disease. Nature Medicine. 2012;**18**(3):363-374

[64] Rui L, Aguirre V, Kim JK, Shulman GI, Lee A, Corbould A, et al. Insulin/IGF-1 and TNF-alpha stimulate phosphorylation of IRS-1 at inhibitory Ser307 via distinct pathways. The Journal of Clinical Investigation. 2001;**107**(2):181-189

[65] De Nardo D, Latz E. NLRP3 inflammasomes link inflammation and metabolic disease. Trends in Immunology. 2011;**32**(8):373-379

[66] Tosti V, Bertozzi B, Fontana L. Health benefits of the Mediterranean diet: Metabolic and molecular mechanisms. The Journals of Gerontology. Series A, Biological Sciences and Medical Sciences. 2018;**73**(3):318-326

[67] Rogero MM, Calder PC. Obesity, inflammation, toll-like receptor 4 and fatty acids. Nutrients. 2018;**10**(4):432

[68] Yahfoufi N, Alsadi N, Jambi M, Matar C. The immunomodulatory and anti-inflammatory role of polyphenols. Nutrients. 2018;**10**(11):1618

[69] Haro C, Montes-Borrego M, Rangel-Zuniga OA, Alcala-Diaz JF, Gomez-Delgado F, Perez-Martinez P, et al. Two healthy diets modulate gut microbial community improving insulin sensitivity in a human obese population. The Journal of Clinical Endocrinology and Metabolism. 2016;**101**(1):233-242

Chapter 6

Immunology of *Helicobacter pylori* Infection

Darmadi Darmadi and Riska Habriel Ruslie

Abstract

Helicobacter pylori (*H. pylori*) is the most common infecting microorganism in humans. *H. pylori* had coexisted with humans for 30,000 years ago and developed extensive survival adaptations. The infection is both active and chronic and leads to several disorders from chronic gastritis to gastric adenocarcinoma. The prevalence of *H. pylori* infection is still high in developing countries. The burden of disease due to infection is also heavy. The persistence of infection is the basis of diseases. H. infection activates innate and adaptive immune responses but the immune response fails to eradicate the infection. *H. pylori* is able to evade both innate and adaptive immune responses. It can neutralize gastric acid, elicit autoimmunity toward parietal cells, prevent phagocytosis, induce apoptosis of immune cells, inhibit lymphocyte proliferation, disrupt imbalance between humoral and cellular adaptive immune responses, promote regulatory T cell activity, and trigger genetic rearrangement. Host factor is involved in the incidence of *H. pylori* infection and its complications. Reinfection after eradication is common. Multiple drug resistance has also emerged. Vaccination is a promising management approach to eradicate *H. pylori* and prevent diseases it caused. The development of the vaccine itself needs to consider the immune escape mechanism of *H. pylori*.

Keywords: adaptive, *Helicobacter pylori*, immune, innate, evasion, vaccine

1. Introduction

Helicobacter pylori (*H. pylori*) is one of the most common infections in humans [1–3]. The microorganism had infected humans for 30,000 years ago and has developed extensive adaptations to survive [2, 4, 5]. Approximately, *H. pylori* infects the stomach of half of the human population globally [3, 6, 7]. Besides residing in the stomach, abundant *H. pylori* are also detected in the oral cavity [8]. The colonization is suspected to be started in the childhood period [2, 4, 7] and maybe persisted for decades or life [2, 4]. The presence of spiral microorganisms resembling *H. pylori* had been identified in the stomachs of the animal during the late nineteenth and early twentieth centuries. Similar spiral bacteria were then isolated in humans, particularly those suffering from peptic ulcer disease or gastric cancer. Previously, the microorganism was named 'Campylobacter-like organism', 'gastric Campylobacter-like organism', '*Campylobacter pyloridis*', or '*Campylobacter pylori*'. The fact that this microorganism is different from members of the genus Campylobacter changes the name to *H. pylori* [9]. The relationship between this microorganism and peptic ulcer disease was established in 1983 [10]. This microorganism causes a wide spectrum of diseases, such as chronic gastritis, peptic ulcer, gastric adenocarcinoma, and mucosa-associated lymphoid tissue lymphoma [6, 7]. In initial reports from all over the world in the first decade after the discovery of *H. pylori*, approximately 95% of duodenal ulcers and 85% of gastric ulcers occurred in the presence of *H. pylori* infection [9]. The World Health Organization has classified H. pylori as a class I carcinogen due to its epidemiological link with gastric malignancy [2, 3, 10]. However in some cases, the presence of *H. pylori* infection is asymptomatic [3, 7, 11]. There is a hypothesis suggesting the role of immune response in the pathogenesis of infection. The immune response toward the infection is ineffective, causing persistent microorganisms and inflammation [6]. In this review, we discuss the host immune response toward *H. pylori* infection in association with disease chronicity and vaccine development.

2. Characteristics of *H. pylori*

H. pylori belongs to the Proteobacteria subdivision, Campylobacterales order, and Helicobacteraceae family. Helicobacter species are divided into two major lineages: the gastric Helicobacter species and the enterohepatic (nongastric) Helicobacter species [9]. *H. pylori* is a spiral, gram-negative, flagellated, microaerophilic, and facultative acidophilic bacterium [1, 5, 11, 12]. Its envelope consists of an inner (cytoplasmic) membrane, periplasm with peptidoglycan, and an outer membrane that consists of phospholipids and lipopolysaccharide [9]. This microorganism is very sensitive to drying and usual disinfectants [12]. It is transmitted via oral–oral and fecal–oral routes [5, 8, 9, 13]. Contamination of water sources is one major cause of transmission [8, 9, 14]. It is reported that 40% of samples of drinking water in Pakistan are contaminated with *H. pylori* [8]. Contamination of drinking water is also reported in 20.3% of samples in Peru. Recent transmission hypothesis has suggested that blowflies and houseflies are responsible as they feed with and breed in fecal material [14]. *H. pylori* extracts nutrients from blood and host cells [5]. The microorganism has extensive genetic diversity resulting in high mutation rates and high recombinant frequency. The virulence factors of *H. pylori* are also affected by this phenomenon and contribute to immune escape and chronic infection [2, 12]. Several methods of DNA rearrangement along with the introduction and deletion of foreign sequences are responsible for genetic diversity [9].

Some factors contributing to *H. pylori* infection are younger age, [4] low socioeconomic status, limited living space, sharing of beds, low parental education level, pollution of daily used water, and history of *H. pylori* infection in family members [4, 8, 13]. Genetic predisposition may play role in the infection of *H. pylori*. People from African and Pacific Islander ancestries have a higher risk for infection despite adjustment for other risk factors [8]. This is supported by another study which reported a higher prevalence in non-whites compared to non-Hispanic whites in the United States. Higher prevalence was even observed in Alaskan natives [10]. A diet containing less vegetables and fruits along with high consumption of fried food is increasing the risk for infection [14]. Another literature states that age and gender are not related to increased risk of infection [13]. Additionally, the effect of smoking and alcohol is uncertain on the incidence of *H. pylori* infection [13, 14].

3. Epidemiology and burden of *H. pylori* infection

The prevalence of *H. pylori* infection is low in childhood and has begun to increase to 20% in adults younger than 40 years of age, and to 50% at the age of 60 [9]. In 2015, approximately 4.4 billion individuals were infected with *H. pylori* [13]. The prevalence rate comprises roughly 4.3% [14]. A systematic review and meta-analysis by Hooi, et al reported that Africa has the highest pooled prevalence of *H. pylori* infection while Oceania has the lowest rate [10]. The prevalence tends to decrease recently due to improvements in sanitation [9, 10]. Similar result is reported by Sjomina et al. a few years later, showing the minimal change in the epidemiology of *H. pylori* infection [14]. In Europe, Northern countries report lower prevalence compared to Southern and Eastern countries. The highest prevalence in Europe is reported in Portugal, reaching 84.2%. In the American continent, Mexico holds the highest prevalence (52.2%) similar to Bhutan in the Asia continent (86%). A study from Nigeria reported a very high prevalence of *H. pylori* infection, which is 93.6% [13]. In the Australia and Oceania region, the highest prevalence is detected on Pacific Island (49%). Minor differences are observed regarding the epidemiology of H. infection in several studies and it is due to the different diagnostic methods utilized from one study to another [8].

H. pylori is associated with 92% of duodenal ulcers and 70% of gastric ulcers. It is also related to 50% of gastric cancer and raises the risk for gastric cancer six times higher compared to those without *H. pylori* infection [11]. The odd for ulcer disease is even higher, reaching 10 times higher than *H. pylori*-negative subjects [9]. A study by Plummer et al. reported that 6.2% of estimated 12.7 million new malignancy cases in 2008 are attributed to *H. pylori* infection [15]. The incidence of gastric cancer is associated with geographical factors, strain diversity, and host immunological responses. The highest incidence of gastric cancer is reported in East Asian countries [16]. In 2017, there were 1.22 million new cases of gastric cancer with 865.000 deaths and 19.1 million disability-adjusted life-years. Not all, but the majority of gastric cancers are related to *H. pylori* infection therefore the microorganism is responsible for the burden of the disease [17]. Eradication of *H. pylori* will give a significant impact from an economic perspective [8, 17]. Eradication leads to decreased consultation with medical practitioners and is proven as an effective cost-saving method [8]. Screening and eradication of *H. pylori* infection in China might prevent one gastric cancer in every four to six cases [10].

4. Immunology at a glance

There are two major groups of immunity in the human body—innate and adaptive immune responses. Innate immunity comprises immune responses which do not require the previous contact with immune triggers [6, 18]. The response is rapid but not specific and has no memory [18]. Innate immunity acts as the first line of defense against harmful substances. Activation of the innate immune system may eliminate the substance and trigger inflammation by releasing mediators such as cytokines, reactive oxygen species, and nitric oxide [6]. Elimination may also be carried on by cell-dependent mechanisms, such as phagocytosis and cytotoxicity [18]. In the gastrointestinal tract, mucosal defense is classified as an innate immune system that consists of mucosal epithelium, gastric acid, and immune cells (macrophage and dendritic cell) [1, 18]. The innate cellular immune may sense the presence of antigen via pattern recognition receptors (PRR), such as toll-like receptors (TLR) [18].

Adaptive immunity is an immune response toward previously contacted immune triggers. This immune system is specific and has immunologic memory. Activation of adaptive immunity is related to innate immunity. For example, antigen-presenting cells (macrophages and dendritic cells), as a part of the innate immune system, trigger activation and differentiation of T-helper (Th) cells, which marks the initiation of the adaptive immune response [6]. Th cells differentiate into Th1, Th2, Th17, and regulatory T (Treg) cells. Th1 plays role in cell-mediated immunity while Th2 in humoral immunity [4–7]. The balance between Th1 and Th2 is important in maintaining a normal host's immune response [6]. Th1 cells secrete tumor necrosis factor (TNF) and interferon (IFN)γ. Th2 cells secrete interleukin (IL)-4, IL-5, and IL-10 which act in suppressing the inflammatory effect of Th1 and in producing antibodies by lymphocyte B cells [3–5, 7]. Th17 plays role in the immune response toward extracellular bacterial infection by secreting IL-17A, IL-17F, IL21, and IL-22. Treg itself has activity in suppressing effector T cells proliferation and cytokine production, therefore moderating inflammation and preventing autoimmunity. Some cytokines secreted by Th17 are IL-10 and transforming growth factor (TGF)-β [3–5].

5. Immune response toward *H. pylori* infection

Many diseases, including infection due to *H. pylori*, involve dysregulation of the immune system. Infection is both active, marked by neutrophilic accumulation, and chronic, marked by lymphocytic deposition [1, 5, 6, 9]. These findings are positive 2 weeks after infection. Anti-*H. pylori* antibodies are also detected at 4 weeks after initial infection, marked by the high levels of immunoglobulin (Ig) M, IgG, and IgA in gastric mucosa of infected patients [1, 3, 11]. A study in mice demonstrated that transient infiltration of macrophages and neutrophils into the glandular stomach is observed in the first 2 days after infection. By day 10 after infection, the numbers of macrophages and neutrophils are decreased to baseline levels. The adaptive immune response is started to appear in the 3rd week, marked by infiltration of T lymphocytes in paragastric lymph nodes and elevated expression of TNFα and IFNγ [19]. The levels of IgM and IgA anti-*H. pylori* in biopsy specimens from the gastric antral region of patients infected with *H. pylori* are 40- to 50-fold higher compared to non-infected subjects [3]. However, the presence of *H. pylori* in the stomach for a long period of time supports the suspected ineffective immune response [2, 6, 9]. The presence of this microorganism causes a persistent and chronic infection [9]. Chronic infection leads to chronic inflammation, gastritis, peptic ulcer, gastric mucosa-associated lymphoid tissue lymphoma, and ultimately, gastric cancer [1, 6].

H. pylori infection activates innate and adaptive immunity, along with humoral and cellular immunity as the parts of the adaptive immune system. There are cytotoxin-associated gene pathogenicity islands (cag PAI) and vacuolating toxin A (VacA) which act as major virulence factors in *H. pylori* infection. Cag PAI encodes a type IV bacterial secretion system that injects bacterial products into gastric epithelial cells resulting in inflammation and increased risk of malignancy [1, 6, 7, 9]. Cag PAI is a protein with a molecular mass of 140 kDa. It is highly immunogenic and present in approximately 50–70% of *H. pylori* strains [9]. VacA, on the other hand, is associated with cellular damage and inflammation [6]. VacA is a protein-sized 95 kDa and secreted from approximately 50% of all *H. pylori* strains. It damages cells by inducing massive vacuolization. The process ends with apoptosis and immune modulation [9]. *H. pylori* enters the gastrointestinal tract, penetrates the mucus gastric layer, and interacts with macrophages, dendritic cells, and monocytes [1, 6, 7]. *H. pylori* adhere to the gastric epithelial cell with the assistance of outer membrane proteins such as BabA, SabA, AlpA, AlpB, and HopZ [1, 5, 20]. After adherence, cag PAI and VacA disrupt gastric epithelial cell polarity, acid secretion (via control of gastrin and H^+/K^+-ATPase), and induce inflammation [1]. TLR on epithelial cells also recognizes bacterial products, such as flagella and lipopolysaccharide. The interaction elicits inflammation and supports the activation of the adaptive immune response [9]. *H. pylori* which have been ingested by antigen-presenting cells activate the adaptive immune response [2, 5]. Macrophages and neutrophils may also eliminate *H. pylori* through nitric oxide (NO)-dependent phagocytosis or reactive oxygen species production [5, 6]. They release cytokines such as IL-12, IL-10, and IL-23 which in turn stimulate naïve Th cells [2, 6]. In the other way, dendritic cells present *H. pylori* antigen to naïve Th cells. Naïve Th cells then differentiate into Th1 or Th2/Treg [1, 6]. However, Th1 is more prominent compared to Th2/Treg cells. Th1 then produces IFNγ, TNFα, and IL-2 [1, 2, 4, 6]. Elevation of pro-inflammatory cytokines, such as IL-1β, TNFα, IL-8, and IL-6, is observed. The release of cytokines promotes inflammation in the stomach and leads to gastritis [1, 6]. In contrast, the role of lymphocyte B cells in *H. pylori* infection is indeterminate. Studies reported that antibodies against *H. pylori* are produced but they might be counterproductive [1, 2, 6]. It is suspected that the immunoglobulins against *H. pylori* are easily degraded and unstable in structure [1, 6]. Other literature states that the presence of IgA anti-*H. pylori* gives a protective effect against infection and gastric malignancy [11]. Further investigation is mandatory regarding the role of humoral antibodies in *H. pylori* infection [6].

6. Mechanism of immune evasion of *H. pylori*

The outer membrane proteins in *H. pylori* are found to be less immunogenic compared to proteins from other pathogens; therefore, the immune response elicited by the innate immune system is less powerful [7, 9]. It is known that *H. pylori*'s lipopolysaccharide has a 500- to 1000-fold lower endotoxic activity than lipopolysaccharide from S. typhimurium and *E. coli* [3]. The presence of arginase enzyme coded by rocF gene in *H. pylori* may decrease L-arginine, the substrate for NO, level. Decreased NO level will impair phagocytosis by macrophage and prevent *H. pylori* elimination. Additionally, this will promote the apoptosis of macrophages [1–3, 6, 7, 20]. *H. pylori* are also capable of producing urea from L-arginine and further ammonia from urea. The process is mediated by the urease enzyme and α-carbonic anhydrase. Ammonia is known for its ability to neutralize gastric HCl and sustain the survival of *H. pylori* [1, 5, 6, 20]. Glucosylation of cholesterol also aids *H. pylori's* survival. This process protects the microorganism from macrophage phagocytosis [5, 6]. *H. pylori* may also evade innate immune recognition by avoiding PRR, a subset of pathogen-associated molecular patterns. It modulates its surface molecules including lipopolysaccharide and flagellin to avoid recognition by toll-like receptors on antigen-presenting cells. The molecules are recognized as self molecules and thus do not trigger the immune response [2, 3, 5]. Even after being phagocytized, *H. pylori* may survive from killing by the aid of cag PAI and VacA. Both delay actin polymerization and phagosome formation inside macrophages [2].

Chronic exposure of dendritic cells to *H. pylori* decreases the ability of dendritic cells to induce Th1 response and support the persistence of infection [1–6]. The malfunction of dendritic cells is due to *H. pylori*-controlled maturation. *H. pylori* restore transcription factor in dendritic cells and inhibit their maturation [2]. Lewis antigen form *H. pylori* may also bind dendritic cell-specific ICAM-3-grabbing nonitegrin (DC-SIGN) and blocked Th1 cell recruitment [4]. Examination of patients with chronic *H. pylori* infection also shows elevation of Treg cells in the gastric tissue compared to healthy subjects [2, 4–6, 20]. *H. pylori* are suggested to promote the expansion of the Treg population and their recruitment to the site of infection [4]. As we know that Treg suppresses the activity of memory T cells, it will relieve inflammation and gastritis severity [1, 2, 4, 6] but at the same time hamper the ability of the host to eliminate pathogens, including *H. pylori* [2, 4, 6, 20]. The condition is hypothesized from the increased level of TGF-β and IL-10 independent of VacA and cag PAI [2, 20]. *H. pylori* inhibits lymphocyte proliferation via IL-2 inhibitory effect from VacA and induction of cell cycle arrest from VacA-independent produced protein [1–6]. The process is made possible via an interference signaling pathway at the level of calcium-calmodulin-dependent phosphatase calcineurin [4]. Gamma-glutamyl transpeptidase (GGT) is another low-molecular-weight protein of *H. pylori* that is capable of inhibiting the proliferation of lymphocytes. The mechanism involves extracellular cleavage of glutathione and the production of reactive oxygen species. The depletion of L-arginine level due to arginase activity of *H. pylori* is also hampering lymphocyte T cell proliferation [4, 5]. Furthermore, VacA may induce T cell apoptosis by reducing Bcl-2, an anti-apoptotic protein [2, 5].

Studies from chronic gastritis found that *H. pylori* may induce autoimmunity which affects gastric parietal cells. Both cellular and humoral antigens damage the cell in patients with gastritis due to *H. pylori* infection [1, 3, 6, 11]. The origin of autoantibody is suspected from the presence of Lewis x and Lewis y antigens which are similar to the H^+/K^+-ATPase β subunit of parietal cells. Parietal cell loss occurs via IFNγ-mediated inflammation and Fas-mediated apoptosis or cytotoxicity [2, 3, 6, 7, 9, 11]. The presence of pro-inflammatory cytokines also inhibits acid secretion from parietal cells. IL-1β and TNFα are the most potent inhibitors [6]. The resulting hypochlorhydria situation allows *H. pylori* to persist and cause prolonged infection [3, 6]. In contrast, those pro-inflammatory cytokines stimulate gastrin secretion by disrupting the negative feedback signal of somatostatin [6].

Coinfection between *H. pylori* and parasitic helminths will cause an imbalance in Th1 and Th2 responses with predominantly Th2 activity [6, 9, 10]. This situation is clearly observed in the African population and referred to as 'African enigma'. 'African enigma' is marked by low gastric cancer despite a high prevalence of *H. pylori* in Africa. Lately, it is known that high rate of helminth coinfection is high in the corresponding population [10]. The variation in Lewis antigen also moderates Th1 response and favors Th2 activity [7]. This condition is supported by a study in mice infected with *H. pylori* showing dysfunctional Th1 response [4]. The imbalance will alleviate inflammation in gastritis but hamper Th1-mediated *H. pylori* elimination [2, 6, 7].

Host factor also contributes to immune response toward H. pylori infection. Host genetic polymorphisms affecting the IL-1 gene cluster elevate the level of IL-1 and lead to the reduction of gastric acid secretion. Low gastric acid secretion promotes infection and colonization of *H. pylori*. A similar situation is induced by a polymorphism in the TNF-α gene. In contrast with the IL-10 gene, the polymorphism causes higher expression of IL-10 and favors anti-inflammatory activity [9]. Defects in cytokine coding genes are involved in the persistence of *H. pylori* infection. Defects in gene coding IL-1 and TNF are associated with decreased cytokines production and increased risk for gastric cancer [7]. Single nucleotide polymorphism in gene coding IL-10 which resulted in increased IL-10 will promote Th2 activity and resulted in prolonged *H. pylori* infection and an elevated risk for recurrent gastric cancer [16]. The presence of *H. pylori* in the macrophages alters the expression of miRNA. The alteration in miRNA, particularly miR-4270 causes increased expression of CD300E, a surface protein on macrophages that affects the antigen presentation capacity of macrophages. Increased CD300E expression is negatively correlated with antigen presentation capacity [21]. Shakhatreh, et al conducted a study on the Jordanian population to determine the association between IL-1 gene polymorphism and *H. pylori* infection. -31T/C polymorphism was found significantly associated with *H. pylori* infection, particularly the TT genotype [22]. Those statements are reinforced by a meta-analysis by Ma et al. They focused on polymorphism in genes that code IL-1. Increased risk for *H. pylori* infection is observed in IL-1B-31C/T polymorphism with an odds ratio of 1.134. Furthermore, IL-1B-511C/T and IL-8-251A/T polymorphisms increase the risk for gastric cancer with odds ratios of 1.784 and 1.810, respectively [23]. Zeyaullah et al. also proposed the role of gene polymorphism in gastric cancer. IL-10-592A/C, IL-10-819T/C, and IL-17-197G/A are all found to be related to gastric cancer. Besides polymorphism in cytokine genes, toll-like receptor genes are also involved. TLR4+ 1196C/T polymorphism is one genetic rearrangement that increases the risk of gastric cancer in *H. pylori*-infected individuals [24].

7. Vaccination and immune response toward *H. pylori*

The trend of antibiotic resistance in *H. pylori* is increasing recently [1, 6, 10, 11]. Resistance rates for metronidazole, clarithromycin, and levofloxacin are the highest, surpassing 50%. This condition is worsening in previously treated subjects [25]. Besides, reinfection may occur even after the complete eradication of the previous infection [6]. This condition urges the development of a vaccine against *H. pylori* [1, 6, 10]. In the past decade, much effort has been devoted to the development of a vaccine as an alternative treatment for *H. pylori* infection [9]. There are two types of vaccine which are potentially possible: prophylactic vaccine and therapeutic vaccine. A therapeutic vaccine that can both eradicate infection and stimulate long-lasting immunity is the most desired one [10, 11]. Vaccine will significantly cut the economic burden from *H. pylori* infection even if the vaccine's efficacy is only 55% [2]. The first report on *H. pylori* vaccine development was submitted in 2011 by Moss et al., followed by Iankov et al. They conducted trials in mice and showed promising results. Cellular immunity, particularly Th1 response, is able to sterilize the microorganism [20]. The humoral immune response also gains the spotlight for the vaccine platform. The induction of Th2 response is proposed to be the basis of effective vaccination. A trial in mice showed high neutralizing specific salivary IgA and serum IgG after oral immunization. Besides preventing infection, the vaccine was also shown to have therapeutic properties. Gastric inflammation of mice in the trial was alleviated after vaccination [9]. In line with previous literature, Espinosa-Ramos conducted a vaccination trial in mice and observed that plasma secretory IgA and IgG are elevated post-vaccination. The presence of antibodies also protected 100% of mice in the study from virulent *H. pylori* [26]. The utilization of the vaccine seems promising, but this option still needs further development, especially in humans, considering the immune evasion ability of *H. pylori* [6, 9]. Public health intervention is still a major concern since preventing is better than treating the infection. Improvement in socioeconomic status together with hygiene and sanitation may decrease the rate of infection as seen in developed countries [9].

8. Conclusion

H. pylori is the most common infection in humans and has infected humans since 30,000 years ago. The prevalence of *H. pylori* infection is still high particularly in developing countries. The highest prevalence is reported in Africa, namely Nigeria. *H. pylori* infection causes a wide spectrum of diseases from chronic gastritis to gastric adenocarcinoma. The disease has high morbidity and mortality rates. The burden from diseases caused by *H. pylori* is also heavy. The transmission of *H. pylori* is via oral-oral or fecal-oral routes. Contamination of water sources for drinking is a significant mode of transmission. The transmission is closely related to socioeconomic status, hygiene, and sanitation. *H. pylori* infection activates both innate and adaptive immunity. In adaptive immunity, Th1 response is dominant compared to Th2. Despite activating the immune system, *H. pylori* eradication by immune response is ineffective. *H. pylori* has abilities to escape from the host's immune system. In the innate immune system, *H. pylori* can neutralize gastric acid via urease enzyme activity and autoimmune-induced parietal cell loss. *H. pylori* prevents phagocytosis and promotes apoptosis of macrophages. Its LPS is less immunogenic compared to other gram-negative bacteria. Chronic infection hampers dendritic cell ability and disturbs activation of the adaptive immune response. In the adaptive immune system, *H. pylori* inhibit lymphocyte proliferation, induces T cell apoptosis, promotes Th2 activity and suppresses Th1 activity via Lewis antigen, and promotes Treg expansion thus dampens inflammation. External factors, such as coinfection with helminths, support the activity of Th2 and hamper *H. pylori* eradication. Genetic rearrangement is induced by *H. pylori* or by the host itself. The rearrangement alters immune response and causes ineffective eradication of *H. pylori*. Multiple antibiotic resistance is observed in *H. pylori,* particularly against metronidazole, clarithromycin, and levofloxacin. This contributes to persistent *H. pylori* infection. Vaccination becomes promising alternative management for preventing infection. Additionally, the vaccine may also have a therapeutic effect. However, the development of a vaccine should pay attention to the immune escape mechanism of *H. pylori*. Public health intervention is still important to holistically manage the infection.

Conflict of interest

The authors declare no conflict of interest.

Author details

Darmadi Darmadi[1*] and Riska Habriel Ruslie[2]

1 Faculty of Medicine, Department of Internal Medicine, Universitas Sumatera Utara, Medan, Indonesia

2 Faculty of Medicine, Department of Child Health, Universitas Prima Indonesia, Medan, Indonesia

*Address all correspondence to: darmadi@usu.ac.id

References

[1] Moyat M, Velin D. Immune responses to *Helicobacter pylori* infection. World Journal of Gastroenterology. 2014;**20**:5583-5593. DOI: 10.3748/wjg.v20.i19.5583

[2] Lina TT, Alzahrani S, Gonzalez J, Pinchuk IV, Beswick EJ, Reyes VE. Immune evasion strategies used by *Helicobacter pylori*. World Journal of Gastroenterology. 2014;**20**:12753-12766. DOI: 10.3748/wjg.v20.i36.12753

[3] Suarez G, Reyes VE, Beswick EJ. Immune response to *H pylori*. World Journal of Gastroenterology. 2006;**12**:5593-5598. DOI: 10.3748/wjg.v12.i35.5593

[4] Larussa T, Leone I, Suraci E, Imeneo M, Luzza F. *Helicobacter pylori* and T helper cells: Mechanisms of immune escape and tolerance. Journal of Immunology Research. 2015;**2015**:981328. DOI: 10.1155/2015/981328

[5] Yang H, Hu B. Immunological perspective: *Helicobacter pylori* infection and gastritis. Mediators of Inflammation. 2022;**2022**:2944156. DOI: 10.1155/2022/2944156

[6] Wilson KT, Crabtree JE. Immunology of *Helicobacter pylori*: Insights into the failure of the immune response and perspectives on vaccine studies. Gastroenterology. 2007;**133**:288-308. DOI: 10.1053/j.gastro.2007.05.008

[7] Bergman M, Del Prete G, van Kooyk Y, Appelmelk B. *Helicobacter pylori* phase variation, immune modulation and gastric autoimmunity. Nature Reviews. Microbiology. 2006;**4**:151-159. DOI: 10.1038/nrmicro1344

[8] Goh K, Chan W, Shiota S, Yamaoka Y. Epidemiology of *Helicobacter pylori* infection and public health implications. Helicobacter. 2011;**16**:1-9. DOI: 10.1111/j.1523-5378.2011.00874.x

[9] Kusters JG, van Vliet AHM, Kuipers EJ. Pathogenesis of *Helicobacter pylori* infection. Clinical Microbiology Reviews. 2016;**19**:449-490. DOI: 10.1128/CMR.00054-05

[10] Hooi JKY, Lai WY, Ng WK, Suen MMY, Underwood FE, Tanyingoh D, et al. Global prevalence of *Helicobacter pylori* infection: Systematic review and meta-analysis. Gastroenterology. 2017;**153**:420-429. DOI: 10.1053/j.gastro.2017.04.022

[11] Zevering Y, Jacob L, Meyer TF. Naturally acquired human immune responses against *Helicobacter pylori* and implications for vaccine development. Gut. 1999;**45**:465-474. DOI: 10.1136/gut.45.3.465

[12] Kalenic S, Plecko V, Presecki V, Katitic M, Dominis M. *Helicobacter pylori*—Bacterial characteristics. Liječnički Vjesnik. 2002;**124**(Suppl. 1):10-13

[13] Eusebi LH, Zagari RM, Bazzoli F. Epidemiology of *Helicobacter pylori* infection. Helicobacter. 2014;**19**(Suppl. 1):1-5. DOI: 10.1111/hel.12165

[14] Sjomina O, Pavlova J, Niv Y, Leja M. Epidemiology of *Helicobacter pylori* infection. Helicobacter. 2018;**23**(Suppl. 1):e12514. DOI: 10.1111/hel.12514

[15] Plummer M, Franceschi S, Vignat J, Forman D, de Martel C. Global burden of gastric cancer attributable to *Helicobacter pylori*. International Journal of Cancer. 2015;**136**:487-490. DOI: 10.1002/ijc.28999

[16] Sato M, Miura K, Kageyama C, Sakae H, Obayashi Y, Kawahara Y, et al. Association of host immunity with *Helicobacter pylori* infection in recurrent gastric cancer. Infect Agents Cancer. 2019;**14**:4. DOI: 10.1186/s13027-019-0221-1

[17] Etemadi A, Safiri S, Sepanlou SG, Ikuta K, Bisignano C, Shakeri R, et al. The global, regional, and national burden of stomach cancer in 195 counties, 1990-2017: A systematic analysis for the global burden of disease study 2017. The Lancet Gastroenterology & Hepatology. 2020;**5**:42-54. DOI: 10.1016/S2468-1253(19)30328-0

[18] Gasteiger G, D'Osualdo A, Schubert DA, Weber A, Bruscia EM, Hartl D. Cellular innate immunity: An old game with new players. Journal of Innate Immunity. 2017;**9**:111-125. DOI: 10.1159/000453397

[19] Algood HMS, Gallo-Romero J, Wilson KT, Peek RM Jr, Cover TL. Host response to *Helicobacter pylori* infection before initiation of the adaptive immune response. FEMS Immunology and Medical Microbiology. 2007;**51**:577-586. DOI: 10.1111/j.1574-695X.2007.00338.x

[20] Ihan A, Pinchuk IV, Beswick EJ. Inflammation, immunology and vaccines for *Helicobacter pylori* infection. Helicobacter. 2012;**17**:16-21. DOI: 10.1111/j.1523-5378.2012.00977.x

[21] Pagliari M, Munari F, Toffoletto M, Lonardi S, Chemello F, Codolo G, et al. *Helicobacter pylori* affects the antigen presentation activity of macrophages modulating the expression of the immune receptor CD300E through miR-4270. Frontiers in Immunology. 2017;**8**:1288. DOI: 10.3389/fimmu.2017.01288

[22] Shakhatreh MAK, Khabour OF, Alzoubi KH, BaniHani MN, Abi-Siniyeh A, Bashir NA, et al. The influence of IL-1B gene polymorphisms on *H. pylori* infection and triple treatment response among Jordanian population. Applied Clinical Genetics. 2020;**13**:139-145. DOI: 10.2147/TACG.S253778

[23] Ma J, Wu D, Hu X, Li J, Cao M, Dong W. Associations between cytokine gene polymorphisms and susceptibility to *Helicobacter pylori* infection and helicobacter pylori related gastric cancer, peptic ulcer disease: A meta-analysis. PLoS One. 2017;**12**:e0176463. DOI: 10.1371/journal.pone.0176463

[24] Zeyaullah M, AlShahrani AM, Ahmad I. Association of *Helicobacter pylori* infection and host cytokine gene polymorphism with gastric cancer. Canadian Journal of Gastroenterology & Hepatology. 2021;**2021**:8810620. DOI: 10.1155/2021/8810620

[25] Liu D, Wang Y, Zhu Z, Zhang S, Zhu X, Wan J, et al. Characteristics of *Helicobacter pylori* antibiotic resistance: Data from four different populations. Antimicrobial Resistance and Infection Control. 2019;**8**:192. DOI: 10.1186/s13756-019-0632-1

[26] Espinosa-Ramos D, Caballero-Hernandez D, Gomez-Flores R, Trejo-Chaves A, Perez-Limon LJ, de la Garza-Ramos MA, et al. Immunization with synthetic *Helicobacter pylori* peptide induces secretory IgA antibodies and protects mice against infection. Canadian Journal of Infectious Diseases and Medical Microbiology. 2019;**2019**:8595487. DOI: 10.1155/2019/8595487

Section 3

Gut Microbiota and Associated Diseases

Chapter 7

Large Association of GI Tract Microbial Community with Immune and Nervous Systems

Alireza Kazempour

Abstract

The gut microbial community has amazing effects on our immune system and nervous system through three pathways: cell signaling, electron transfer, and biological cycles. However, this relationship is two-way and has its own risks or benefits. Except for the brain, there is no place in the body that does not have cytokines (but not all of them). Cytokines are one of the most important immune molecules that play an important role in maintaining homeostasis in our body and the connection between the central nervous system and our immune system. So it is clear that many beneficial microbes in the gut are stimulated when we are hungry or when our nervous system is under pressure from external stimuli. These microbes die or damage intestinal epithelial tissues and stimulate immune molecules such as interleukins or IFNs upstream.

Keywords: GI-tract, microbial community, immune system, nervous systems

1. Introduction

The digestive system includes all the structures between the mouth and the anus. The gastrointestinal tract (GI-tract) begins at the end of the esophagus and ends at the anus, and includes the stomach, duodenum, small intestine, large intestine, and rectum. The weight of microbial communities living in the human gut is about the same as the weight of the human brain. The brain weight of an adult human is between 1 and 1.3 kg, in contrast to about 1 to 1.5 kg of human body weight forms the intestinal microbial community [1, 2]. This microbial community consists of more than 1000 different and heterogeneous bacteria that provide environmental factors to the digestive system and play an important role in the maturation of the host immune system [3].

This microbial colonization in intestinal mucosal tissues plays an important role in promoting host innate- immunity [4]. The diverse and resident microbial populations in the gut promote the growth and maturation of the host immune system through a variety of methods, including the development of lymphatic structures, differentiation and maturation of B and T immune cells, intestinal immune tolerance, and response to T-cell CD4 receptors [5]. Interactions and metabolism by intestinal microbes directly affect the activity of the intestine; How? This is very simple, most of the microorganisms who live in the gut have anaerobic respiration (e.g. citric-acid cycle, oxidative phosphorylation, amino acid, and fatty acid metabolism, etc.).

These respirations systems can stimulate, activate, or regulate many immune molecules called cytokines [6–8].

Cytokines are commonly known as inflammatory mediators and immune responses that have very low molecular weight and function similarly to hormones. Also, cytokines can affect the secretory cells and other cells that receive them [9]. In fact, they regulate all the mechanisms of the vertebrate body and respond to external stimuli. Some cytokines play critical roles in our bodies and transmit immune messages (e.g. IL-1, IL-6, TNF-α, and IFNs), which we see as fever, inflammation, pain, and fatigue in the presence of injury or complication; but this is not all their function, even they can affect the hypothalamic–pituitary–adrenal axis (HPA-axis) pathways and most of the biomarkers [10]. The network of cytokine activity is such that it communicates between all cells and the immune system. A cytokine can also stimulate its target cell to produce more cytokines or completely disrupt their production [11]. Cytokines perform their functions by binding to specific receptors on the target cell membrane, four receptor proteins for cytokines have been identified that are classified into five families, including immunoglobulin receptors, class I cytokine receptors (hematopoietin), class II cytokine receptors (interferons), TNF receptors, and chemokine family receptors [12].

Many observations suggest that the intestinal microbiome interacts with inflammation of the brain and CNS function. The nervous system and GI-tract communicate with each other through a two-way network of signaling pathways consisting of several connections including the vagus nerve, immune system, metabolites, and bacterial products [13]. The gut microbiota and the brain can affect each other directly CNS and indirectly autonomic nervous system (ANS). The vagus nerve is the most important part of the sympathetic and parasympathetic system (dependent on the ANS) that controls many of our essential functions and daily activities (e.g., mood control, immune response, digestion, and heart rate) [14, 15]. In direct signaling, endocrine secretion by the central nervous system (CNS) can stimulate intestinal bacteria. This direct signaling usually involves the concentration of catecholamine, which is also effective in physical and psychological stress. But in the indirect signaling method in addition to CNS. The ANS is also involved. So the ANS plays an important role in maintaining the integrity, modulating, and regulating the permeability of epithelial surfaces, intestinal physiology, and microbial function [15].

2. GI-tract microbiome

The gut microbiota is a complex collection of bacteria, archaea, viruses, and fungi that enter our digestive system daily through swallowing foods or swallowing saliva, so they can be colonized in our GI tract. The classification composition of the intestinal microbiome varies greatly from person to person due to the internal microbiome and external microbiome agents. The first factor (microbiome-intrinsic) depends on the condition of the microbiome after puberty during life and through species interactions [16]. The second factor (microbiome-external) refers to the various layers of the environment that affect or interact with the gut microbiome. Experimentally, they overlap into three categories: external hosting factors, intrinsic hosting factors, and environmental factors [16].

The intestinal microbial community contains 1000–1500 species of bacteria. However, about 160 species of bacteria can be present in a person's gut. That is why there is a fundamental difference in the composition of the microbiome between individuals, which indicates the dependence of the microbiome on environmental changes and genetic inheritance [17].

Studies of human dietary changes in the intestinal microbiome and gene expression patterns in adults are associated with changes in the diversity, structure, and function of the intestinal microbiome. In fact, the same dietary changes in the gut microbiome are associated with some changes in brain function or activity [15–17]. The symbiotic relationship between the gut microbial community and humans is beneficial to both parties. As human hosts, we provide important habitat and nutrients for our intestinal microbiome, and the gut microbiome supports the development of our metabolic system and the maturation of our intestinal immune system by providing beneficial nutrients. Each intestinal microbial community regulates a number of homeostatic mechanisms, including immune function and protection of the intestinal barrier in a healthy host [17, 18].

The composition of the gut microbiome is influenced by factors such as diet, antibiotic use, disease status ways of being born, and many other elements throughout human life. However, microbes form a complex symbiotic relationship with the host, where the host provides the microbiota with a nutrient-dense environment, and the microbiota, in turn, provides metabolic, protective, and structural functions that are not encoded or produced by the host genome [19].

Each person's gastrointestinal microbiome has six major phylum of bacteria and approximately 15 predominant species, as shown below [19]:

i. Firmicutes {Ruminococcus, Clostridium, Lactobacillus, Eubacterium, Faecalibacterium, Roseburia

ii. Bacteroidetes {Bacteriodes, Prevotella, Xylanibacter

iii. Actinobacteria {Collinsella, Bifidobacterium

iv. Verrucomicrobia {Akkermansia

v. Proteobacteria {Escherichia, Desulfovibrio

vi. Euryarchaeota {Methanobrevibacter

In addition to bacteria, studies have shown that 101 species belonging to 85 fungal genera isolated from the oral cavity of healthy volunteers, which represent three dominant phyla (*Ascomycota*, *Basidiomycota,* and *Zygomycota*) and more than ten classes of fungi which accounted for 99% of the population in all of the studies. Yeasts of the genus *Saccharomyces*, *Malassezia*, and *Candida* are also the predominant fungi found in fecal samples in most studies [20–22]. So the intestinal fungal populations can be called "silent populations". This is because the population of fungal species, also known as "microbiome" occupies a very small volume of our GI-tract [23].

Most commensal fungi that live in our gut are uncultivable, but many of these fungi are pathogenic and under normal circumstances are not harmful to our bodies. The amount of fungus that lives in a person's gut is related to that person's eating habits and intestinal pH level. Also, their presence in the GI-tract of monogastric animals is only 0.1% of the total intestinal microbiome. According to observations, *Candida* and *Phialemonium* can survive in the acidic environment of the stomach, there are also many types of fungi that survive in the acidic environment and grow in the human GI-tract. However, the most common phylum and predominant species of fungi who live in the GI-tract based on their morphological and reproductive traits are as shown below [22–26]:

i. Ascomycota
- **Paecilomyces**
- **Penicillium**
- **Candida**
- **Aspergillus**
- **Fonsecaea**
- **Geotrichum**
- **Saccharomyces**

ii. Basidiomycota
- **Trichosporon**
- **Rhodotorula**

iii. Zygomycota
- **Rhizopus**
- **Mucor**

GI-tract viruses after bacteria and fungi constitute the predominant population of the intestinal microbiome. It can be expected that more than 10^{12} viruses can live in the human gut and play an important role in regulating complex microbial networks active in the gut habitat [27]. Viruses, like the other microbes in the GI-tract, have a significant variation in their species among other people. However, not enough information is available on the functional role of most intestinal viruses, but they appear to be effective in some bacterial functions, such as generating or transmitting resistance and protection against other intestinal pathogens [28]. Also, about more than 90% of intestinal viruses communities are composed mainly of bacteriophages, while eukaryotic viruses are less than 10%. Now, two types of virus variants and the most common phylum have been identified in the human gut, which are as shown below [28, 29]:

i. Bacteriophages
- ***Siphoviridae***
- ***Podoviridae***
- ***Myoviridae***
- ***Microviridae***
- ***inoviridae***

ii. Eukaryotic Viruses
- ***adenoviridae***
- ***alphaflavoviridae***
- ***astroviridae***
- ***Arenaviridae***
- ***circoviridae***
- ***Geminiviridae***
- ***Genemoviridae***
- ***papilomaviridae***
- ***picornaviridae***
- ***polyomaviridae***
- ***parvoviridae***
- ***Virgaviridae***
- ***Rudiviridae***

Although archaea have a very small fraction of the microbiota, but some of them (e.g. *Methanobrevibacter*) play a very important role in intestinal methanogens. The archaea domain contains a wide range of organisms that share properties with prokaryotic and eukaryotic domains [28, 30]. Methanogens are the unique and specific metabolism of some archaea species that are widespread in environments (e.g., freshwater, marine sediments, soils and intestines of humans, and many animal species). The archea that lives in our body is found in our mouth, esophagus, and intestines. But each of them is colonized in a specific part of our digestive system. However, archaea extracted from the human body are classified into three kingdoms and more than ten phyla as shown below [30]:

i. Thaumarchaeota {**Nitrososphaerales**

ii. Crenarchaeota {**Sulfolobales**

iii. Euryarchaeota
- **Archaeoglobales**
- **Halobacteriales**
- **Methanopyrales**
- **Methanobacteriales**
- **Methanococcales**
- **Methanomassiliicoccales**
- **Methanomicrobiales**
- **Methanocellales**
- **Methanosarcinales**
- **Thermococcales**
- **Thermoplasmatales**

The microbial ecology of the GI tract is composed of chemically and physically diverse micro-environment habitats stretching from the esophagus to the rectum; colonization or transient occupation by microbes is about 150–200 m^2 of the gut surface [28]. The symbiotic relationship between the gut microbiota and the host, mediated by a complex metabolic network, includes immunity, nerves, and glands. These symbiotic relationships can lead to severe interference with synthesized microbial metabolites. The predominant functions of the gut microbiota and key metabolites are associated with host health control, reflecting the multifaceted function of the host microbiome, immune system, nerves, and vital organs [31].

2.1 Encounter and interaction of microbial ecology of the GI-tract

Thousands of microbial species inhabit the GI-tract, and observations show that microbial communities such as bacteria and fungi interact with each other, in such a way that targeting bacteria or fungi can inadvertently lead to fungal or bacterial dysbiosis. Many of these studies have shown that some fungi have a strong effect on the reassembly of intestinal bacterial communities after antibiotic treatment (e.g., *Candida albicans*) [24]. Studies on the colonization of *C. albicans* in animal models showed that the fungus partially increased the host's immunity against pathogenic agents (e.g., ***Clostridium difficile***) by increasing the level of IL-17, a pro-inflammatory cytokine. It has also been shown that dysbiosis of intestinal microbial agents can reduce the abundance of anti-inflammatory bacteria (e.g., *Lacticaseibacillus*) and increase pro-inflammatory bacteria (e.g., *Escherichia* and *Shigella*) [24, 32].

The physical structure between the microbiome and the epithelial cells is one of the most important factors in enhancing the selective acceptance of the intestinal microbiome, as the secretion of moderate amounts by the intestinal epithelium causes a complete change in the growing strains at the epithelial level [33]. In distinct intestinal habitats, environmental and competitive microbial filters are the driving force behind the removal and formation of microbial diversity, these factors during colonization and evolution probably explain the diversity of species [34]. *Actinobacteria* can be considered as a keystone phylum, because they are rare and have many connections between bacteria species outside and inside the host body. The number of *Bacteroidetes* is large and they are very widespread in GI-tract, so we can consider them as the predominant phyla [35]. However, the level of intestinal bacterial microbiome phyla can be considered relatively stable over time. Many factors may affect their sustainability (e.g., microbial energy and metabolites produce) [28].

Unhealthy nutrition or poor diets can alter intestinal microbial interactions and dietary diversity, resulting in changes in the availability of microbial nutrients and/or ligands that carry information from the gut to the brain in response to food intake [36]. As a result, they disrupt energy homeostasis, host energy, and metabolites interactions with intestinal microbiota have a significant effect on overall human health, including energy reabsorption and immune system regulation [28, 36]. In humans, digestible carbohydrates are digested by enzymes secreted by the dominant members of the large intestinal microbiota, most of these microbiota are located in the colon (e.g., *Bacteroides* and *Prevotella* species), but healthy nutrition and proper diet can induce beneficial and proper functions by human gastrointestinal microbes (e.g., breakdown of food, synthesis of vitamins and biomolecules, and interaction with the immune system) [37, 38].

Gastrointestinal diseases have been shown to be directly detectable by changes in the microbiome as well as an increase in invasive microbial strains or a decrease in intestinal regulatory microbiome species [39]. Host genetics and horizontal transmission of microbial genes are important factors that play a key role in the composition of the gut microbiome and the frequent replacement of gut microbes, although the horizontal transmission of peripheral microbes can lead to the development of common microbes in the intestinal microbiome ecosystem or alter their colonization patterns by altering the horizontal transfer of interspecific genes, which in turn diversifies the gut microbiota [40].

If the two microbes are positively correlated, they are more likely to facilitate each other but this approach increases colonization-resistant Bacteroides species, whereby the invasive microbial strain cannot colonize the host unless the same microbial species has already been colonized from a common microbial phylum in the GI-tract [41]. The dimensions of this issue can be expressed as: Some important members of the class *Enterobacteriaceae* are responsible for many gastrointestinal complications and significant mortality rates (e.g. *Salmonella*, *Shigella*, and *Yersinia*) [42], as mentioned earlier the *Escherichia* is also a species of *Enterobacteriaceae*, but the interesting and important thing is that *E. coli* and *Shigella* have genetically similarity to each other (about 80 to 90%), and both of them carry the virulence plasmid (pINV) as extra genome; so, it can be considered that *Shigella* and *E. coli* transmit their potentials to bind to intestinal epithelial surfaces, pathogenesis, and even antibiotic resistance by horizontal gene transfer of their plasmid together [42, 43].

It is clear that any pathogen that enters the GI-tract can attach to epithelial surfaces and colonize itself through similar groups, emphasizing and using mechanisms of microbial agents that are genetically similar to them. Conversely, some probiotic microbial groups extracted from the human gut environment (e.g., *Lactobacillus* and *Bifidobacterium*) compete for nutrients and growth medium with this group of pathogen gut microbial colonies, which can act as controlling or killing agents for these bacteria [42–44].

2.2 GI-tract microbiome products

The food we eat throughout the day is the main source of precursors for the production of GI-tract microbial metabolites [44, 45]. Our diet modulates the gut microbiome because the food we eat is also consumed by the gut microbiome and causes changes in the ecosystem and the microbial metabolic properties of the gut [45]. The intestinal microbial ecosystems can change their function in response to changes in our diet. In fact, the type of diet we eat over a long period of time affects the microbial activity of other microbial species in our gut [45], bacteria produce a large number of metabolites that contain structural components and act as signaling molecules for a number of types of our mucosal cells [46].

Enteroendocrine (EE) cells respond differently to many nutrients and intestinal conditions. The intestinal microbiome affects the hormonal secretion of enteroendocrine (EE) cells downstream and facilitates host metabolism or pathogenic metabolites [46]. The gut microbiota plays an important role in human metabolism by enzymes that are not encoded in the human genome (e.g., breakdown of polysaccharides or polyphenols and the synthesis of vitamins) [47]. In the composition of intestinal microbiome metabolites, the processing and absorption of several nutrients and metabolites, including bile acids, lipids, amino acids, vitamins, and short-chain fatty acids (SCFA) derived from intestinal bacteria, are directly related to diet and digestion, and can facilitate or modulate immune cells through direct and indirect mechanisms [45]. The product of microbial degradation of food sources in the gut are bioactive metabolites that bind to target receptors, activate signaling cascades, and modulate several metabolic pathways with local and systemic effects [48].

2.2.1 SCFA metabolite

SCFAs are the main metabolite and the end products of food fiber fermentation by intestinal anaerobic microbiota and have several beneficial effects on mammalian energy metabolism [49]. Acetate, propionate, and butyrate act as post-biological molecules and are present in the large intestine, all three of which SCFAs that are produced by bacterial species consume lactate and succinate [48, 49]. For the microbial community, SCFAs are an essential extra end products that is needed to balance the production of equivalent redox in the anaerobic environment of the intestine [49]; the SCFAs, which are produced in the colon, are absorbed into the tissues through the circulatory system (e.g. Acetate), metabolized in the liver (e.g. Propionate), and consumed by local colonocytes as their primary fuel source (e.g. Butyrate) [46]. Past studies have shown that some bacterial strains excreted from the gut (e.g. *E. coli*) can metabolize acetate by converting acetate to acetyl coenzyme-A (acCoA) by using the reversible pathway of acetate kinase (AckA)-phosphotransacetylase (Pta) pathway [50]. In addition to modulating redox stress, the SCFAs increased the colon defense barrier and can be involved in many of the intestinal activities as shown below:

i. Triggering of Foxp3+ T regulatory (Treg) cells and tolerance

ii. Induction of IgA secretion from B cells

iii. Bacterial competitive exclusion

iv. Promotion of mucus secretion by gut epithelial cells

v. Contribution to the intestinal barrier integrity

vi. Inhibition of the pro-inflammatory transcription factor (NF-κB) and decreasing of oxidative stress

Observations have shown that butyrate, a molecule of SCFAs, can modulate neuronal functions by gene expression of neurotransmitters as well as gastrointestinal stimulation, and also shown that butyrate increases the proportion of choline acetyltransferase by the Src-kinase signaling pathway and the acetylation of histone H3K9 in enteric neurons [45].

2.2.2 Amino-acids metabolite

Another metabolite that is a product by the colon microbiome is amino acids. Some of these amino acids (e.g., Serotonin and tryptophan) have a direct impact on host cell metabolism. Disorders caused by these two bacterial amino acids, can have several effects on the gut-brain axis and vice versa [51, 52]. The GI-tract has three main pathways for tryptophan (Trp) metabolism, which lead to serotonin (5-hydroxytryptamine), kynurenine (Kyn), and indole derivatives, which are directly or indirectly controlled by microbiota [53]. Also, the GI-tract contains large amounts of serotonin (5-hydroxytryptamine) and its receptors (5-HT). Some of the spore-forming (SP) bacteria (e.g. *Bacillus* and *Clostridium* species) have been shown to enhance the level of serotonin receptor biosynthesis by intestinal enterochromaffin cells (ECs) [54].

2.2.3 Bile-acids metabolite

Bile acids are the end products of cholesterol catabolism. They are also signaling molecules that regulate metabolic systems that activate nuclear receptors and G protein-coupled receptors (GPCRs) to regulate hepatic lipid, glucose, and energy homeostasis and impound metabolic homeostasis [55]. To convert cholesterol to bile acids, there are 17 separate enzymes located in the cytosol, endoplasmic reticulum, mitochondria, and peroxisomes. These enzymes can catalyze steroid chain changes and oxidative cleavage of three carbons from the cholesterol side-chain to form C24 bile acids. There are two main pathways of bile acid biosynthesis [55].

Primary bile acids (BAs) are produced inside the liver cells and then released into the duodenum to facilitate the absorption of lipids or fat-soluble vitamins. Both nutritional and microbial factors have been shown to affect the composition of the intestinal BA pool and modulate an important population of FOXP3 + regulatory T (T reg) cells that express transcription factor RORγ [56].

Secondary bile acid is produced by the microbial biotransformation of cholate, deoxycholate enhances gastrointestinal motility by activating TGR5 G-protein-coupled receptors on ECs, Sp-induced metabolites increase 5-HT levels in ECs, and Sp colonization improves GI-tract motility [54].

2.2.4 Lipid metabolite

Some intestinal microbiome bacteria, by consuming lipids, can act both as a substrate for bacterial metabolic processes and as a factor to inhibit bacterial growth in the structural and ecological changes of gut microbiota [57]. Several potential lipid mediators have been identified that act as metabolic messengers to communicate energy status and regulate substrate use between tissues. Also, these mediators can be exogenously distributed in the intestine and effect glucose and lipid metabolism [58]. It has been shown that some intestinal bacteria (e.g., *Lactobacillus*, *Butyrivibrio*, and *Megasphaera*) can react with fatty acid double bonds to produce metabolites that we are unable to synthesize. Many of these metabolites may affect the physiological functions and health of the host. The conjugated linoleic acid (CLA) is one of these metabolites that exerts opposite or different effects [57].

The gut microbiota processes lipids and other digestion nutrient factors to produce metabolites with impacts on host lipid homeostasis and putative effects on pathophysiological functions [57], lipogenesis is controlled by several rate-limiting enzymes that convert acetyl-CoA to palmitate, palmitoleate, stearate, and oleate [58]. The effect of butyrate on vagal inputs to NPY neurons has been identified. Butyrate can also promote the oxidation of fatty acids by consuming carbohydrates, especially in conditions of reduced nutrition throughout the day [36, 59].

Also, lipids play a protective role in the structure of intestinal gram-negative bacteria. Gram-negative bacteria have lipopolysaccharide in their structure, which consists of lipids and polysaccharides. The important point is that this structure acts as a pathogen for this group of bacteria. What we need to know is that LPS is a large glycolipid composed of three structural domains: lipid A, core oligosaccharide, and O antigen [60, 61].

Lipoproteins are absorbed by fat cells with or without LPS. However, LPS are directly and indirectly involved in the inflammatory response in adipose tissue. The LPS is also involved in the transfer of macrophages from the M2 phenotype to M1; in addition, LPS within adipocytes may activate the caspase [62]. The exact structure of LPS varies from bacteria to bacteria and is highly regulated in host cells and is closely related to bacterial virulence. It should be noted that additional enzymes and gene products can modify the basic structure of LPS in some bacteria (especially pathogenic bacteria) [63].

3. GI-tract communications anatomy

Before describing signaling pathways, we need to get a little familiar with the anatomy of the GI-tract and involved systems. Depending on the physiological structure of males and females, the structure of the pelvis will be different. Actually, the outlined subdivision of the pelvic connective tissue is identical in the male and female. The only difference is that in women the uterus is located between the bladder and the rectum and divides the pelvic peritoneum into two sacks, but this is not the case in the male pelvic cavity [64].

3.1 GI-tract neuroanatomy

A complex set of nerve masses and fibers extending from the brainstem to the sacrum, together with neurons in the sympathetic and parasympathetic systems, control a variety of functions, including swallowing, digestion, and excretion. Intestinal-associated neurons and neural networks are generally classified as belonging to the enteric nervous system (ENS) [65], which is described in terms of function and action as follows:

Functions:

- Brainstem
  - **DRG → (afferent nerves) → nodose ganglia**
    - **stomach**
    - **small intestine**
  - **(efferent nerves) → vagal nerve**
    - **stomach**
    - **small intestine**.
- Spinalcord
  - **Celiac ganglia**
    - **DRG → (afferent nerves) → {small intestine**
    - **(efferent nerves) →**
      - **stomach**
      - **small intestine**
  - **Superior Mesenteric ganglia { (efferent nerves) →**
    - **small intestine**
    - **large intestine**
  - **Inferior Mesenteric ganglia { (efferent nerves) → large intestine**
  - **pelvic nerve**
    - **DRG → (afferent nerves) → { large intestine**
    - **(efferent nerves) → { large intestine**

Actions:

$$\text{Sympathetic} \xrightarrow{\textit{catecholamines}} \boldsymbol{\alpha}\text{- or } \beta\text{-adrenergic receptors} \rightarrow$$

$$\begin{cases} \textit{postganglionic vasoconstrictor neurons} \\ \quad \textit{secretion inhibitory neurons} \\ \quad \textit{motility inhibitory neurons} \end{cases}$$

$$\text{Parasympathetic} \begin{cases} \xRightarrow{\textbf{vagus nerve}} \textbf{Intestinal nerve innervation} \\ \xRightarrow{\textbf{neurotransmitter acetylcholine}} \begin{cases} \textbf{motility} \\ \textbf{digestion} \\ \textbf{secretory function} \end{cases} \end{cases}$$

The GI-tract is innervated through its connections to the CNS and by the ENS in the wall of the GI-tract, ENS works in coordination with the CNS reflex to the command center and in the neural pathways that pass through the sympathetic ganglia to control gastrointestinal function [66].

3.2 GI-tract lymphatic system anatomy

The anatomy of the lymphatic system include the thymus, GI-tract, lymph nodes, spleen, and tonsils, and is very similar to the circulatory system expansion. In many organs of the body (e.g., neck, chest, pelvis, etc.), this system is seen in the form of lymph vessels in cooperation with these organs. The lymphatic vascular system consists of a network of vessels that extends to every part of the body except the brain and spinal cord. Of course, lymphatic vessels are found only in the hard palate [67, 68].

Even though the body fluids can move between blood vessels and tissues through very small pores. So in this system, lymphoid organs and lymph nodes monitor and control the composition of body fluids (ie. blood and hemolymph), which includes the following activities [67, 68]:

i. Absorption of pathogens

ii. Strengthening the immune response

iii. Treatment of infection

Also, many endocrine functions require the lymphatic system and even the absorption and transfer of fats and fat-soluble vitamins from the digestive system to the lymph. From there, most organs with the lymphatic system drain their by-products into the lymphatic system and enter the circulatory system from there. That is why this system has a one-way function [67]. There is an extensive network of lymph nodes and lymph vessels in the pelvis that are connected to the tissues and organs of the pelvis, especially the intestines. The gut-associated lymphoid tissues (GALT) perform many functions including monitoring the proliferation and regeneration of gut epithelial cells, Peyer's patches in the small intestine, controlling water absorption from the intestine, and intestinal health conditions [69].

3.3 GI-tract tissues anatomy

If we want to look at the mechanism of action of the digestive system under a magnifying glass, it is necessary to know which organs and which tissues we should examine.

The GI-tract, which begins in the esophagus and ends in the anus (i.e., esophagus, stomach, liver, gallbladder, pancreas, small-intestine, appendix, large intestine, rectum, and anus), has different tissues, biochemical and biophysical functions, and mechanisms [70]. Each of which must be examined separately and their functions considered together. After swallowing, food enters the stomach through the esophagus (passes through the muscular cuff) and by mechanical and chemical activity in the stomach, food enters the duodenum (beginning of the small intestine) entirely as a concentrated liquid containing digestive acids and enzymes (i.e., gastric juice) then, due to physical and chemical activities, the intestine is fully digested to provide the materials and compounds needed by the body. To make it easier, two important GI-tract organs (i.e., stomach and intestines), tissues, and their physical and chemical functions are summarized below [70–72]:

Gaster

- **Tissue =**
  - **mucosa =**
    - **gastric glands**
    - **muscularis mucosae**
  - **submucosa = {blood vessels**
  - **muscularis =**
    - **oblique muscle**
    - **circular muscle**
    - **longitudinal muscle**
  - **serosa = {connective tissu layer**
- **physical function =**
  - **1.fundus = fundus muscular = increase gastric accommodation**
  - **2.corpus = Peristaltic muscular = contraction waves**
  - **3.antrum = ring – shaped muscular = peristaltic food pump**
- **chemichal function =**
  - ***Decomposition = reduces the size of food particles***
  - ***Destruction = tenderize the food matrix***

Intestine

- **Tissue =**
  - **1 – Mesentery**
  - **2 – Vesseles =**
    - **Vein**
    - **Artery**
    - **Nerve =**
      - **Submucosal plexus**
      - **Myenteric plexus**
  - **3 – Muscularis =**
    - **Circular muscle**
    - **Longitudinal muscle**
  - **4 – Serosa =**
    - **Areolar connective tissue**
    - **Epitheulium**
  - **5 – Submucosa =**
    - **Lymphatic tissue**
    - **Glands**
    - **Lumen**
  - **6 – Mucosa =**
    - **Epithelium**
    - **Lamina propia**
    - **Muscularis mucosae**
- **Physical function =**
  - **Small intestine / (Ileum, Duodenu, jejunum) =**
    - ***Peristalsis***
    - ***Segmentation***
    - ***Absorption***
  - **Large intestine / (*ascending, descending and sigmoid colon*) =**
    - ***Peristalsis***
    - ***Segmentation***
    - ***Absorption***
- **Chemichal function =**
  - **Small intestine =**
    - **Lipid Breackdown**
      - **Pancreatic lipases**
      - **Phospholipase**
    - **protein Breakdown**
      - **Trypsin**
      - **Chymotrypsin**
      - **Carbopeptidase**
      - **Elastase**
    - **Strach Breakdown**
      - **Pancreatic amylase**
      - **Dextrinase**
      - **Sucrose**
      - **Maltase**
      - **Lactase**
      - **Amyloglucosidase**
    - **Nucleic Acid Breakdown**
      - **Nucleases**
      - **Nucleosidases**
      - **Phosphatases**
  - **Large intestine = {Fermentation = fermentation products**

3.4 GI-tract immune system anatomy

The immune system can be defined as a complex system that protects the body against microorganisms, infectious agents, and a variety of autoimmune diseases and carcinogens, immune system can respond to any antigen in both specific and non-specific forms. The immune system function can also be seen in both innate and adaptive forms in all systems, infrastructures, and various cellular and molecular mechanisms to stop or eliminate invasive antigens [73].

Apart from the lymphoid cell, various organs and cells are also involved in intestinal immunity (i.e., goblet cells, entero-endocrine cells, macrophages, mast cells), these appropriate subsets of lymphoid cells are usually found in the epithelium (e.g., T suppressor) or in the lamina propria (e.g., T helper), IgA is also mostly produced by plasma cells [74]. The inductive sites are organized into specialized aggregations of lymphoid follicles called Peyer's patches, are demonstrated as typical organized lymphatic structures of the intestine. They are present and found in large numbers from before birth to the senescence and also present in the ileum, duodenum, and jejunum [74, 75]. In superior vertebrates, such as mammalians, the immune system is made up of primary and secondary lymphatic organs that are organized in an almost identical morphology. The thymus and bone marrow are the major organs of the primary lymphatic system, and the spleen, lymph nodes, and mucosal- associate lymphoid tissue (MALT) are the secondary lymphatic system. Innate immunity is found in all living things and can detect protected and common molecular structures in pathogens and microorganisms. These include the identification of polysaccharides, lipopolysaccharides (LPS), peptidoglycans, bacterial and viral DNA and RNA through the interaction of specific receptors (e.g., toll-like receptor TLR) [73].

The GI-tract has the largest volume of microbes in the human body, maintaining an elegant balance between immunity against pathogens and tolerance toward commensal microbiome, such as immune balance, or intestinal homeostasis, is accomplished by fine-tuning and cooperating with various branches of the immune system, including the innate and adaptive immune system [76]. The gastrointestinal mucosa separates the digestive fluid inside the duct, which contains a large number of antigens from various sources, and prevents the antigens from freely reaching the body, it also allows for some vital host and peripheral intestinal interactions. The mucosal immunity is related to secretory IgA; The IgA is derived from mucosal plasma cells after the proliferation of its precursors in antigen-induced Peyer patches. In fact, IgA is transported to the intestinal tract after binding to the secretory component (SC) as a dimer. However, the induction of local immunity and intestinal systemic tolerance may be a specific immune response to the gut-associated lymphoid tissue (GALT) [77].

Therefore, the immune system can deal with any pathogen in different conditions, depending on the location, amount and type of damage, and all this is due to the chemical structures at the cellular and molecular levels of organisms. Chemical structures help identify the invasive agent and the type of response to them. These structures, which are generally proteins, are produced and secreted by epithelial, endothelial, dendritic cells (DCs), and lymph nodes and are commonly known as cytokines [73].

In total, the number of proteins that have cytokine activity reaches more than 200, their secretion depends on the effective concentrations of cytokines that are created in the vicinity of target cells [78]. Cytokines are involved in the interaction of lymph cells, hematopoietic cells, and inflammatory cells. They are usually having a short half-life but the network of cytokine activity is such that it communicates between all cells and factors involved in the immune system. Also, the inflammatory responses, regulation of hematopoiesis, proliferation control, and cellular differentiation are different biological responses that can induce by cytokines [11, 79].

Cytokines are a general name for a complex of proteins that are involved in our immunity in the form of structural molecules. This complex including of lymphokine (cytokines produced by lymphocytes), chemokines (cytokines with chemical activity), interleukins (cytokines produced by leukocytes that affect other leukocytes), and monokine (cytokines produced by monocytes). All of these cytokines can work together and can even counteract the effects of each other. Also, cytokines stimulate B- and T-cell-dependent responses. In the immune system, T-cells respond well to the activation of B-cells in response to antigens, the proliferation and the activation of eosinophils, neutrophils, and basophils by cytokines. The cytokines act by binding to specific receptors on the target cell membrane. So far, four types of receptor proteins for cytokines have been identified that are classified into five families including immunoglobulin receptors, class I cytokine receptors (hematopoietins), class II cytokine receptors (interferons), TNF receptors, and chemokine family receptors [11, 12].

As mentioned earlier, the gut contains the largest immune system and intestinal mucus is considered as the primary site of interaction with common and pathogenic organisms. The innate immune system acts to restrict the passage of microbiota through the mucosal barrier, so intestinal epithelial cells, in coordination with antigen-presenting cells (APCs), form the first line of defense in the intestine. Cytokine binding to the T-cell receptor promotes T-cell expansion or expression of distinct Th subsets or to regulatory T cells (Tregs). Th1 cells produce proinflammatory cytokines, including IFN-γ and TNF-α, which are important for cell-mediated immunity against most bacteria. In contrast, Th2 cells produce anti-inflammatory cytokines, including IL-4 and IL-13, which are critical for humoral mediated immunity against extracellular pathogens. Cytokines bind to cell surface receptors in immune and non-immune cells, activating the JAK–STAT signaling pathway and positively regulating intestinal function by regulating the expression of specific target genes [80].

4. GI-tract signaling pathway

The gut-brain axis (GBA) is a two-way communication between the CNS and the intestines that connects the emotional and cognitive centers of the brain to the functions of the peripheral intestine. The interaction between the microbiota and the GBA is two-way, meaning that they can communicate with each other through signaling from the gut microbiota to the brain and from the brain to the gut microbiota using neural, endocrine, immune, and humoral connections. This communication from brain to gut includes the CNS, autonomic nervous system (ANS), enteric nervous system (ENS), hypothalamic–pituitary–adrenal (HPA) axis, and vice versa from gut to brain pathway including the ascending pain pathways, cytokines (e.g. TNF-α, IL) and entero-endocrine cells (e.g. serotonin) [81, 82].

Evidence suggests that gut microorganisms can stimulate the vagus nerve and play an important role in mediating effects on the brain and behavior. The vagus nerves distinguish between non-pathogenic and potentially pathogenic bacteria, and can even mediate signals in the absence of overt inflammation and vagal pathways that, depending on the nature of the stimulus, can induce anxiolytic and anti-anxiety effects. By interacting with immune cells, mediators are released that reduce inflammation. This role of modulating vagal nerve immunity has consequences for modulating brain function and even a variety of moods [83]. Also, the response to HPA with the initial modification of the gastrointestinal flora, and the effects of the initial stress on the barrier function of the GI-tract and the flora, demonstrates the ability of both systems to prepare each other for future problems [82].

All responses to food stimuli occur in the small intestine and also, especially the colon. The colon is an essential part of the GI-tract and acts as a filter and facilitates the absorption of nutrients from food, water, electrolytes, and vitamins through the intestinal tract. Within these, "macro" environments are several "micro" environments where bacteria can live, such as the lumen of the bowel, the mucus layer overlying the epithelium, mucus within intestinal crypts, and the surface of mucosal epithelial cells. The intestinal epithelial cells (IECs) produce multiple tubular injections that form crypts that increase tissue uptake levels. In the crypt domain, the intestinal stem cell (ISC) niche enables continuous regeneration of the intestinal lining (e.g., enterocytes, neuroendocrine cells, tuft cells, Paneth cells, M cells, and goblet cells), IECs can proliferate, differentiate, and move upward (mucus) until they are replaced in the human colon five to seven days later. IECs also communicate with microbiota, coordinate innate and adaptive effector cell functions. The IECs form a continuous epithelium of cells that are tightly linked by different types of cell–cell junctions that assist in maintaining the integrity of the barrier [84, 85].

The RAS superfamily of small GTPase including RAS, Rho/Rac, Arf, and Rab subfamilies are critical regulators of intestinal epithelial homeostasis and barrier function. At the molecular level, RAS proteins cycle between an inactive state, where they are bound to guanosine diphosphate (GDP), and an active STAT, bound to guanosine triphosphate (GTP) [84].

To better understand the signaling pathways from gut to brain and brain to gut, we need to examine these signaling pathways in two structures (prokaryote, eukaryote), Since intestinal bacteria are the most active in terms of communication, in this section, the focus will be on bacteria, which we will discuss below:

4.1 GI-tract prokaryotes signaling

Bacteria constantly monitor and interpret the conditions inside their cells and their environment to maintain their survival to be able to adjust and provide appropriate responses to the environmental changes around them. Therefore, they use a variety of small molecules for extracellular and intracellular signaling. Hence, these bacterial signals, which are seen in both intracellular and extracellular forms, play an important role in creating or responding to environmental changes in establishing communication between bacteria with other members of their community or other living bacteria that share environmental conditions [86]. Bacterial signaling systems located on their cell membranes are complex and is recognized in three major types (i.e. one-component system, two-component system, extra-cytoplasmic sigma factors), they can also communicate with each other and transmit functional signals as a cell-to-cell signaling mechanism called Quorum Sensing(QS) [85, 87].

The adaptive responses to peripheral signals are mainly generated by transcriptional regulators through two systems, one-component, and two-component signal transmission systems. These systems scan small molecular proteins inside and outside the cell and modulate gene expression to provide the appropriate physiological response to the prevailing conditions [88].

One-component signaling systems include members of the ToxR family and they do not contain a phosphoryl acceptor domain, therefore, representing the simplest form of bacterial transmembrane signaling systems. In two-component systems, integrated membrane histidine kinase generally acts as a sensor for various stimuli and is also responsible for transmitting information across the membrane. The number of systems regulating the histidine kinase reaction varies widely between bacterial species. But the signaling system of the ECF sigma factors is small regulatory proteins that bind to RNA polymerase and stimulate transcription of specific genes. Many bacteria, particularly those with more complex genomes, contain multiple ECF sigma factors, and these regulators often outnumber all other types of sigma factors [87].

Quorum Sensing (QS) may be used as a system for bacteria to prevent the population from growing to levels that are unsustainable in their environment. If all the nutrients are depleted and waste products are not removed from their environment, it will be deleterious for the community as a whole. In fact, QS is used to determine the fitness of a bacterial population. The QS is found in three major forms in bacteria: one is used primarily by gram-negative bacteria, one is used primarily by gram-positive bacteria, and one has been proposed to be universal. The paradigm for QS in gram-negative bacteria is the LuxIR system. The LuxIR system uses the LuxI protein, or a homolog of this protein, to synthesize an autoinducer (AIs) and LuxR (or a homolog of LuxR) as a regulator that binds to the AIs and modulates gene expression. The QS system used by gram-positive bacteria utilizes peptides as AIs signaling molecules. These autoinducing polypeptides (AIPs) are produced in the cytoplasm as precursor peptides and then cleaved, modified, and exported. The extracellular AIPs are detected via two-component systems in which the external portion of a membrane-bound sensor kinase protein detects the AIP and then phosphorylates and activates a response regulator that binds to DNA and modulates transcription. And the third QS system present in bacteria is found in a wide numbers of bacteria, including both gram-negative and gram-positive species. This system, called the LuxS or autoinducer-2 (AI-2) system, has been detected in more than 55 species by sequence analysis or functional assays. This system is called LuxS/AI-2 system, which is effective in communication between bacterial species [85].

4.2 GI-tract eukaryote cell signaling

The first gut signaling system is related to cell regulation. As mentioned earlier, the RAS superfamily is critical regulator of intestinal epithelial homeostasis and barrier function cells, the RAS superfamily has nine main effectors for several pathways which are briefly described below:

$$
\text{RAS.GTP} \overset{\textbf{Effectors}}{\Longrightarrow}
\begin{cases}
\textbf{PIK3C2A} = \textbf{Pathway} \begin{cases} \textbf{PIP3} \\ \textbf{AKT} \\ \textbf{BAD} \\ \textbf{P53} \\ \textbf{NFkB} \\ \textbf{mTORC1} \end{cases} \\
\langle \textbf{RGL2} | \textbf{RALGDS} | \textbf{RGL1} \rangle = \textbf{Pathway} \begin{cases} \textbf{Ral} \\ \textbf{PLD} \\ \textbf{Sec5} \\ \textbf{RalPBP} \\ \textbf{Arf6} \end{cases} \\
\langle \textbf{RASSF5} | \textbf{RASSF2} | \textbf{RASSF4} \rangle = \textbf{Pathway} \begin{cases} \textbf{MST1} \\ \textbf{MST2} \\ \textbf{LATS} \end{cases} \\
\langle \textbf{ARAF} | \textbf{BRAF} | \textbf{RAF1} \rangle = \textbf{Pathway} \begin{cases} \textbf{MEK} \\ \textbf{ERK} \\ \textbf{ETS} \end{cases} \\
\textbf{ARP1} = \textbf{pathway} \begin{cases} \textbf{RAC} \\ \textbf{PAC} \\ \textbf{RHO} \end{cases} \\
\langle \textbf{RIN2} | \textbf{SNX27} \rangle = \textbf{Pathway} \begin{cases} \textbf{ABL} \\ \textbf{RAB4} \\ \textbf{RAB5} \end{cases} \\
\textbf{MLLT4} = \textbf{Pathway} \begin{cases} \textbf{Actin} \\ \textbf{Nectin} \\ \textbf{Cadherin} \end{cases} \\
\langle \textbf{RAPGEF5} | \textbf{RASIP1} \rangle = \textbf{Pathway} \begin{cases} \textbf{RAP} \\ \textbf{Signaling/adhesion} \end{cases} \\
\textbf{MYO9B} = \textbf{Pathway} \{ \textbf{Myosin/actin adhesion}
\end{cases}
$$

All the effector pathways had responses and effects on colon physiology (e.g., actin or nectin and cadherin or RAP, signaling/adhesion can respond to cell–cell junctions).

RalGDS effector and the activation of Ral GTPases are critical for the regeneration of intestinal stem cells, and also the RASSF-MST-LATS pathway coordinates intestinal regeneration through cell proliferation, apoptosis, and differentiation functions. AFDN is involved in the formation of cell–cell junctions and thereby controls adhesion between different IECs [84].

The second gut signaling system is related to immune regulation, which is regulated by cytokines. As mentioned earlier the cytokine can be present in many tissue or cells as regulator immune molecules, they are essential mediators of the interactions between activated immune cells and non-immune cells, including epithelial and mesenchymal cells [89]. So, the cytokines regulate the intensity and duration of the immune response by stimulating or inhibiting proliferation, differentiation, trafficking, or emigration of lymphocytes all the while acting as a messenger for both the arms of the immune system [90]. Cytokine production by Peyer's patch (PP) cells was examined in response to probiotic and pathogenic bacteria, some probiotics bacteria (e.g., *Lacticaseibacillus casei)* have the ability to induced (e.g. IL-6, IL-8, IL-12) or reduced (e.g. Th1 cells by IFN-γ secretion in PP cells) other cytokines as well [91].

The third gut signaling system is related to hormones. The gut hormones (e.g., cholecystokinin and glucagon-like peptide-1) released following a meal and act on local receptors to regulate glycemia via a neuronal gut-brain axis and provide feedback via nutrient sensing and local hormonal signaling. The small intestine contains a variety of regulatory signals including:

i. Proximal hormones within the duodenum and jejunum, cholecystokinin (CCK) in I cells, and glucose-dependent insulinotropic polypeptide (GIP) in K cells.

ii. Distal hormones in the ileum and large intestine within L cells, glucagon-like peptide-1/2 (GLP-1/2), oxyntomodulin (OXN), and peptide YY (PYY).

The secretion of these hormones is stimulated by nutrients within the intestine that then act on their respective receptors either centrally, or locally on vagal afferents that are in close proximity to enteroendocrine cells, to regulate metabolic homeostasis through various changes in food intake, gastric emptying, intestinal motility, and/or energy expenditure [92].

5. Interaction and regulation between microbiota and the CNS and immune system

The human immune system has evolved to maintain a symbiotic relationship between the host and the microbiota, and disruption of the dynamic immune-microbial interaction leads to profound effects on human health (e.g., interaction between resident microbiota and immune signals, CNS development) as described in below [94].

5.1 Inflammasome signaling pathway

Inflammasome is an innate immune signaling complex, which is activated in response to diverse microbial and endogenous danger signals. Also, the various pattern-recognition receptors (PRRs) in different families, including NLRP1, NLRC3, NLRP6, NLRP7, NLRC4, and AIM2, have been identified to effect in inflammasome activity. Inflammasomes activation recruits ACS (apoptosis-associated speck-like protein containing a caspase recruitment domain) and the cysteine protease caspase 1 through caspase activation and recruitment domain (CARD) to induce the proteolytic cleavage of pro-caspase1 to generate mature and active caspase 1, which further process pro-IL-1β and pro-IL-18 to the final production of bioactive IL-1β and IL-18 proteins [93].

5.2 IFN-I signaling pathway

IFN-I is a pleiotropic and ubiquitous cytokine that plays an essential role in both innate and adaptive immunity and maintenance of host homeostasis. IFN-I is induced by pathogen-associated molecular patterns (PAMPs). Secretion of endogenous IFN-I depends on activation of several classes of PRRs. They play a significant role in priming the host to various viral, bacterial, or tumor components. Effects of IFN-I on inflammation and host hemostasis have been linked to the recruitment of Tregs. Also, the commensal lactic acid bacteria have been shown to trigger TLR3-mediated IFN-β secretion by DCs in the intestine [93].

5.3 NF-κB signaling pathway

The interaction between microbiota and NF-κB signaling is also responsible for CNS inflammation. NF-κB family of transcription factors contribute to both innate and adaptive immune responses and maintenance of the immune system. So, the NF-κB family of transcription factors contribute to both innate and adaptive immune responses and maintenance of the immune system [93].

6. Conclusion

In summary, the gut microbiome binds to intestinal epithelial cells and uses cell signaling and junctions to communicate with each other and with the host CNS. A complex diverse of microorganisms live in the GI-tract which is called gut microbiome, profoundly affect many aspects of host physiology, including nutrient metabolism, infection resistance, and immune system development. The GI-tract is strongly innervated by a complex network of neurons that coordinate vital physiological functions. In addition to CNS; ENS senses and response to the dynamic ecosystem of the GI-tract by converting chemical signals from the environment into nerve impulses that propagate throughout the intestine and other organs of the body, also the local axonal reflexes and autonomic long-range sensory reflexes in GI-tract play an important role in the regulation of immunity by parasympathetic or sympathetic nerves. As a result, the interactions between the nervous system and the immune system enable the gut to respond to the variety of food products it absorbs, and the wide variety of pathogens and microbiomes it holds. Gut microbiota can promote different subsets of immune cells through antigen stimulation and activation of immune signaling pathways. All the interactions that the gut microbiome creates reflect both on our mental states and in our immune system, and vice versa.

We can say that we and our gut microbes talk to each other through these signal pathways, solve each other's needs, and ensure each other's safety.

Author details

Alireza Kazempour
Islamic Azad University, Lahijan Branch, Gilan, Iran

*Address all correspondence to: ali42153@gmail.com

References

[1] Hartmann P, Ramseier A, Gudat F, Mihatsch MJ, Polasek W. Das Normgewicht des Gehirns beim Erwachsenen in Abhängigkeit von Alter, Geschlecht, Körpergrösse und Gewicht [Normal weight of the brain in adults in relation to age, sex, body height and weight]. Pathologe. 1994;**15**(3):165-170. DOI: 10.1007/s002920050040

[2] Angelakis E, Armougom F, Million M, Raoult D. The relationship between gut microbiota and weight gain in humans. Future Microbiology. 2012;**7**(1):91-109. DOI: 10.2217/fmb.11.142

[3] Sobhani I, Amiot A, Le Baleur Y, Levy M, Auriault ML, Van Nhieu JT, et al. Microbial dysbiosis and colon carcinogenesis: Could colon cancer be considered a bacteria-related disease? Therapeutic Advances in Gastroenterology. 2013;**6**(3):215-229. DOI: 10.1177/1756283X12473674

[4] Gensollen T, Iyer SS, Kasper DL, Blumberg RS. How colonization by microbiota in early life shapes the immune system. Science (New York, N. Y.). 2016;**352**(6285):539-544. DOI: 10.1126/science.aad9378

[5] Zhao Q, Elson CO. Adaptive immune education by gut microbiota antigens. Immunology. 2018;**154**(1):28-37. DOI: 10.1111/imm.12896

[6] Vernocchi P, Del Chierico F, Putignani L. Gut microbiota metabolism and interaction with food components. International Journal of Molecular Sciences. 2020;**21**(10):3688. DOI: 10.3390/ijms21103688

[7] Belizário JE, Faintuch J, Garay-Malpartida M. Gut microbiome dysbiosis and immunometabolism: New Frontiers for treatment of metabolic diseases. Mediators of Inflammation. 2018;**2018**: 2037838. DOI: 10.1155/2018/2037838

[8] Schirmer M, Smeekens SP, Vlamakis H, Jaeger M, Oosting M, Franzosa EA, et al. Linking the human gut microbiome to inflammatory cytokine production capacity. Cell. 2016; **167**(4):1125-1136.e8. DOI: 10.1016/j.cell.2016.10.020

[9] Cohen MC, Cohen S. Cytokine function: A study in biologic diversity. American Journal of Clinical Pathology. 1996;**105**(5):589-598. DOI: 10.1093/ajcp/105.5.589

[10] Bennett JM, Reeves G, Billman GE, Sturmberg JP. Inflammation-Nature's way to efficiently respond to all types of challenges: Implications for understanding and managing "the epidemic" of chronic diseases. Frontiers in Medicine. 2018;**5**:316. DOI: 10.3389/fmed.2018.00316

[11] Zhang JM, An J. Cytokines, inflammation, and pain. International Anesthesiology Clinics. 2007;**45**(2): 27-37. DOI: 10.1097/AIA.0b013e318034194e

[12] Foxwell BM, Barrett K, Feldmann M. Cytokine receptors: Structure and signal transduction. Clinical and Experimental Immunology. 1992;**90**(2):161-169. DOI: 10.1111/j.1365-2249.1992.tb07922.x

[13] Rutsch A, Kantsjö JB, Ronchi F. The gut-brain Axis: How microbiota and host inflammasome influence brain physiology and pathology. Frontiers in Immunology. 2020;**11**:604179. DOI: 10.3389/fimmu.2020.604179

[14] Breit S, Kupferberg A, Rogler G, Hasler G. Vagus nerve as modulator of the brain-gut Axis in psychiatric and

inflammatory disorders. Frontiers in Psychiatry. 2018;**9**:44. DOI: 10.3389/fpsyt.2018.00044

[15] Osadchiy V, Martin CR, Mayer EA. The gut-brain Axis and the microbiome: Mechanisms and clinical implications. Clinical Gastroenterology and Hepatology: The Official Clinical Practice Journal of the American Gastroenterological Association. 2019;**17**(2):322-332. DOI: 10.1016/j.cgh.2018.10.002

[16] Schmidt T, Raes J, Bork P. The human gut microbiome: From association to modulation. Cell. 2018; **172**(6):1198-1215. DOI: 10.1016/j.cell.2018.02.044

[17] Shi N, Li N, Duan X, Niu H. Interaction between the gut microbiome and mucosal immune system. Military Medical Research. 2017;**4**:14. DOI: 10.1186/s40779-017-0122-9

[18] Adelman MW, Woodworth MH, Langelier C, Busch LM, Kempker JA, Kraft CS, et al. The gut microbiome's role in the development, maintenance, and outcomes of sepsis. Critical Care (London, England). 2020;**24**(1):278. DOI: 10.1186/s13054-020-02989-1

[19] Bliss ES, Whiteside E. The gut-brain Axis, the human gut microbiota and their integration in the development of obesity. Frontiers in Physiology. 2018;**9**: 900. DOI: 10.3389/fphys.2018.00900

[20] Hallen-Adams HE, Suhr MJ. Fungi in the healthy human gastrointestinal tract. Virulence. 2017;**8**(3):352-358. DOI: 10.1080/21505594.2016.1247140

[21] Qiu X, Ma J, Jiao C, Mao X, Zhao X, Lu M, et al. Alterations in the mucosa-associated fungal microbiota in patients with ulcerative colitis. Oncotarget. 2017; **8**(64):107577-107588. DOI: 10.18632/oncotarget.22534

[22] Naranjo-Ortiz MA, Gabaldón T. Fungal evolution: Diversity, taxonomy and phylogeny of the fungi. Biological Reviews of the Cambridge Philosophical Society. 2019;**94**(6):2101-2137. DOI: 10.1111/brv.12550

[23] Sam QH, Chang MW, Chai LY. The fungal mycobiome and its interaction with gut bacteria in the host. International Journal of Molecular Sciences. 2017;**18**(2):330. DOI: 10.3390/ijms18020330

[24] Pérez JC. Fungi of the human gut microbiota: Roles and significance. International Journal of Medical Microbiology: IJMM. 2021;**311**(3): 151490. DOI: 10.1016/j.ijmm.2021.151490

[25] Li J, Chen D, Yu B, He J, Zheng P, Mao X, et al. Fungi in gastrointestinal tracts of human and mice: From community to functions. Microbial Ecology. 2018;**75**(4):821-829. DOI: 10.1007/s00248-017-1105-9

[26] Borges FM, de Paula TO, Sarmiento M, de Oliveira MG, Pereira M, Toledo IV, et al. Fungal diversity of human gut microbiota among eutrophic, overweight, and obese individuals based on aerobic culture-dependent approach. Current Microbiology. 2018;**75**(6): 726-735. DOI: 10.1007/s00284-018-1438-8

[27] Shkoporov AN, Hill C. Bacteriophages of the human gut: The "known unknown" of the microbiome. Cell Host & Microbe. 2019;**25**(2): 195-209. DOI: 10.1016/j.chom.2019.01.017

[28] Hillman ET, Lu H, Yao T, Nakatsu CH. Microbial ecology along the gastrointestinal tract. Microbes and Environments. 2017;**32**(4):300-313. DOI: 10.1264/jsme2.ME17017

[29] Gregory AC, Zablocki O, Zayed AA, Howell A, Bolduc B, Sullivan MB. The gut Virome database reveals age-dependent patterns of Virome diversity in the human gut. Cell Host & Microbe. 2020;**28**(5):724-740.e8. DOI: 10.1016/j.chom.2020.08.003

[30] Gaci N, Borrel G, Tottey W, O'Toole PW, Brugère JF. Archaea and the human gut: New beginning of an old story. World Journal of Gastroenterology. 2014;**20**(43): 16062-16078. DOI: 10.3748/wjg.v20.i43.16062

[31] Kho ZY, Lal SK. The human gut microbiome - a potential controller of wellness and disease. Frontiers in Microbiology. 2018;**9**:1835. DOI: 10.3389/fmicb.2018.01835

[32] Chen Y, Zhou J, Wang L. Role and mechanism of gut microbiota in human disease. Frontiers in Cellular and Infection Microbiology. 2021;**11**:625913. DOI: 10.3389/fcimb.2021.625913

[33] Schluter J, Foster KR. The evolution of mutualism in gut microbiota via host epithelial selection. PLoS Biology. 2012; **10**(11):e1001424. DOI: 10.1371/journal.pbio.1001424

[34] Walter J, Ley R. The human gut microbiome: Ecology and recent evolutionary changes. Annual Review of Microbiology. 2011;**65**:411-429. DOI: 10.1146/annurev-micro-090110-102830

[35] Trosvik P, de Muinck EJ. Ecology of bacteria in the human gastrointestinal tract–identification of keystone and foundation taxa. Microbiome. 2015;**3**:44. DOI: 10.1186/s40168-015-0107-4

[36] Romaní-Pérez M, Bullich-Vilarrubias C, López-Almela I, Liébana-García R, Olivares M, Sanz Y. The microbiota and the gut-brain Axis in controlling food intake and energy homeostasis. International Journal of Molecular Sciences. 2021;**22**(11):5830. DOI: 10.3390/ijms22115830

[37] Moran NA, Ochman H, Hammer TJ. Evolutionary and ecological consequences of gut microbial communities. Annual Review of Ecology, Evolution, and Systematics. 2019;**50**(1):451-475. DOI: 10.1146/annurev-ecolsys-110617-062453

[38] Selber-Hnatiw S, Rukundo B, Ahmadi M, Akoubi H, Al-Bizri H, Aliu AF, et al. Human gut microbiota: Toward an ecology of disease. Frontiers in Microbiology. 2017;**8**:1265. DOI: 10.3389/fmicb.2017.01265

[39] Zhang CX, Wang HY, Chen TX. Interactions between intestinal microflora/probiotics and the immune system. BioMed Research International. 2019;**2019**:6764919. DOI: 10.1155/2019/6764919

[40] Tasnim N, Abulizi N, Pither J, Hart MM, Gibson DL. Linking the gut microbial ecosystem with the environment: Does gut health depend on where we live? Frontiers in Microbiology. 2017;**8**:1935. DOI: 10.3389/fmicb.2017.01935

[41] Coyte KZ, Rakoff-Nahoum S. Understanding competition and cooperation within the mammalian gut microbiome. Current Biology: CB. 2019; **29**(11):R538-R544. DOI: 10.1016/j.cub.2019.04.017

[42] Dekker JP, Frank KM. Salmonella, shigella, and yersinia. Clinics in Laboratory Medicine. 2015;**35**(2): 225-246. DOI: 10.1016/j.cll.2015.02.002

[43] Lan R, Stevenson G, Reeves PR. Comparison of two major forms of the shigella virulence plasmid pINV: Positive

selection is a major force driving the divergence. Infection and Immunity. 2003;**71**(11):6298-6306. DOI: 10.1128/IAI.71.11.6298-6306.2003

[44] Behnsen J, Deriu E, Sassone-Corsi M, Raffatellu M. Probiotics: Properties, examples, and specific applications. Cold Spring Harbor Perspectives in Medicine. 2013;**3**(3):a010074. DOI: 10.1101/cshperspect.a010074

[45] Vernocchi P, Del Chierico F, Putignani L. Gut microbiota metabolism and interaction with food components. International Journal of Molecular Sciences. 2020;**21**(10):3688. DOI: 10.3390/ijms21103688

[46] Martin AM, Sun EW, Rogers GB, Keating DJ. The influence of the gut microbiome on host metabolism through the regulation of gut hormone release. Frontiers in Physiology. 2019;**10**:428. DOI: 10.3389/fphys.2019.00428

[47] Rowland I, Gibson G, Heinken A, Scott K, Swann J, Thiele I, et al. Gut microbiota functions: Metabolism of nutrients and other food components. European Journal of Nutrition. 2018;**57**(1):1-24. DOI: 10.1007/s00394-017-1445-8

[48] Portincasa P, Bonfrate L, Vacca M, De Angelis M, Farella I, Lanza E, et al. Gut microbiota and short chain fatty acids: Implications in glucose homeostasis. International Journal of Molecular Sciences. 2022;**23**(3):1105. DOI: 10.3390/ijms23031105

[49] den Besten G, van Eunen K, Groen AK, Venema K, Reijngoud DJ, Bakker BM. The role of short-chain fatty acids in the interplay between diet, gut microbiota, and host energy metabolism. Journal of Lipid Research. 2013;**54**(9):2325-2340. DOI: 10.1194/jlr.R036012

[50] Wolfe AJ. Glycolysis for microbiome generation. Microbiology Spectrum. 2015;**3**(3). DOI: 10.1128/microbiolspec.MBP-0014-2014

[51] O'Mahony SM, Clarke G, Borre YE, Dinan TG, Cryan JF. Serotonin, tryptophan metabolism and the brain-gut-microbiome axis. Behavioural Brain Research. 2015;**277**:32-48. DOI: 10.1016/j.bbr.2014.07.027

[52] Jenkins TA, Nguyen JC, Polglaze KE, Bertrand PP. Influence of tryptophan and serotonin on mood and cognition with a possible role of the gut-brain Axis. Nutrients. 2016;**8**(1):56. DOI: 10.3390/nu8010056

[53] Agus A, Planchais J, Sokol H. Gut microbiota regulation of tryptophan metabolism in health and disease. Cell Host & Microbe. 2018;**23**(6):716-724. DOI: 10.1016/j.chom.2018.05.003

[54] Yano JM, Yu K, Donaldson GP, Shastri GG, Ann P, Ma L, et al. Indigenous bacteria from the gut microbiota regulate host serotonin biosynthesis. Cell. 2015;**161**(2):264-276. DOI: 10.1016/j.cell.2015.02.047

[55] Chiang JY. Bile acid metabolism and signaling. Comprehensive Physiology. 2013;**3**(3):1191-1212. DOI: 10.1002/cphy.c120023

[56] Song X, Sun X, Oh SF, Wu M, Zhang Y, Zheng W, et al. Microbial bile acid metabolites modulate gut RORγ^+ regulatory T cell homeostasis. Nature. 2020;**577**(7790):410-415. DOI: 10.1038/s41586-019-1865-0

[57] Schoeler M, Caesar R. Dietary lipids, gut microbiota and lipid metabolism. Reviews in Endocrine & Metabolic Disorders. 2019;**20**(4):461-472. DOI: 10.1007/s11154-019-09512-0

[58] Liu S, Alexander RK, Lee CH. Lipid metabolites as metabolic messengers in inter-organ communication. Trends in Endocrinology and Metabolism: TEM. 2014;**25**(7):356-363. DOI: 10.1016/j.tem.2014.05.002

[59] Li Z, Yi CX, Katiraei S, Kooijman S, Zhou E, Chung CK, et al. Butyrate reduces appetite and activates brown adipose tissue via the gut-brain neural circuit. Gut. 2018;**67**(7):1269-1279. DOI: 10.1136/gutjnl-2017-314050

[60] d'Hennezel E, Abubucker S, Murphy LO, Cullen TW. Total lipopolysaccharide from the human gut microbiome silences toll-like receptor Signaling. mSystems. 2017;**2**(6):e00046-e00017. DOI: 10.1128/mSystems.00046-17

[61] Bertani B, Ruiz N. Function and biogenesis of lipopolysaccharides. EcoSal Plus. 2018;**8**(1). DOI: 10.1128/ecosalplus.ESP-0001-2018

[62] Hersoug LG, Møller P, Loft S. Role of microbiota-derived lipopolysaccharide in adipose tissue inflammation, adipocyte size and pyroptosis during obesity. Nutrition Research Reviews. 2018;**31**(2):153-163. DOI: 10.1017/S0954422417000269

[63] Wang X, Quinn PJ. Lipopolysaccharide: Biosynthetic pathway and structure modification. Progress in Lipid Research. 2010;**49**(2):97-107. DOI: 10.1016/j.plipres.2009.06.002

[64] Fritsch H, Lienemann A, Brenner E, Ludwikowski B. Clinical anatomy of the pelvic floor. Advances in Anatomy, Embryology, and Cell Biology. 2004;**175**:III–64. DOI: 10.1007/978-3-642-18548-9

[65] Jacobson A, Yang D, Vella M, Chiu IM. The intestinal neuro-immune axis: Crosstalk between neurons, immune cells, and microbes. Mucosal Immunology. 2021;**14**(3):555-565. DOI: 10.1038/s41385-020-00368-1

[66] Furness JB, Callaghan BP, Rivera LR, Cho HJ. The enteric nervous system and gastrointestinal innervation: Integrated local and central control. Advances in Experimental Medicine and Biology. 2014;**817**:39-71. DOI: 10.1007/978-1-4939-0897-4_3

[67] Null M, Agarwal M. Anatomy, lymphatic system. [updated 2022 Feb 10]. In: StatPearls [Internet]. Treasure Island (FL): StatPearls Publishing; 2022. Available from: https://www.ncbi.nlm.nih.gov/books/NBK513247/

[68] Moore JE Jr, Bertram CD. Lymphatic system flows. Annual Review of Fluid Mechanics. 2018;**50**:459-482. DOI: 10.1146/annurev-fluid-122316-045259

[69] Breslin JW, Yang Y, Scallan JP, Sweat RS, Adderley SP, Murfee WL. Lymphatic vessel network structure and physiology. Comprehensive Physiology. 2018;**9**(1):207-299. DOI: 10.1002/cphy.c180015

[70] Brandstaeter S, Fuchs SL, Aydin RC, Cyron CJ. Mechanics of the stomach: A review of an emerging field of biomechanics. GAMM-Mitteilungen. 2019;**42**(3):e201900001

[71] Kulkarni S, Ganz J, Bayrer J, Becker L, Bogunovic M, Rao M. Advances in enteric neurobiology: The “brain” in the gut in health and disease. The Journal of Neuroscience: The Official Journal of the Society for Neuroscience. 2018;**38**(44):9346-9354. DOI: 10.1523/JNEUROSCI.1663-18.2018

[72] Reboredo-Rodríguez P, González-Barreiro C, Martínez-Carballo E, Cambeiro-Pérez N, Rial-Otero R,

Figueiredo-González M, et al. Applicability of an In-vitro digestion model to assess the bioaccessibility of phenolic compounds from olive-related products. Molecules (Basel, Switzerland). 2021;**26**(21):6667. DOI: 10.3390/molecules26216667

[73] Reyes-Cerpa S, Maisey K, Reyes-López F, Toro-Ascuy D, Sandino AM, Imarai M. Fish cytokines and immune response. In: Türker H, editor. New Advances and Contributions to Fish Biology [Internet]. London: IntechOpen; 2012 [cited 2022 Feb 26]. Available from: https://www.intechopen.com/chapters/41563doi:10.5772/53504

[74] Pabst R. The anatomical basis for the immune function of the gut. Anatomy and Embryology. 1987;**176**(2):135-144. DOI: 10.1007/BF00310046

[75] Mason KL, Huffnagle GB, Noverr MC, Kao JY. Overview of gut immunology. Advances in Experimental Medicine and Biology. 2008;**635**:1-14. DOI: 10.1007/978-0-387-09550-9_1

[76] Montalban-Arques A, Chaparro M, Gisbert JP, Bernardo D. The innate immune system in the gastrointestinal tract: Role of intraepithelial lymphocytes and lamina propria innate lymphoid cells in intestinal inflammation. Inflammatory Bowel Diseases. 2018; **24**(8):1649-1659. DOI: 10.1093/ibd/izy177

[77] Börsch G. Der Gastrointestinaltrakt als Immunorgan: Das darmassoziierte Immunsystem [the gastrointestinal tract as an immunologic organ: The gut-associated immune system]. Klinische Wochenschrift. 1984;**62**(15):699-709. DOI: 10.1007/BF01725702

[78] Gonçalves A, Machado R, Gomes AC, Da CA. Nanotechnology solutions for controlled cytokine delivery: An applied perspective. Applied Sciences. 2020;**10**(20):7098. DOI: 10.3390/app10207098

[79] Mirantes C, Passegué E, Pietras EM. Pro-inflammatory cytokines: Emerging players regulating HSC function in normal and diseased hematopoiesis. Experimental Cell Research. 2014; **329**(2):248-254. DOI: 10.1016/j.yexcr.2014.08.017

[80] Shea-Donohue T, Fasano A, Smith A, Zhao A. Enteric pathogens and gut function: Role of cytokines and STATs. Gut Microbes. 2010;**1**(5): 316-324. DOI: 10.4161/gmic.1.5.13329

[81] Carabotti M, Scirocco A, Maselli MA, Severi C. The gut-brain axis: Interactions between enteric microbiota, central and enteric nervous systems. Annals of Gastroenterology. 2015;**28**(2):203-209

[82] Keightley PC, Koloski NA, Talley NJ. Pathways in gut-brain communication: Evidence for distinct gut-to-brain and brain-to-gut syndromes. The Australian and New Zealand Journal of Psychiatry. 2015;**49**(3):207-214. DOI: 10.1177/0004867415569801

[83] Forsythe P, Bienenstock J, Kunze WA. Vagal pathways for microbiome-brain-gut axis communication. Advances in Experimental Medicine and Biology. 2014;**817**:115-133. DOI: 10.1007/978-1-4939-0897-4_5

[84] Ternet C, Kiel C. Signaling pathways in intestinal homeostasis and colorectal cancer: KRAS at Centre stage. Cell Communication and Signaling: CCS. 2021;**19**:31. DOI: 10.1186/s12964-021-00712-3

[85] Kaper JB, Sperandio V. Bacterial cell-to-cell signaling in the gastrointestinal tract. Infection and Immunity. 2005;

73(6):3197-3209. DOI: 10.1128/IAI.73.6.3197-3209.2005

[86] Camilli A, Bassler BL. Bacterial small-molecule signaling pathways. Science (New York, N.Y.). 2006; **311**(5764):1113-1116. DOI: 10.1126/science.1121357

[87] Jung K, Fabiani F, Hoyer E, Lassak J. Bacterial transmembrane signalling systems and their engineering for biosensing. Open Biology. 2018;**8**(4): 180023. DOI: 10.1098/rsob.180023

[88] Darnell RL, Nakatani Y, Knottenbelt MK, Gebhard S, Cook GM. Functional characterization of BcrR: A one-component transmembrane signal transduction system for bacitracin resistance. Microbiology (Reading, England). 2019;**165**(4):475-487. DOI: 10.1099/mic.0.000781

[89] Andoh A, Yagi Y, Shioya M, Nishida A, Tsujikawa T, Fujiyama Y. Mucosal cytokine network in inflammatory bowel disease. World Journal of Gastroenterology. 2008; **14**(33):5154-5161. DOI: 10.3748/wjg.14.5154

[90] Dhama K, Mahendran M, Chauhan RS, Tomar S. Cytokines-their functional roles and prospective applications in veterinary practice: A review. Journal of Immunology and Immunopathology. 2008;**10**(2):79-89

[91] Chiba Y, Shida K, Nagata S, Wada M, Bian L, Wang C, et al. Well-controlled proinflammatory cytokine responses of Peyer's patch cells to probiotic lactobacillus casei. Immunology. 2010;**130**(3):352-362. DOI: 10.1111/j.1365-2567.2009.03204.x

[92] Côté CD, Zadeh-Tahmasebi M, Rasmussen BA, Duca FA, Lam T. Hormonal signaling in the gut. The Journal of Biological Chemistry. 2014; **289**(17):11642-11649. DOI: 10.1074/jbc.O114.556068

[93] Ma Q, Xing C, Long W, et al. Impact of microbiota on central nervous system and neurological diseases: The gut-brain axis. Journal of Neuroinflammation. 2019;**16**:53. DOI: 10.1186/s12974-019-1434-3

[94] Belkaid Y, Hand TW. Role of the microbiota in immunity and inflammation. Cell. 2014;**157**(1):121-141. DOI: 10.1016/j.cell.2014.03.011

Chapter 8

Role of Gut Microbiome and Enteric Bacteria in Gallbladder Cancer

Jyoti Sharma, Farhanul Huda, Manisha Naithani, Sudhir Kumar Singh, Navin Kumar and Somprakas Basu

Abstract

Gallbladder cancer (GBC) is associated with a sinister prognosis, a short survival time, and early metastasis to distant sites. Chronic inflammation of the gallbladder due to gallstone disease and biliary bacteria remain key factors in the pathogenesis of GBC. The association of chronic bacterial infections with the development of GBC has provided a new perspective on the causation of GBC. A strong link between chronic Salmonella infection and enterohepatic strains of Helicobacter species with GBC has been suggested. It is believed that many other enteric bacterial strains, predominantly the Enterobacteriaceae species, are associated with the development of GBC. However, the available literature mainly comprises observational studies and small meta-analyses necessitating the requirement of a higher level of evidence. This chapter discusses the role of the gut microbiome, dysbiosis and its association with carcinogenesis, and the organisms associated with the causation of GBC.

Keywords: gallbladder neoplasm, dysbiosis, gut microbiome, brain-gut axis, gastrointestinal microbiome

1. Introduction

Gallbladder cancer (GBC) is the most common malignancy of the biliary tract with an aggressive clinical course and short median survival [1]. While being a rarity in the western world, GBC is one of the major causes of cancer-related morbidity and mortality in South Asian and Southeast Asian countries [2]. Females are more commonly affected than males. According to the cancer statistics of 2020, GBC accounts for 0.6% of the total cancer cases and is associated with 0.9% of total cancer-related deaths [3]. Around 10% of the global GBC burden is contributed by India, with the Northern, Central, and North-eastern parts as the highest contributors [4]. Only 10% of cases present at an early stage which can be owed to the aggressive tumor biology of this cancer and the lack of effective screening techniques for its early detection [5]. Chronic inflammation of the gallbladder remains a major factor in the pathogenesis of GBC, although the causes are multi-factorial. Gall stones, heavy metals, environmental toxins, and carcinogens have all been implicated in chronic irritation of the gallbladder mucosa, thereby leading to dysplasia and subsequent development of neoplasia.

The landscape of the microbiome populating our digestive tract has received a lot of scientific attention in recent years [5]. There is ample evidence linking the human microbiome and its metabolites to carcinogenesis. It is proven that balanced flora or microbial eubiosis is related to health while dysbiosis or unbalanced flora can lead to various diseases, including cancers [6, 7]. There can be multiple triggers causing dysbiosis, including fluctuations in the environment, inflammation, infection, medications, dietary changes, or genetic predisposition. The International Agency for Research on Cancer labeled ten microbial species as carcinogens [8]. Around 15–20% of cancers are linked to microbial pathogens, with *Helicobacter pylori* (*H. pylori*), human papillomavirus, Hepatitis B virus, and hepatitis C virus being the four predominant species, driving 90% of infection-associated cancers [6, 9, 10]. However, there is very limited information available on the microbial species inhabiting the human gall bladder, except for a few cultivable species of bacteria associated with cholelithiasis [11, 12]. It was seen that the biliary tract has an abundance of Enterobacteriaceae [13]. Microorganisms in the common bile duct of patients having gallstones were more commonly those that inhabited the human respiratory tract and oral cavity rather than intestinal microbes [14]. Very recently, culture-negative bile samples acquired from normal gallbladders were evaluated using 6S ribosome gene analysis. A very simple and less diverse bacterial flora was found comprising the Firmicutes, Proteobacteria, and Actinobacteria phyla [15].

Detection of some bacteria does not indicate its causality in inflammation or cancer. However, recent amassing evidence indicates that microbiota dysbiosis and chronic inflammation contribute to carcinogenesis [16]. Several reports point towards strains of Salmonella and Helicobacter colonizing the gall bladder and are linked to an escalated risk of developing GBC [17, 18]. Premalignant lesions were found to be coexisting with chronic Salmonella infestation, despite the absence of gallstones [19]. Various experimental studies and epidemiologic data support the induction of carcinogenesis due to dysbiosis of the gallbladder microbiome. However, results indicating only cultivable species limit these claims. Also, despite the proximity of a large diverse microflora reservoir in the gut, little is known about its impact on the human bile microbiome. In this chapter, we aim to provide a comprehensive review of all the available literature on the gut and biliary microbiome and their association with GBC.

2. Understanding the human gut microbiome

The term microbiome has been derived from two words, "micro" and "biome", meaning, a specific microbial community with distinct physiological and chemical properties, residing in a well-defined habitat which is their "theatre of activity". This definition was proposed by Whipps while working on mycoparasites [20]. The term "gut microbiome" or "human microbiome" was coined by Joshua Lederberg in 2001 and since then it has been a topic for debate among researchers [21]. The human microbiome can be defined as a specific community of commensal, symbiotic and pathogenic micro-organisms that reside within our body spaces [22]. These include gut bacteria, eukaryotes, archaea, and specific viruses [23, 24]. In a healthy individual, these bacteria are responsible for various synthetic and metabolic functions and detoxification of various xenobiotics [25]. They form an integral part of the "gut-brain axis" which is bidirectional communication between the gut and the

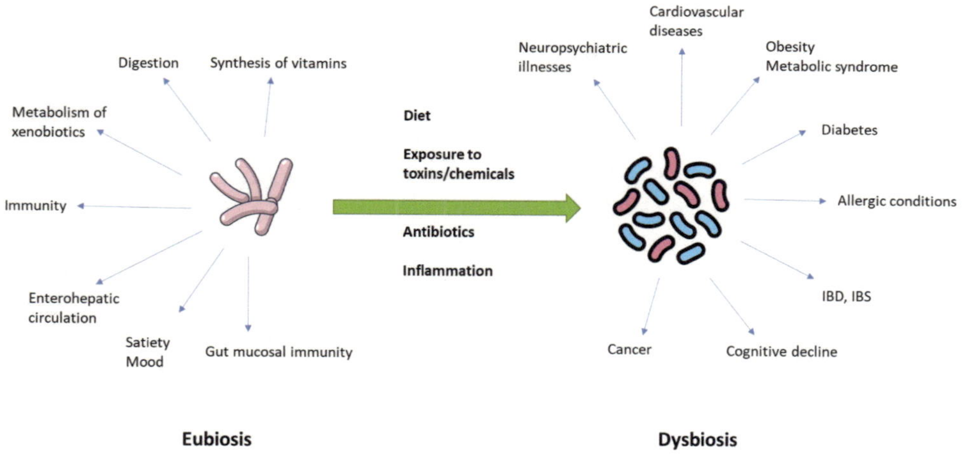

Figure 1.
Importance of gut bacterial microflora. The figure illustrates the role of enteric bacteria in the maintenance of homeostasis. The "eubiotic" bacteria display a complex interaction with the various synthetic and metabolic functions of our body as well as in the "gut-brain crosstalk". Alteration or "dysbiosis" due to any factor (diet, chemicals, antibiotics, inflammatory conditions) may lead to "blooms" of harmful bacteria. The dysbiotic bacteria have now been linked to various cardiovascular, metabolic, neuropsychiatric diseases, including cancer.

cognitive and emotional centers of the brain. This link is responsible for satiety and appetite regulation, elevation of mood, cognitive development, and neuroprotection [26]. Studies have also found a link between the gut microbiome and immune homeostasis. The complex and bidirectional interaction between the gut microbiota and the host immune system is responsible for the development of both innate and adaptive immunity, thus preventing the body from pathogenic organisms [27, 28]. Moreover, the microbiota is also responsible for the maintenance of gut mucosal integrity and prevents the overgrowth of pathogenic organisms, thus maintaining the first line of defense against the pathogens [29]. Therefore, any imbalance in the gut microbiota may lead to the development of various autoimmune diseases. This concept of "dysbiosis" or "imbalance" in the gut microbiota may result in relative "blooms" of harmful bacteria, especially Enterobacteriaceae [30, 31]. Dysbiosis can be caused by a variety of factors, namely, dietary changes, inflammatory conditions, exposure to drugs, and toxins [32, 33] (**Figure 1**). The gut bacteria have been linked to a wide variety of cardiovascular diseases [34, 35], obesity [36, 37], inflammatory bowel disease [38], irritable bowel syndrome [39], and some neuropsychiatric diseases like depression [40]. But what has intrigued the researchers is the role of gut microbiota in the development of cancer.

3. Mechanism of carcinogenesis

To ascertain the role of gut microbiota in the development of cancer, we need to look at the mechanisms responsible for carcinogenesis. The normal cells get altered into cancerous cells, by changes at the cellular, genetic or epigenetic levels. This process is known as "cell transformation" [41]. TP53 is a tumor suppressor gene that encodes the protein P53. P53 acts as a tumor suppressor which causes a transient cell cycle arrest, allowing the cells to repair the damage caused to the DNA before the cell divides. The cells that are unable to repair the damage undergo apoptosis. This ensures that the potentially oncogenic mutations are not propagated [42]. Chronic inflammation causes alteration in the TP53 gene, leading to its inactivation. This results in an unregulated cell cycle and cell division, leading to the accumulation of mutations and uncontrolled cellular proliferation. TP53 alterations were seen in biliary epithelia of patients with gallstone disease with an increased frequency with the disease progression from metaplasia to carcinoma [43]. TP53 was the most commonly mutated gene, followed by PIK3CA, SMAD4, ARID1A, KRAS [44–46] and amplification of ERBB2 [47].

The study of these genetic alterations and mechanisms of carcinogenesis has been made possible with the development of various ex vivo and in vivo animal models. These models have been used extensively to decipher the etiopathogenesis of GBC and to develop and test the treatment protocols [48]. Ex vivo models use cell lines to study the tumor characteristics and cellular and genetic abnormalities. But it was seen that different cell lines yielded different tumor characteristics for the same type of tumor, thus complicating the interpretation [49]. In vivo models were superior to the cell lines as they used genetically engineered animals that could retain the genetic mutations and could undergo cellular differentiation. The tumor cells with mutated or amplified genes were inoculated in them and studied for the development of cancer [49–51]. The drawback of these models was the lack of innate immunity which led to altered results as the cancer was not strictly recapitulated [51].

Although these models indicate a causal relationship between the risk factors and carcinogenesis, the human body reacts quite differently as compared to an animal model, thus necessitating the need for the development of an ideal human model.

4. Role of gut microbiota in cancer causation

There has been an ongoing debate among researchers on the role of gut microbiota in the causation of cancer as cancer is neither a contagious nor an infectious disease [16]. The first proposition of the possible role of gut microbiota in cancer causation was given by Russel in 1890, which was supported by positive results over the subsequent years [52–54]. However, in 1963, a group of scientists from NCI, USA claimed that the bacteria found in the cancer tissues were probably contaminants [16]. This subject remained controversial until Marshall, in his study, proved the association between *H. pylori* and gastric adenocarcinoma [55]. This was a breakthrough study in this aspect and since then, a number of bacteria have been linked to a variety of cancers [56–58]. However, the mechanism by which the microbiota cause cancer is still unclear. While there is no concrete evidence supporting the causation of cancer, there may be a role of the bacteria in its progression [59].

Microbiota may act as a carcinogen in two ways, either by inducing a chronic inflammatory state or direct injury by material toxins and metabolites [16, 60–62]. Release of pro-inflammatory mediators like TNF-α and IL-1 and generation of reactive oxygen species (ROS) stimulates lymphoepithelial proliferation and cell division. This leads to immune dysregulation, thereby leading to tumorigenesis [27]. It causes alteration in the cell cycle leading to immunosuppression [63]. It also results in genetic and cellular damage and genomic instability which preclude carcinogenesis [64]. The bacterial toxins are genotoxins that cause DNA damage and may lead to the development of cancer [65, 66]. Thus, chronic bacterial infections demonstrate a dual role in carcinogenesis by both stimulating and inhibiting the immune system.

5. The biliary microbiome

According to traditional thinking, the biliary tract has always been considered sterile. This is because of the anti-microbial properties possessed by bile which affects the bacterial membrane and DNA [67]. However, inflammatory conditions of the biliary tract, like acute cholecystitis and cholangitis have frequently cultured bacterial colonies commonly found in the human gut; the common organisms being, *Escherichia coli*, Enterobacter, Pseudomonas, and Citrobacter spp. [68]. This can be explained by the pathophysiology of these diseases, which is, biliary obstruction and gut bacterial translocation. However, recent studies have indicated that even under nonpathogenic circumstances, the human bile comprises a rich diversity of microbial flora which is actively involved in the regulation of the size and composition of the bile acid pool as well as the metabolism of bile acids [69, 70]. However, this normal biliary microbiome mainly included Firmicutes, Actinobacteria, Proteobacteria, and Bacteroidetes which were also found in the human gut [70, 71]. This can partly be explained by the close association of the human gut and the biliary tract and the involvement of enteric bacteria in enterohepatic circulation but the evidence is largely limited to animal models and an ideal human experimental model is required [72].

Dysbiosis of the gut bacteria has been implicated in the development and progression of various cancers, including gastric [73, 74], colorectal [75], and oral cancers [76, 77], however, their association with causation or progression of the hepatobiliary cancers is still in question. The natural synergy that exists between the bile acid metabolism and the biliary microbiome reaffirms the proposition that biliary microbial dysbiosis may lead to various biliary tract diseases including gallstone formation and the development of cancer.

6. Mechanism of carcinogenesis in the biliary tract

Cholangiocytes are considered the potential cells of origin for biliary tract cancers, including gallbladder cancer [78]. Any insult to the cholangiocytes leads to the release of pro-inflammatory mediators like IL-6 and IL-1β which results in the differentiation of T helper cells (Th-17 cells). The cholangiocytes interact with Th-17 cells leading to their activation and proliferation, in order to compensate for the cell loss [79]. Moreover, the bacteria and their products are recognized by the cholangiocytes through the Pathogen Associated Molecular Patterns (PAMPs) present in the bile, which interacts with the pattern recognition receptors, that are, the Toll-like receptors (TLRs) and the NOD-like receptors (NLRs), leading to their activation [80]. This results in collagen deposition and fibrosis. The resultant cholangiopathy may lead to ductopenia, dysplasia, and malignant transformation [81]. Chronic inflammation leads to the release of mediators like IL-17, TNF-α, and TGF-β which cause genetic alterations in the tumor suppressor genes and the proto-oncogenes resulting in cell transformation [82]. These mediators are among the few which have been implicated in the causation of carcinogenesis [83–85].

7. Enteric bacteria and gallbladder cancer

Gallbladder cancer is the most common biliary tract cancer and the etiopathogenesis is multifactorial [86]. However, chronic inflammation [87] and gallstone disease represent the most important aetiologies in the development of GBC and are supported by Level II evidence [88]. The recent development of culture and culture-independent techniques have identified various organisms which are associated with the formation of both pigmented as well as cholesterol gallstones [89, 90]. These dysbiotic organisms are mainly enteric bacteria that have the ability to form a biofilm, thereby resisting cellular and DNA damage caused by bile. They are namely, *E. coli*, *Pseudomonas aeruginosa*, *Klebsiella pneumoniae*, *Enterococcus spp*, *Acinetobacter spp* which were associated with the patients presenting with gallstones [91–93]. *Clostridium*, *Bifidobacterium*, *Peptostreptococcus*, *Bacteroides* were among the other bacteria leading to the formation of gallstones by interfering with the enterohepatic circulation [94, 95]. With the development of Polymerase chain reaction—denaturing gradient gel electrophoresis (PCR-DGGE), various uncultivable bacteria like *Staphylococcus hemolyticus*, *Enterobacter or Citrobacter spp*, *Morganella spp*, *Salmonella spp.*, *Capnocytophaga spp*, *Lactococcus species*, *Bacillus spp*, and *H. Pylori* have been isolated in different compositions [72, 96]. Some pathogens of the oral cavity have also been implicated in the formation of gallstones by affecting the motility of the gallbladder and the production of mucin [97]. These bacteria can be indirectly linked with GBC. Recent studies have demonstrated positive cultures of enteric bacteria in GBC patients projecting a direct association of the gut flora with GBC, however, the level of evidence is low [91].

Although various bacteria have now been identified and linked with the development of gallstones and their theoretical association with GBC, *Helicobacter species* and *Salmonella typhi* have been extensively studied and are strongly implicated in the development of GBC [98, 99].

8. The Helicobacter species and gallbladder cancer

The *Helicobacter species*, especially *H. pylori*, have been largely implicated in the causation of gastric as well as intestinal cancers by the mechanism of inflammation-induced tumorigenesis and are now being associated with the development of various hepatobiliary cancers [100].

H. pylori induces a chronic inflammatory state by resulting in the release of various pro-inflammatory mediators like TNF-α, IL-1, IL-6, and other vasoactive substances [101]. They also prevent cell adhesion and lead to the migration of the mutated epithelial cells [102]. Free radicals produced cause oxidative damage to the biliary epithelium [103]. IL-8 production may also promote inflammation and alter cellular proliferation and apoptosis [87]. The Cag- A protein secreted by *H. pylori* is one of the most extensively studied virulence factors responsible for producing a chronic inflammatory state as seen in gastric epithelial cells and increasing the risk of gallstones [90]. Some strains possess pathogenicity islands which produce a type IV secretion system and also result in a "hummingbird" epithelial phenotype of epithelial cells which are implicated in rival cell death, thus resulting in the death of the normal biliary microbiota and producing a Helicobacter "bloom" [104, 105]. This "bloom" of Helicobacter results in chronic inflammation of the biliary epithelium by the various mechanisms mentioned, leading to dysplasia and subsequent neoplasia.

Since the first evidence of *H. pylori* in gallbladder mucosa in a patient with cholecystitis was detected by Kawaguchi et al. [106] in 1996, ample studies have shown an association of Helicobacter species, especially *H. pylori*, with the formation of gallstones [107–109] which have also been reiterated by recent meta-analyses [110]. Another study by Kuroki et al. [111] reported a higher biliary epithelial proliferation rate in patients infected with *Helicobacter species* as compared to the control group.

The isolation of bacteria in these studies was done using various next-generation sequencing techniques. These techniques are being utilized to establish an association between the *Helicobacter species* and GBC. While some have shown promising results [112, 113], others have negated this theory [114, 115]. Apart from *H. pylori*, attempt at isolating other enterohepatic strains like *H. bilis*, *H. hepaticus,* and *H. pullorum* have been done in a number of studies [116]. Dewhirst and Fox [117] identified 5 strains of *H. bilis*, 2 strains of *Flexisipira rappini*, and one strain of *H. pullorum* using PCR analysis in patients with gallbladder diseases and GBC. Various studies have demonstrated high positivity of *H. bilis* in patients with biliary tract and gallbladder cancer, suggesting an association of *H. bilis* with GBC [118–120].

There have been several meta-analyses suggesting an association between Helicobacter infection and cancer of the biliary tract with conflicting results [121, 122]. However, a recent meta-analysis has suggested a significant association between *H. pylori* infection and biliary tract cancer [98]. The available research has suggested a strong association between *Helicobacter species* and GBC; however, these studies are largely limited to observational studies or small meta-analyses necessitating the need for a higher level of evidence in order to establish a general consensus.

9. *Salmonella typhi* and gallbladder cancer

Salmonella enterica serovar typhi is a gram-negative, flagellated, rod-shaped bacteria which is the causative agent of typhoid fever. It resides in the gallbladder and results in chronic inflammation of the gall bladder mucosa leading to the formation of gallstones [123]. It also produces a biofilm that prevents it from the anti-bacterial action of the bile and thus results in its persistence in the gallbladder leading to a chronic carrier state.

Salmonella typhi has been strongly associated with the development of gallstone disease and chronic infection with *S. typhi* is now being linked to GBC. The earliest evidence dates back to 1964 when Cargill et al. suggested a probable association between chronic typhoid and paratyphoid carriers and GBC [76]. In 1971, Axelrod et al., also reported a similar association between *S. typhi* and GBC. Since then, several studies have reiterated the results [124–126]. There are certain proposed mechanisms by which Salmonella may result in a chronic inflammatory state and subsequent development of cancer. The typhoid toxin is carcinogenic and causes alterations in the cell cycle and DNA damage [127]. AvrA, is an effector protein synthesized by Salmonella pathogenicity island 1 via Type III secretion system which subdues the host inflammatory response and prevents autophagy, thus resulting in the persistence of the bacteria and the chronic carrier state [128, 129].

Typhoid fever and GBC are endemic in the Gangetic belt and the northern states of India which provides strong evidence to suggest an association between *S. typhi* and GBC [7, 130, 131]. A study conducted in Northern India demonstrated that patients with gallbladder cancer had a higher Vi polysaccharide as compared to the control group and the risk of developing GBC in typhoid carrier patients was 8.47 times higher than the non-carrier group, thus concluding the chronic typhoid carrier state as a risk factor for GBC [130]. This has been reinforced by a number of recent studies [127, 132]. Although there is emerging evidence suggesting a positive association between *S. typhi* and GBC but the data is limited, with conflicting results, thus requiring larger epidemiological studies to establish a consensus [99, 133].

10. Miscellaneous bacteria and gallbladder cancer

Gene fragments of *Collibacillus*, *B. fragilis*, *Klebsiella*, *C. perfringens* and *Clostridium* spp. have been identified in the bile and gallbladder tissue of patients with GBC [134]. A positive correlation between the bacterial species of *E. coli*, *E. fecalis*, *Klebsiella*, and *Enterobacter spp. B10* along with *Peptostreptococcus stomatis*, *Fusobacterium*, *Firmicutes nucleatum*, and *Enterococcus faecium* with the development

Study/year	Sample	Bacterial strain	Isolation technique	Inference
Welton et al. [125]	Deceased typhoid carriers	S. typhi	Record registers	Chronic typhoid carriers are 6 times more likely to die of hepatobiliary cancer than controls ($P < 0.001$)
Caygill et al. [124]	Chronic typhoid carriers	S. typhi	Record registers	167-fold higher risk of GBC in chronic typhoid carriers Chronic, and not acute infection is a risk factor for GBC
Csendes et al. [137]	Tissue, bile	E. coli, E. faecalis, Klebsiella Enterobacter	Culture	Both aerobic and anaerobic gram-negative bacteria were found and may have a role in GBC
Shukla et al. [130]	Serum	S. typhi	IHA Vi antigen	Significantly high Vi positivity in patients with gallbladder carcinoma compared to controls Risk of developing GBC is 8.47 times more in culture-positive typhoid carriers than the noncarriers
Dutta et al. [131]	Serum	S. typhi	ELISA Vi antigen	Chronic typhoid carrier state is a risk factor for GBC
Dewhirst et al. [117]	Multiple sources: animal and human tissue, blood, stool, fetus	H. bilis, Flexisipira rappini, H. pullorum	PCR (16S rRNA)	Correlation of Helicobacter species with GBC and other biliary tract diseases Identified 5 strains of H. bilis, 2 strains of Flexisipira rappini, and one strain of H. pullorum
Matsukura et al. [118]	Bile	H. bilis	PCR (16S rRNA)	H. bilis infection in bile was associated with gallbladder cancer in Japanese and Thai patients
Fukuda et al. [112]	Bile, tissue	Helicobacter	PCR, Histology, IHC	Significantly high positivity of Helicobacter DNAs in 52.6% of patients with hepatobiliary cancer than that in the benign cases ($P = 0.03$)
Lu et al. [134]	Tissue	Colibacillus B. fragilis, Klebsiella C. perfringens Clostridium	PCR 16S rRNA	Possible association of both aerobic and anaerobic bacteria with GBC
Murata et al. [119]	Tissue	H. bilis	Nested PCR (16S rRNA)	4 out of 14 cases with biliary tract cancer were positive for H. bilis which may indicate their role in GBC

Study/year	Sample	Bacterial strain	Isolation technique	Inference
Kobayashi et al. [120]	Bile	H. pylori H. hepaticus H. bilis	PCR	Helicobacter DNA was detected in bile of 86% of malignant biliary diseases DNA fragments of Helicobacter species other than H. pylori, H. hepaticus, and H. bilis were commonly detectable
Bohr et al. [115]	Tissue	Helicobacter spp.	Culture, IHC, PCR (16S rRNA)	Helicobacter species do not play a predominant role in the pathogenesis of GSD and GBC in the German population
Shimoyama et al. [113]	Blood	H. hepaticus	ELISA	H. hepaticus-specific antigen was significantly higher in patients with biliary tract cancer ($P < 0.05$)
Iyer et al. [132]	Tissue	143 HPV S. typhi Ty2 S. typhi CT18 S. typhimurium-LT2 S. choleraesuis-SCB67 S. paratyphi-TCC S. paratyphi SPB7	PCR analysis	Association of non-typhoidal Salmonella species with GBC along with typhoidal strains Chronic carrier state is a risk factor for GBC
Tsuchiya et al. [114]	Blood	H. pylori	ELISA	No significant differences in antibody titers or H. pylori infection positivity rates between cases and controls
Song et al. [135]	Tissue	Peptostreptococcus stomatis Enterococcus faecium	DNA extraction and metagenomic sequencing	Existence of an altered microbiota in GBC

Table 1.
Studies show the association of the gut microbiome with gallbladder cancer.

of GBC has also been found in recent studies [135–137]. These bacteria were commensals of the gut and have been associated with colorectal cancer [138], gastric cancer [139], and metastatic melanoma [140]. Thus, their presence in gallbladder tissue and bile may indicate their association with GBC. **Table 1** summarizes the various studies showing an association of gut microbiota with GBC.

11. Therapeutic perspective: "microbial therapeutics"

There is a complex interplay between the human body and its microbiome. While a normal gut flora is essential for homeostasis, dysbiosis may lead to a multitude of diseases. Several mechanisms associated with carcinogenesis are now being utilized in its prevention. GBC has been associated with chronic inflammation and chronic typhoid carrier state; thus, many animal models have been developed to study the role of antibiotics in the eradication of Salmonella, thereby reducing the chances of development of GBC. But the results have been conflicting [128] and Cholecystectomy remains the only definitive treatment for eradication of the carrier state of Salmonella [128]. There was a rise in the number of prophylactic cholecystectomies owing to this but it also saw an increase in the number of colorectal malignancies due to gut bacterial dysbiosis, thereby emphasizing their role in the development of cancer [141].

The role of the gut microbiome in the maintenance of homeostasis encouraged the researchers to utilize their potential in the therapeutic management of the disease. Microbiome therapeutics consist of additive therapy, subtractive therapy, and modulatory therapy. Additive therapy with genetically engineered or natural probiotic agents has shown some benefit in colorectal cancer and is now being utilized in GBC. There is emerging evidence regarding the association of probiotics and dietary changes with a decreased incidence of gallstone disease, thereby reducing the chances of GBC, thus additive therapy with natural or genetically engineered probiotic organisms may prove beneficial. However, there is still a dearth of evidence in this aspect [142, 143]. Subtractive therapy is being utilized by genetically engineered *E. coli* strains with a cloned antibiofilm protease Deg P gene or a cloned Lysine and Pyosin gene which results in inhibition of growth of pathogenic bacteria in the gut, thereby preventing dysbiosis [144, 145]. *E. coli* strains with a cloned antibiotic Microcin H47 gene may also help in inhibiting and displacing Salmonella from the gut [146]. These two methods can be utilized in patients with a Chronic salmonella carrier state and may be used as an alternative to Cholecystectomy. This may also reduce the incidence of GBC in these patients. Apart from this, genetically engineered bacteria are also being used to test the effect and toxicity profile of chemotherapy [147, 148], develop cancer vaccines and targeted biological therapies [149].

12. Conclusion

The gut microbiome forms an integral part of the human body and is often referred to as the "forgotten organ". Its role in health and disease has been studied extensively over the past two decades but the possibility of its role in cancer causation has caught the eye of researchers. The association between the gut microbiome and cancer has provided new insight into understanding the pathophysiology of cancer and planning the management strategies. There is a strong correlation between gut microbial dysbiosis and the development of colorectal and gastric adenocarcinomas, however, their role in hepatobiliary cancers, especially GBC remains poorly understood. This can be owed to the short survival of GBC resulting in vast unexplored domains of this disease and the difficulty to isolate the bacteria involved via routine culture methods. Moreover, the lack of an ideal animal or a human model has greatly limited the research. The advent of the next-generation sequencing methods has seen emerging evidence linking various bacteria to the etiopathogenesis of GBC, but causality is far from proven. A higher level of evidence either in the form of larger meta-analyses or larger epidemiological studies is needed to establish a consensus.

Author details

Jyoti Sharma, Farhanul Huda, Manisha Naithani, Sudhir Kumar Singh, Navin Kumar and Somprakas Basu*
All India Institute of Medical Sciences, Dehradun, India

*Address all correspondence to: somprakas.surg@aiimsrishikesh.edu.in

References

[1] Okumura K, Gogna S, Gachabayov M, Felsenreich DM, McGuirk M, Rojas A, et al. Gallbladder cancer: Historical treatment and new management options. World Journal of Gastrointestinal Oncology. 2021;**13**(10):1317-1335

[2] Lai CHE, Lau WY. Gallbladder cancer—A comprehensive review. Surgeon. 2008;**6**(2):101-110

[3] Sung H, Ferlay J, Siegel RL, Laversanne M, Soerjomataram I, Jemal A, et al. Global cancer statistics 2020: GLOBOCAN estimates of incidence and mortality worldwide for 36 cancers in 185 countries. CA: A Cancer Journal for Clinicians. 2021;**71**(3):209-249

[4] Mathur P, Sathishkumar K, Chaturvedi M, Das P, Sudarshan KL, Santhappan S, et al. Cancer statistics, 2020: Report from National Cancer Registry Programme, India. JCO Global Oncology. 2020;**6**:1063-1075

[5] Dutta U. Gallbladder cancer: Can newer insights improve the outcome? Journal of Gastroenterology and Hepatology. 2012;**27**(4):642-653

[6] Dzutsev A, Badger JH, Perez-Chanona E, Roy S, Salcedo R, Smith CK, et al. Microbes and cancer. Annual Review of Immunology. 2017;**35**:199-228

[7] Nath G, Singh YK, Kumar K, Gulati AK, Shukla VK, Khanna AK, et al. Association of carcinoma of the gallbladder with typhoid carriage in a typhoid endemic area using nested PCR. Journal of Infection in Developing Countries. 2008;**2**(4):302-307

[8] De Martel C, Ferlay J, Franceschi S, Vignat J, Bray F, Forman D, et al. Global burden of cancers attributable to infections in 2008: A review and synthetic analysis. The Lancet Oncology. 2012;**13**(6)

[9] de Martel C, Georges D, Bray F, Ferlay J, Clifford GM. Global burden of cancer attributable to infections in 2018: A worldwide incidence analysis. The Lancet Global Health. 2020;**8**(2):e180-e190. DOI: 10.1016/S2214-109X(19)30488-7

[10] Bhatt AP, Redinbo MR, Bultman SJ. The role of the microbiome in cancer development and therapy. CA: A Cancer Journal for Clinicians. 2017;**67**(4):326-344

[11] Vitetta L, Sali A, Moritz V, Shaw A, Carson P, Little P, et al. Bacteria and gallstone nucleation. The Australian and New Zealand Journal of Surgery. 1989;**59**(7):571-577

[12] Stewart L, Grifiss JML, Jarvis GA, Way LW. Biliary bacterial factors determine the path of gallstone formation. The American Journal of Surgery. 2006;**192**(5):598-603

[13] Liu J, Yan Q, Luo F, Shang D, Wu D, Zhang H, et al. Acute cholecystitis associated with infection of Enterobacteriaceae from gut microbiota. Clinical Microbiology and Infection. 2015;**21**(9):851.e1-851.e9. DOI: 10.1016/j.cmi.2015.05.017

[14] Shen H, Ye F, Xie L, Yang J, Li Z, Xu P, et al. Metagenomic sequencing of bile from gallstone patients to identify different microbial community patterns and novel biliary bacteria. Scientific Reports. 2015;**5**:1-13. DOI: 10.1038/srep17450

[15] Shinoda T, Quality A, Minagawa M, Toya M, Motooka D, Kirikae T, et al.

Characterization of The Microbiome In Bile From Normal Human Gallbladders. Durham, North Carolina, United States: Research Square. pp. 1-10

[16] Nath G, Gulati AK, Shukla VK. Role of bacteria in carcinogenesis, with special reference to carcinoma of the gallbladder. World Journal of Gastroenterology. 2010;**16**(43):5395

[17] Nath G, Singh H, Shukla V. Chronic typhoid carriage and carcinoma of the gallbladder. European Journal of Cancer Prevention. 1997;**6**:557-559

[18] De Martel C, Plummer M, Parsonnet J, Van Doorn LJ, Franceschi S. Helicobacter species in cancers of the gallbladder and extrahepatic biliary tract. British Journal of Cancer. 2009;**100**(1):194-199

[19] Gonzalez-Escobedo G, La Perle KMD, Gunn JS. Histopathological analysis of Salmonella chronic carriage in the mouse hepatopancreatobiliary system. PLoS One. 2013;**8**(12):1-14

[20] Berg G, Rybakova D, Fischer D, Cernava T, Vergès M-CC, Charles T, et al. Correction to: Microbiome definition re-visited: Old concepts and new challenges. Microbiome. 2020;**8**(1):1-22

[21] Lederberg J, McCray A. Ome sweet omics—A genealogical treasury of words the scientist magazine®. Science. 2001;**15**(3-4):8

[22] Kim Y, Choi D. Microbiome of hepatobiliary diseases. Hanyang Medical Reviews. 2018;**38**(2):80

[23] Marchesi JR. Prokaryotic and eukaryotic diversity of the human gut. In: Advances in Applied Microbiology. Amsterdam, Netherlands: ScienceDirect; 2010

[24] Breitbart M, Haynes M, Kelley S, Angly F, Edwards RA, Felts B, et al. Viral diversity and dynamics in an infant gut. Research in Microbiology. 2008;**159**(5):367-373

[25] Valdes AM, Walter J, Segal E, Spector TD. Role of the gut microbiota in nutrition and health. BMJ. 2018;**361**:36-44

[26] Margolis KG, Cryan JF, Mayer EA. The microbiota-gut-brain Axis: From motility to mood. Gastroenterology. 2021;**160**(5):1486-1501. DOI: 10.1053/j.gastro.2020.10.066

[27] Chen GY, Shaw MH, Redondo G, Núñez G. Innate immune receptor nod1 protects the intestine from inflammation-induced tumorigenesis. Cancer Research. 2008;**68**(24):10060-10067

[28] Zheng D, Liwinski T, Elinav E. Interaction between microbiota and immunity in health and disease. Cell Research. 2020;**30**(6):492-506. DOI: 10.1038/s41422-020-0332-7

[29] Jandhyala SM, Talukdar R, Subramanyam C, Vuyyuru H, Sasikala M, Reddy DN. Role of the normal gut microbiota. World Journal of Gastroenterology. 2015;**21**(29):8836-8847

[30] Lupp C, Robertson ML, Wickham ME, Sekirov I, Champion OL, Gaynor EC, et al. Host-mediated inflammation disrupts the intestinal microbiota and promotes the overgrowth of Enterobacteriaceae. Cell Host & Microbe. 2007;**2**(2):119-129

[31] Arthur JC, Perez-Chanona E, Mühlbauer M, Tomkovich S, Uronis JM, Fan TJ, et al. Intestinal inflammation targets cancer-inducing activity of the microbiota. Science. 2012;**338**(6103): 120-123

[32] Kamada N, Seo SU, Chen GY, Núñez G. Role of the gut microbiota

in immunity and inflammatory disease. Nature Reviews. Immunology. 2013;**13**(5):321-335

[33] Pickard JM, Zeng MY, Caruso R, Núñez G. Gut microbiota: Role in pathogen colonization, immune responses, and inflammatory disease. Immunological Reviews. 2017;**279**(1):70-89

[34] Karlsson F, Tremaroli V, Nielsen J, Bäckhed F. Assessing the human gut microbiota in metabolic diseases. Diabetes. 2013;**62**:3341-3349

[35] Le Chatelier E, Nielsen T, Qin J, Prifti E, Hildebrand F, Falony G, et al. Richness of human gut microbiome correlates with metabolic markers. Nature. 2013;**500**(7464):541-546

[36] Pedersen HK, Gudmundsdottir V, Nielsen HB, Hyotylainen T, Nielsen T, Jensen BAH, et al. Human gut microbes impact host serum metabolome and insulin sensitivity. Nature. 2016;**535**(7612):376-381

[37] Turnbaugh PJ, Hamady M, Yatsunenko T, Cantarel BL, Duncan A, Ley RE, et al. A core gut microbiome in obese and lean twins. Nature. 2009;**457**(7228):480-484

[38] Khan I, Ullah N, Zha L, Bai Y, Khan A, Zhao T, et al. Alteration of gut microbiota in inflammatory bowel disease (IBD): Cause or consequence? IBD treatment targeting the gut microbiome. Pathogens. 2019;**8**(3):1-28

[39] Koloski NA, Jones M, Talley NJ. Aliment Pharmacol Ther—2016—Koloski—Evidence that independent gut-to-brain and brain-to-gut pathways operate in the irritable bowel syndrome and functional dyspepsia: A 1-year population-based prospective study. Alimentary Pharmacology & Therapeutics;**44**(6):592-600

[40] Zheng P, Zeng B, Zhou C, Liu M, Fang Z, Xu X, et al. Gut microbiome remodeling induces depressive-like behaviors through a pathway mediated by the host's metabolism. Molecular Psychiatry. 2016;**21**(6):786-796

[41] Wang-Michelitsch J, Michelitsch TM. Cell transformation in tumor-development: A result of accumulation of Misrepairs of DNA through many generations of cells. 2015;**3**:1-14. Available from: http://arxiv.org/abs/1505.01375

[42] Bieging KT. Unravelling mechanisms of p53-mediated tumour suppression. Nature Reviews. Cancer. 2014;**14**(5):359-370

[43] Jain K, Mohapatra T, Das P, Misra MC, Gupta SD, Ghosh M, et al. Sequential occurrence of preneoplastic lesions and accumulation of loss of heterozygosity in patients with gallbladder stones suggest causal association with gallbladder cancer. Annals of Surgery. 2014;**260**(6):1073-1080

[44] Mehrotra R, Tulsyan S, Hussain S, Mittal B, Singh Saluja S, Singh S, et al. Genetic landscape of gallbladder cancer: Global overview. Mutation Research—Reviews in Mutation Research. 2018;**778**:61-71. DOI: 10.1016/j.mrrev.2018.08.003

[45] Rutering J, Ilmer M, Recio A, Coleman M, Vykoukal J, Alt E, et al. Mutational landscape of metastatic cancer revealed from prospective clinical sequencing of 10,000 patients. Nature Medicine. 2016;**5**(6):1-8

[46] Abdel-Wahab R, Yap TA, Madison R, Pant S, Cooke M, Wang K, et al. Genomic profiling reveals high frequency of DNA repair genetic aberrations in gallbladder cancer. Scientific Reports. 2020;**10**(1):1-8. DOI: 10.1038/s41598-020-77939-6

[47] Javle M, Bekaii-Saab T, Jain A, Wang Y, Kelley RK, Wang K, et al. Biliary cancer: Utility of next-generation sequencing for clinical management. Cancer. 2016;**122**(24):3838-3847

[48] Egberts JH, Schniewind B, Schafmayer C, Kruse ML, Sipos B, Fändrich F, et al. Establishment of a novel orthotopic xenograft model of human gallbladder carcinoma. Clinical & Experimental Metastasis. 2007;**24**(3):141-148

[49] Tumor H. New Human Tumor Cell Lines. Boston, MA: Springer; 1975

[50] Emura F, Kamma H, Ghosh M, Koike N, Kawamoto T, Saijo K, et al. Establishment and characterization of novel xenograft models of human biliary tract carcinomas. International Journal of Oncology. 2003;**23**(5):1293-1300

[51] Kato S, Fushimi K, Yabuki Y, Maru Y, Hasegawa S, Matsuura T, et al. Precision modeling of gall bladder cancer patients in mice based on orthotopic implantation of organoid-derived tumor buds. Oncogenesis. 2021;**10**(4):1-13. DOI: 10.1038/s41389-021-00322-1

[52] Schilsky RL, Nass S, Le Beau MM, Benz EJ. Progress in cancer research, prevention, and care. The New England Journal of Medicine. 2020;**383**(10):897-900

[53] Roossinck MJ. Plants, viruses and the environment: Ecology and mutualism. Virology. 2015;**479**:271-277

[54] Livingston VW, Alexander-Jackson E. An experimental biologic approach to the treatment of neoplastic disease; determination of actinomycin in urine and cultures as an aid to diagnosis and prognosis. Journal of the American Medical Women's Association. 1965;**20**(9):858-866

[55] Marshall BJ, Windsor HM. The relation of *Helicobacter pylori* to gastric adenocarcinoma and lymphoma: Pathophysiology, epidemiology, screening, clinical presentation, treatment, and prevention. Medical Clinics. 2005;**89**(2):313-344

[56] Littman AJ, Jackson LA, Vaughan TL. *Chlamydia pneumoniae* and lung cancer: Epidemiologic evidence. Cancer Epidemiology Biomarkers and Prevention. 2005;**14**:773-778

[57] Ellmerich S, Schöller M, Duranton B, Gossé F, Galluser M, Klein JP, et al. Promotion of intestinal carcinogenesis by *Streptococcus bovis*. Carcinogenesis. 2000;**21**(4):753-756

[58] Dehio C. Bartonella—Host-cell interactions and vascular tumour formation. Nature Reviews Microbiology. 2005;**3**:621-631

[59] Morales-Sánchez A, Fuentes-Pananá EM. Human viruses and cancer. Viruses. 2014;**6**:4047-4079

[60] Lax AJ, Thomas W. How bacteria could cause cancer: One step at a time. Trends in Microbiology. 2002;**10**:293-299

[61] Karin M, Greten FR. NF-κB: Linking inflammation and immunity to cancer development and progression. Nature Reviews. Immunology. 2005;**5**(10):749-759

[62] Travaglione S, Fabbri A, Fiorentini C. The Rho-activating CNF1 toxin from pathogenic *E. coli*: A risk factor for human cancer development? Infectious Agents and Cancer. 2008;**3**:1-9

[63] Vedham V, Divi RL, Starks VL, Verma M. Multiple infections and cancer: Implications in epidemiology. Technology in Cancer Research & Treatment (TCRT). 2014;**13**(2):177-194

[64] Machado AMD, Figueiredo C, Touati E, Máximo V, Sousa S, Michel V, et al. *Helicobacter pylori* infection induces genetic instability of nuclear and mitochondrial DNA in gastric cells. Clinical Cancer Research. 2009;**15**(9):2995-3002

[65] Garrett WS. Cancer and the microbiota. Science. 2015;**348**(6230): 80-86

[66] Schwabe RF, Jobin C. The microbiome and cancer. Nature Reviews Cancer. 2013;**13**:800-812

[67] Merritt ME, Donaldson JR. Effect of bile salts on the DNA and membrane integrity of enteric bacteria. Journal of Medical Microbiology. 2009;**58**:1533-1541

[68] Csendes A, Burdiles P, Maluenda F, Diaz JC, Csendes P, Mitru N. Simultaneous bacteriologic assessment of bile from gallbladder and common bile duct in control subjects and patients with gallstones and common duct stones. Archives of Surgery. 1996;**131**(4):389-394

[69] Keren N, Konikoff FM, Paitan Y, Gabay G, Reshef L, Naftali T, et al. Interactions between the intestinal microbiota and bile acids in gallstones patients. Environmental Microbiology Reports. 2015;**7**(6):874-880

[70] Molinero N, Ruiz L, Milani C, Gutiérrez-Díaz I, Sánchez B, Mangifesta M, et al. The human gallbladder microbiome is related to the physiological state and the biliary metabolic profile. Microbiome. 2019;**7**(1):1-17

[71] Jiménez E, Sánchez B, Farina A, Margolles A, Rodríguez JM. Characterization of the bile and gall bladder microbiota of healthy pigs. Microbiology. 2014;**3**(6):937-949

[72] Binda C, Gibiino G, Coluccio C, Sbrancia M, Dajti E, Sinagra E, et al. Biliary diseases from the microbiome perspective: How microorganisms could change the approach to benign and malignant diseases. Microorganisms. 2022;**10**(2):312

[73] Dias-Jácome E, Libânio D, Borges-Canha M, Galaghar A, Pimentel-Nunes P. Gastric microbiota and carcinogenesis: The role of non-*Helicobacter pylori* bacteria—A systematic review. Revista Espanola de Enfermedades Digestivas. 2016;**108**(9):530-540

[74] Herrera V, Parsonnet J. *Helicobacter pylori* and gastric adenocarcinoma. Clinical Microbiology and Infection. 2009;**15**(11):971-976

[75] Gagnière J, Raisch J, Veziant J, Barnich N, Bonnet R, Buc E, et al. Gut microbiota imbalance and colorectal cancer. World Journal of Gastroenterology. 2016;**22**:501

[76] Chattopadhyay I, Verma M, Panda M. Role of oral microbiome signatures in diagnosis and prognosis of oral cancer. Technology in Cancer Research & Treatment. 2019;**18**:1533033819867354

[77] Zhang WL, Wang SS, Wang HF, Tang YJ, Tang YL, Liang XH. Who is who in oral cancer? Experimental Cell Research. 2019;**384**:111634

[78] Moeini A, Haber PK, Sia D. Cell of origin in biliary tract cancers and clinical implications. JHEP Reports. 2021;**3**(2):1-13

[79] O'Hara SP, Tabibian JH, Splinter PL, Larusso NF. The dynamic biliary epithelia: Molecules, pathways, and disease. The Journal of Hepatology. 2013;**58**(3):575-582. DOI: 10.1016/j.jhep.2012.10.011

[80] Fukata M, Vamadevan AS, Abreu MT. Toll-like receptors (TLRs)

and Nod-like receptors (NLRs) in inflammatory disorders. Seminars in Immunology. 2009;**21**(4):242-253

[81] Yoo KS, Lim WT, Choi HS. Biology of cholangiocytes: From bench to bedside. Gut and Liver. 2016;**10**(5):687-698

[82] Landskron G, De La Fuente M, Thuwajit P, Thuwajit C, Hermoso MA. Chronic inflammation and cytokines in the tumor microenvironment. Journal of Immunology Research. 2014;**2014**:149185

[83] Woo CH, Eom YW, Yoo MH, You HJ, Han HJ, Song WK, et al. Tumor necrosis factor-α generates reactive oxygen species via a cytosolic phospholipase A2-linked cascade. The Journal of Biological Chemistry. 2000;**275**(41):32357-32362

[84] He D, Li H, Yusuf N, Elmets CA, Athar M, Katiyar SK, et al. IL-17 mediated inflammation promotes tumor growth and progression in the skin. PLoS One. 2012;**7**(2):1-9

[85] Zen Y, Harada K, Sasaki M, Chen TC, Chen MF, Yen T. Intrahepatic cholangiocarcinoma escapes from growth inhibitory effect of transforming growth factor-β1 by overexpression of cyclin D1. Laboratory Investigation. 2005;**85**(4):572-581

[86] Sharma A, Sharma KL, Gupta A, Yadav A, Kumar A. Gallbladder cancer epidemiology, pathogenesis and molecular genetics: Recent update. World Journal of Gastroenterology. 2017;**23**(22):3978-3998

[87] Espinoza JA, Bizama C, Ferreccio C, Javle M, Miquel JF. The inflammatory inception of gallbladder cancer. Biochimica et Biophysica Acta (BBA)—Reviews on Cancer. 2018;**1865**(2):245-254

[88] Pilgrim CHC, Groeschl RT, Christians KK, Gamblin TC. Modern perspectives on factors predisposing to the development of gallbladder cancer. HPB. 2013;**15**:839-844

[89] Kawai M, Iwahashi M, Uchiyama K, Ochiai M, Tanimura H, Yamaue H. Gram-positive cocci are associated with the formation of completely pure cholesterol stones. The American Journal of Gastroenterology. 2002;**97**(1):83-88

[90] Maurer KJ, Ihrig MM, Rogers AB, Ng V, Bouchard G, Leonard MR, et al. Identification of cholelithogenic enterohepatic Helicobacter species and their role in murine cholesterol gallstone formation. Gastroenterology. 2005;**128**(4):1023-1033

[91] Hazrah P, Oahn KTH, Tewari M, Pandey AK, Kumar K, Mohapatra TM, et al. The frequency of live bacteria in gallstones. HPB. 2004;**6**(1):28-32

[92] Stewart L, Griffiss JML, Jarvis GA, Way LW. Gallstones containing bacteria are biofilms: Bacterial slime production and ability to form pigment solids determines infection severity and bacteremia. Journal of Gastrointestinal Surgery. 2007;**11**(8):977-984

[93] Tajeddin E, Sherafat SJ, Majidi MRS, Alebouyeh M, Alizadeh AHM, Zali MR. Association of diverse bacterial communities in human bile samples with biliary tract disorders: A survey using culture and polymerase chain reaction-denaturing gradient gel electrophoresis methods. European Journal of Clinical Microbiology & Infectious Diseases. 2016;**35**(8):1331-1339

[94] Cai JS, Chen JH. The mechanism of enterohepatic circulation in the formation of gallstone disease. The Journal of Membrane Biology. 2014;**247**(11):1067-1082

[95] Wu T, Zhang Z, Liu B, Hou D, Liang Y, Zhang J, et al. Gut microbiota

dysbiosis and bacterial community assembly associated with cholesterol gallstones in large-scale study. BMC Genomics. 2013;**14**(1):1-11

[96] Marshall JM, Flechtner AD, La Perle KM, Gunn JS. Visualization of extracellular matrix components within sectioned Salmonella biofilms on the surface of human gallstones. PLoS One. 2014;**9**(2):e89243

[97] Grigor'eva IN, Romanova TI. Gallstone disease and microbiome. Microorganisms. 2020;**8**(6):835

[98] Cherif S, Bouriat K, Rais H, Elantri S, Amine A. *Helicobacter pylori* and biliary tract cancers: A meta-analysis. Canadian Journal of Infectious Diseases and Medical Microbiology. 2020;**2020**:1-7

[99] Nagaraja V, Eslick GD. Systematic review with meta-analysis: The relationship between chronic Salmonella typhi carrier status and gall-bladder cancer. Alimentary Pharmacology & Therapeutics. 2014;**39**(8):745-750

[100] Polk DB, Peek RM. *Helicobacter pylori*: Gastric cancer and beyond. Nature Reviews Cancer. 2010;**10**:403-414

[101] Kasprzak A, Szmyt M, Malkowski W, Przybyszewska W, Helak-Łapaj C, Seraszek-Jaros A, et al. Analysis of immunohistochemical expression of proinfammatory cytokines (IL-1α, IL-6, and TNF-α) in gallbladder mucosa: Comparative study in acute and chronic calculous cholecystitis. Folia Morphologica. 2015;**74**(1):65-72

[102] Zhou D, Guan WB, Wang JD, Zhang Y, Gong W, Quan ZW. A comparative study of clinicopathological features between chronic cholecystitis patients with and without *Helicobacter pylori* infection in gallbladder mucosa. PLoS One. 2013;**8**(7):e70265

[103] Sipos P, Krisztina H, Blázovics A, Fehér J. Cholecystitis, gallstones and free radical reactions in human gallbladder. Medical Science Monitor. 2001;**7**(1):CR84-CR88

[104] Bagnoli F, Buti L, Tompkins L, Covacci A, Amieva MR. *Helicobacter pylori* CagA induces a transition from polarized to invasive phenotypes in MDCK cells. Proceedings of the National Academy of Sciences of the United States of America. 2005;**102**(45):16339-16344

[105] Krisch LM, Posselt G, Hammerl P, Wesslera S. CagA phosphorylation in helicobacter pylori-infected B cells is mediated by the nonreceptor tyrosine kinases of the src and Abl families. Infection and Immunity. 2016;**84**(9):2671-2680

[106] Kawaguchi M, Saito T, Ohno H, Midorikawa S, Sanji T, Handa Y, et al. Bacteria closely resembling *Helicobacter pylori* detected immunohistologically and genetically in resected gallbladder mucosa. Journal of Gastroenterology. 1996;**31**(2):294-298

[107] Monstein HJ, Jonsson Y, Zdolsek J, Svanvik J. Identification of *Helicobacter pylori* DNA in human cholesterol gallstones. Scandinavian Journal of Gastroenterology. 2002;**37**(1):112-119

[108] Bulajic M, Maisonneuve P, Schneider-Brachert W, Müller P, Reischl U, Stimec B, et al. *Helicobacter pylori* and the risk of benign and malignant biliary tract disease. Cancer. 2002;**95**(9):1946-1953

[109] Nilsson I, Shabo I, Svanvik J, Monstein HJ. Multiple displacement amplification of isolated DNA from human gallstones: Molecular identification of Helicobacter DNA by means of 16S rDNA-based pyrosequencing analysis. Helicobacter. 2005;**10**(6):592-600

[110] Cen L, Pan J, Zhou B, Yu C, Li Y, Chen W, et al. *Helicobacter pylori* infection of the gallbladder and the risk of chronic cholecystitis and cholelithiasis: A systematic review and meta-analysis. Helicobacter. 2018;**23**(1):1-10

[111] Kuroki T, Fukuda K, Yamanouchi K, Kitajima T, Matsuzaki S, Tajima Y, et al. *Helicobacter pylori* accelerates the biliary epithelial cell proliferation activity in hepatolithiasis. Hepatogastroenterology. 2002;**49**(45):648-651

[112] Fukuda K, Kuroki T, Tajima Y, Tsuneoka N, Kitajima T, Matsuzaki S, et al. Comparative analysis of Helicobacter DNAs and biliary pathology in patients with and without hepatobiliary cancer. Carcinogenesis. 2002;**23**(11):1927-1931

[113] Shimoyama T, Takahashi R, Abe D, Mizuki I, Endo T, Fukuda S. Serological analysis of Helicobacter hepaticus infection in patients with biliary and pancreatic diseases. Journal of Gastroenterology and Hepatology. 2010;**25**(SUPPL. 1):1-4

[114] Tsuchiya Y, Mishra K, Kapoor VK, Vishwakarma R, Behari A, Ikoma T, et al. Plasma *Helicobacter pylori* antibody titers and *Helicobacter pylori* infection positivity rates in patients with gallbladder cancer or cholelithiasis: A hospital-based case-control study. The Asian Pacific Journal of Cancer Prevention (APJCP). 2018;**19**(7):1911-1915

[115] Bohr URM, Kuester D, Meyer F, Wex T, Stillert M, Csepregi A, et al. Low prevalence of Helicobacteraceae in gall-stone disease and gall-bladder carcinoma in the German population. Clinical Microbiology and Infection. 2007;**13**(5):525-531. DOI: 10.1111/j.1469-0691.2007.01690.x

[116] Fatemi SM, Doosti A, Shokri D, Ghorbani-Dalini S, Molazadeh M, Tavakoli H, et al. Is there a correlation between *Helicobacter pylori* and enterohepatic helicobacter species and gallstone cholecystitis? The Middle East Journal of Digestive Diseases. 2018;**10**(1):24

[117] Dewhirst FE, Fox JG, Mendes EN, Paster BJ, Gates CE, Kirkbride CA, et al. Flexispira rappini strains represent at least 10 Helicobacter taxa. International Journal of Systematic and Evolutionary Microbiology. 2000;**50**(5):1781-1787

[118] Matsukura N, Yokomuro S, Yamada S, Tajiri T, Sundo T, Hadama T, et al. Association between *Helicobacter bilis* in bile and biliary tract malignancies: *H. bilis* in bile from Japanese and Thai patients with benign and malignant diseases in the biliary tract. Japanese Journal of Cancer Research. 2002;**93**(7):842-847

[119] Murata H, Tsuji S, Tsujii M, Fu HY, Imura HTAN, Oto MT, et al. *Helicobacter bilis* infection in biliary tract cancer. Alimentary Pharmacology & Therapeutics. 2004;**20**:90-94

[120] Kobayashi T, Harada K, Miwa K, Nakanuma Y. Helicobacter genus DNA fragments are commonly detectable in bile from patients with extrahepatic biliary diseases and associated with their pathogenesis. Digestive Diseases and Sciences. 2005;**50**(5):862-867

[121] Zhou D, Wang JD, Weng MZ, Zhang Y, Wang XF, Gong W, et al. Infections of Helicobacter spp. in the biliary system are associated with biliary tract cancer: A meta-analysis. European Journal of Gastroenterology & Hepatology. 2013;**25**(4):447-454

[122] Xiao M, Gao Y, Wang Y. Helicobacter species infection may be associated with cholangiocarcinoma: A meta-analysis. International Journal of Clinical Practice. 2014;**68**(2):262-270

[123] Young D, Hussell T, Dougan G. Chronic bacterial infections: Living with unwanted guests. Molecular Microbiology. 2002;**3**(11):1026-1032

[124] Caygill CPJ, Hill MJ, Braddick M, Sharp JCM. Cancer mortality in chronic typhoid and paratyphoid carriers. Lancet. 1994;**343**(8889):83-84

[125] Welton JC, Marr JS, Friedman SM. Association between hepatobiliary cancer and typhoid carrier state. Lancet. 1979;**313**(8120):791-794

[126] Strom BL, Soloway RD, Rios-Dalenz JL, Rodriguez-Martinez HA, West SL, Kinman JL, et al. Risk factors for gallbladder cancer. An international collaborative case–control study. Cancer. 1995;**76**(10):1747-1756

[127] Di Domenico EG, Cavallo I, Pontone M, Toma L, Ensoli F. Biofilm producing *Salmonella typhi*: Chronic colonization and development of gallbladder cancer. International Journal of Molecular Sciences. 2017;**18**(9):1887

[128] Scanu T, Spaapen RM, Bakker JM, Pratap CB, Wu L. Salmonella manipulation of host signaling pathways provokes cellular Transformation associated with gallbladder carcinoma. Cell Host & Microbe. 2015;**17**(6):55104

[129] Lu R, Bosland M, Xia Y, Zhang YG, Kato I, Sun J. Presence of Salmonella AvrA in colorectal tumor and its precursor lesions in mouse intestine and human specimens. Oncotarget. 2017;**8**(33):55104

[130] Shukla VK, Singh H, Pandey M, Upadhyay SK, Nath G. Carcinoma of the gallbladder—Is it a sequel of typhoid? Digestive Diseases and Sciences. 2000;**45**(5):900-903

[131] Dutta U, Garg PK, Kumar R, Tandon RK. Typhoid carriers among patients with gallstones are at increased risk for carcinoma of the gallbladder. The American Journal of Gastroenterology. 2000;**95**(3):784-787

[132] Iyer P, Barreto SG, Sahoo B, Chandrani P, Ramadwar MR, Shrikhande SV, et al. Non-typhoidal Salmonella DNA traces in gallbladder cancer. Infectious Agents and Cancer. 2016;**11**(1):10-13. DOI: 10.1186/s13027-016-0057-x

[133] Koshiol J, Wozniak A, Cook P, Adaniel C, Acevedo J, Azócar L, et al. Salmonella enterica serovar typhi and gallbladder cancer: A case–control study and meta-analysis. Cancer Medicine. 2016;**5**(11):3235-3310

[134] Lu Y, Zhang BY, Sen SJ, Wu LQ. Expression of the bacterial gene in gallbladder carcinoma tissue and bile. Hepatobiliary & Pancreatic Diseases International: HBPD INT. 2004;**3**(1):133-135

[135] Song X, Wang X, Hu Y, Li H, Ren T, Li Y, et al. A metagenomic study of biliary microbiome change along the cholecystitis-carcinoma sequence. Clinical and Translational Medicine. 2020;**10**(2):1-12

[136] Kumar S, Kumar S, Kumar S. Infection as a risk factor for gallbladder cancer. Journal of Surgical Oncology. 2006;**93**(8):633-639

[137] Csendes A, Becerra M, Burdiles P, Demian I, Bancalari K, Csendes P. Bacteriological studies of bile from the gallbladder in patients with carcinoma of the gallbladder, cholelithiasis, common bile duct stones and no gallstones disease. European Journal of Surgery, Acta Chirurgica. 1994;**160**(6-7):363-367

[138] Yu J, Feng Q, Wong SH, Zhang D, Yi Liang Q, Qin Y, et al. Metagenomic

analysis of faecal microbiome as a tool towards targeted non-invasive biomarkers for colorectal cancer. Gut Microbiota. 2017;**66**(1):70-78

[139] Coker OO, Dai Z, Nie Y, Zhao G, Cao L, Nakatsu G, et al. Mucosal microbiome dysbiosis in gastric carcinogenesis. Gut Microbiota. 2018;**67**(6):1024-1032

[140] Matson V, Fessler J, Bao R, Chongsuwat T, Zha Y, Alegre ML, et al. The commensal microbiome is associated with anti-PD-1 efficacy in metastatic melanoma patients. Science. 2018;**359**(6371):104-108

[141] Sarashina-Kida H, Negishi H, Nishio J, Suda W, Nakajima Y, Yasui-Kato M, et al. Gallbladder-derived surfactant protein D regulates gut commensal bacteria for maintaining intestinal homeostasis. Proceedings of the National Academy of Sciences of the United States of America. 2017;**114**(38):10178-10183

[142] Wang L, Guo M-J, Gao Q, Yang J-F, Yang L, Pang X-L, et al. The effects of probiotics on total cholesterol: A meta-analysis of randomized controlled trials. Medicine (Baltimore). 2018;**97**(5):e9679

[143] Jones ML, Martoni CJ, Parent M, Prakash S. Cholesterol-lowering efficacy of a microencapsulated bile salt hydrolase-active *Lactobacillus reuteri* NCIMB 30242 yoghurt formulation in hypercholesterolaemic adults. British Journal of Nutrition. 2012;**107**(10):1505-1513

[144] Fang K, Jin X, Hong SH. Probiotic *Escherichia coli* inhibits biofilm formation of pathogenic *E. coli* via extracellular activity of DegP. Scientific Reports. 2018;**8**(1):1-12. DOI: 10.1038/s41598-018-23180-1

[145] Gupta S, Bram EE, Weiss R. Genetically programmable pathogen sense and destroy. ACS Synthetic Biology. 2013;**2**(12):715-723

[146] Sassone-Corsi M, Nuccio SP, Liu H, Hernandez D, Vu CT, Takahashi AA, et al. Microcins mediate competition among Enterobacteriaceae in the inflamed gut. Nature. 2016;**540**(7632):280-283. DOI: 10.1038/nature20557

[147] Kumari P, Beniwal S, Khatri P, Singhal M, Saugat S. P-112 Weekly gemcitabine for the treatment of biliary tract and gallbladder cancer. Annals of Oncology. 2016;**27**(Supplement 2):ii33

[148] Park JS, Oh SY, Kim SH, Kwon HC, Kim JS, Kim HJ, et al. Single-agent gemcitabine in the treatment of advanced biliary tract cancers: A phase II study. Japanese Journal of Clinical Oncology. 2005;**35**(2):68-73

[149] Sarotra P, Medhi B. Use of bacteria in cancer therapy. Recent Results in Cancer Research. 2016;**209**:111-121

Chapter 9

Celiac Disease, Management, and Follow-Up

Ángela Ruiz-Carnicer, Verónica Segura, Carolina Sousa and Isabel Comino

Abstract

Celiac disease (CD) is a systemic immune-mediated disorder characterized by a specific serological and histological profile triggered by gluten ingestion, which is given in genetically predisposed subjects. Heterogeneous clinical presentation is characteristic in CD, affecting any organ or tissue with gastrointestinal, extraintestinal, seronegative, or nonresponsive manifestations. CD diagnosis is based on several criteria, including genetic and serological tests, clinical symptoms and/or risk conditions, and duodenal biopsy. Currently, the available treatment for CD is a strict gluten-free diet (GFD) that essentially relies on the consumption of naturally gluten-free foods, such as animal-based products, fruits, vegetables, legumes, and nuts, as well as gluten-free dietary products that may not contain more than 20 mg of gluten per kg of food according to Codex Alimentarius. However, it is difficult to maintain a strict oral diet for life and at least one-third of patients with CD are exposed to gluten. Difficulties adhering to a GFD have led to new tools to monitor the correct adherence to GFD and alternative forms of treatment.

Keywords: celiac disease, gluten-free diet, gluten immunogenic peptides, dietary adherence, non-dietary therapies

1. Introduction

Celiac disease (CD) is a chronic immune-mediated enteropathy triggered by exposure to dietary gluten in genetically predisposed individuals [1]. The diagnosis rate of this pathology has increased in the last 10 years [2], so worldwide epidemiologic data are now available showing that CD is ubiquitous, with a prevalence of 1.4% [3], higher in female than male individuals [2–7].

Clinically, CD presents with a wide variety of gastrointestinal and extraintestinal symptoms that differ considerably according to the age of presentation [8] or even be an asymptomatic disease. Digestive symptoms and growth retardation are frequent in the pediatric population diagnosed within the first years of life [9]. However, in adults, symptoms can be nonspecific gastrointestinal or extraintestinal of various kinds.

Currently, the only available treatment for CD is a strict, lifelong gluten-free diet (GFD), which requires significant patient education, motivation, and follow-up [10]. Adherence to a GFD is not easy, with the ubiquitous nature of gluten, cross-contamination of foods, inadequate food-labeling regulations, and social constraints [11]. Current methods to evaluate adherence to a GFD include the use of a dietary questionnaire and monitoring of serological findings or clinical symptoms; however, neither of these methods generates a direct nor an accurate measurement of dietary adherence [1, 11, 12]. Small bowel biopsy is the "gold standard" for CD diagnosis, but according to most clinical guidelines, its role in the follow-up of patients with CD is limited to cases involving a lack of clinical response or symptom recurrence [13–17]. Nonresponsive CD occurs frequently, particularly in those diagnosed in adulthood. Persistent or recurring symptoms should lead to a review of patients to exclude alternative diagnoses and a review of GFD to ensure there is no obvious gluten contamination and confirm adherence to GFD [1]. Possible causes include age at diagnosis, follow-up time, the existence of social differences, the intake of certain drugs (PPIs, NSAIDs), severe clinical symptoms at diagnosis, inadequate adherence to diet, or the presence of inadvertent contamination of the diet [18, 19].

In this chapter, we synthesized the latest research findings and evidence related to the management of CD and GFD, including emerging tools to monitor the correct adherence to GFD and the development of non-dietary therapies.

2. Pathogeny

Pathogeny development of CD is due to a combination of environmental (gluten and other factors), genetic (HLA system), and immunological factors (response of intestinal T lymphocytes).

2.1 Gluten

The major environmental factor responsible for the development of CD is gluten, which is a complex mixture of prolamin and glutelin storage proteins of certain cereals, such as wheat, barley, rye, oats, and their derivates. These common dietary proteins have unusual biochemical properties that include a high abundance of glutamine and proline residues, which render them resistant to degradation by gastrointestinal proteases [20], leaving large peptides. These peptides enter the lamina propria of the small intestine via transcellular or paracellular routes where, in affected individuals, an immune reaction occurs.

2.2 Immunological factors

The most accepted model for explaining CD immunopathogenesis is the two-signal model mediated by a first innate immune response (direct toxic effect of gluten on the epithelium) followed by a secondary antigen-specific adaptive response (through CD4+ T lymphocytes of the lamina propria) [20, 21]. Some peptides, such as, the 19-mer gliadin peptide, trigger an innate immune response mainly characterized by the production of IL-15 by epithelial cells. Result is the disruption of the epithelial barrier, by increasing the permeability and inducing enterocyte apoptosis [20]. These peptides enter the lamina propria of the small intestine via transcellular or paracellular routes [20] where, in affected individuals, an adaptive immune reaction occurs that is facilitated by increased intestinal permeability that allows the passage of immunogenic peptides, such as 33-mer, to the lamina propria. At the same time, some glutamine residues of these peptides are catalytically deaminated by tissue transglutaminase (tTG). This deamination, in turn, increases the immunogenicity of peptides due to high-affinity interactions between modified residues and ligand binding sites of HLA-DQ2 and HLA-DQ8 molecules [22] expressed by dendritic cells. Gliadin peptides are then presented to gliadin-reactive CD4+ T cells. During this process, antibodies against tTG, gliadin, and actin are made through unclear mechanisms. These antibodies might contribute to extra-intestinal manifestations of CD, such as dermatitis herpetiformis and gluten ataxia. Moreover, the immune response initiates a cascade of reactions that degenerate into crypt hyperplasia and flattening of the intestinal villi.

2.3 Genetic factors

The importance of a genetic component for the development of CD is evident, based on the familial occurrence and the high concordance among identical twins [23, 24]. Almost 100% of patients with CD possess specific variants of the HLA class II genes HLA-DQA1 and HLA-DQB1 that, together, encode the two chains (α and β) of CD-associated heterodimer proteins DQ2 and DQ8 that are expressed on the surface of antigen-presenting cells [25]. More than 90% of patients with CD are DQ2 positive and most of the others are DQ8 positive [26]. HLA-DQ2 and HLA-DQ8 risk heterodimers are present in approximately 30–40% of the general population, and of these, approximately 1% develop the disease, so HLA DQ2/8 seems necessary, but not sufficient for the development of CD [27].

Several studies have been carried out to identify non-HLA susceptibility genes. Among these are a large number of CD-associated genes basically encode interleukins, interleukin receptors, and tumor necrosis factors or receptors that are involved in innate immunity and epithelial stress signals (COELIAC2, COELIAC 3, CTLA4, and COELIAC4) [28].

2.4 Other environmental factors

Other environmental factors that could contribute to the development of CD have also been studied, such as the time and manner of introduction of gluten, the type of delivery, the start and duration of breastfeeding, the microbiome or early exposure to antibiotics, among others [29, 30]. However, the studies carried out to date do not confirm the different hypotheses proposed. Recently, the link between viral infections and loss of oral gluten tolerance has been investigated, since infections caused by rotavirus, reovirus, astrovirus, enterovirus, and adenovirus are very common in childhood. This opens the door to a new field of knowledge that could allow the design of preventive strategies in the future of CD [31, 32].

3. Clinical manifestations

Clinical characteristics of CD differ considerably depending on the age of presentation, and it can also be profuse or simply present analytical abnormalities [33–36]. It can manifest clinically with a wide variety of symptoms that affect multiple organs and systems and that can be both gastrointestinal (diarrhea, vomiting, abdominal pain, bloating, bloating, constipation, gastroesophageal reflux, among others) and extra-intestinal (tiredness, dermatitis herpetiformis, anemia, osteoporosis, infertility, growth retardation, neuropathy, ataxia, delayed puberty, etc.) [8, 25]. Symptomatic

CD can be classified into classic and non-classic. Any case presenting with malabsorption is classified as a classic CD. Although the clinical presentation is changing toward an affectation of older individuals with milder symptoms. The symptomatic classical disease was previously the most common presentation, and although it remains a prominent mode of presentation, subclinical and nonclassical cases now make up roughly 30% and 40–60% of new cases, respectively [37, 38].

4. Diagnosis

The diagnosis of CD may require genetic and serological tests and a duodenal biopsy.

4.1 Genetic risk markers

The main genetic risk factor for CD is the presence of HLA-DQ2 and -DQ8 heterodimers, which are identified in 90% and 5–7% of patients with CD, respectively [7]. Since these alleles are found in 30–40% of the general population (HLA-DQ2 being the most common) [39], their absence is important due to their negative predictive value (NPV). Therefore, the HLA-DQ2/HLA-DQ8 test plays an important role in CD diagnosis and is recommended in the following situations [40]—(a) exclusion of the disease, especially in patients who have started GFD; (b) in situations of uncertain diagnosis due to negative serology, but histology suggestive of CD; (c) to differentiate siblings in whom it is intended to ensure that it is unlikely that they will develop the disease from those who will need monitoring; (d) in subjects with autoimmune diseases and other diseases in which CD should be investigated.

A negative result for HLA-DQ2/HLA-DQ8 means a very low probability of developing the disease. Therefore, this test can be used to support the diagnosis of CD, since it has a high NPV, allowing exclusion with 99% certainty [41]. However, it has little positive predictive value (PPV) (only around 12%), so its determination has no diagnostic value in situations with elevated antibodies directed against tTG and should be reserved as second-line in patients with diagnostic doubt [42, 43].

4.2 Specific serum antibodies

Various serological tests have been developed to detect CD—antigliadin antibodies (anti-AGA), antibodies against deaminated gliadin peptides (anti-DGP), anti-endomysia antibodies (anti-EMA), and anti-transglutaminase antibodies (anti-tTG). Serological tests are important for two reasons—(1) they select patients in whom duodenal biopsy should be indicated to confirm clinical suspicion, and (2) they confirm the diagnosis in cases in which enteropathy has been observed [43].

Anti-AGA has been used for decades and is reasonably safe when the probability of suffering from CD is very high. However, it has been shown that these antibodies present variability in their diagnostic precision, due to the fact that they have low sensitivity and specificity; therefore, they should not be included in routine tests for the diagnosis of CD [41, 44].

Anti-EMA has a relatively low sensitivity (80–90%), but its specificity is close to 100%. However, they require more complex laboratory techniques and depend on the experience of the laboratory staff, remaining as a second-line test adequate to confirm clinical suspicion [1].

Anti-tTG IgA has a sensitivity and specificity of 95 and 90%, respectively [41, 45]. Anti-DGP has shown good precision, although lower than anti-tTG IgA, so an isolated positive result for IgA and/or IgG-DGP in patients at low risk for CD, predicts the disease only in 15%, being in the rest of the cases false positives. Therefore, in a first approximation, anti-tTG are the preferred antibodies for the diagnosis of CD according to the ESPGHAN diagnostic criteria [46, 47]. Anti-DGP is considered less sensitive or specific for the detection of CD compared to anti-tTG and anti-EMA. However, these last two antibodies are less sensitive in children under 2 years of age. It should also be taken into account that anti-tTG can be negative in 5–16% of patients with histologically confirmed CD [48]. Therefore, there is no serological test with perfect sensitivity and specificity [44]. In case of general IgA deficiency, which is observed in 2–3% of patients with CD, the IgG-based test (anti-DGP IgG and anti-tTG IgG) should be performed. IgG anti-tTG has diagnostic utility in patients with selective IgA deficiency (IgA < 0.07 mg/dl). Regarding anti-DGP IgG, there is no evidence of greater efficacy compared to anti-tTG IgG or anti-EMA IgG [41].

4.3 Intestinal biopsy

Duodenal biopsy of the small intestine is a key point in the diagnosis of CD. A distinctive pattern of histological abnormalities has been identified in this disease, including partial or total villous atrophy, elongated crypts, decreased villus/crypt ratio, increased crypt mitotic index, increased crypt density of intraepithelial lymphocytes (IELs), and infiltration of plasma cells in the lamina propria. An increase in IELs tends to be located at the tips of the villi and are usually CD8+ [37]. The presence of a diffuse and uniform infiltrate of these lymphocytes is the most sensitive finding, but it is not specific to CD. A count of at least 25 IELs/100 enterocytes represents a definitive increase in IELs [49, 50]. Immunohistochemical studies have shown that the increase in IELs represents an expansion of cytotoxic T cells alpha-beta and gamma-delta. Gamma-delta T cells are observed in 1–10% of the normal small intestinal mucosa but increase in patients with CD, where they may represent 15–30% of all IELs [1]. In addition, the absence of the brush border can be identified, as well as alterations in epithelial cells.

There are three grading systems to establish the severity of histological damage proposed by Marsh, Oberhuber [51], and Corazza-Villanaci [52]. Marsh system, with three types of grades, was replaced in 1999 by Oberhuber [51], which proposes a better standardization with six types [51]. In 2007, a new, simpler classification was published by Corazza-Villanacci [52]. These classifications are qualitative and subjective [1, 37]. Marsh-Oberhuber classification is used by most pathologists both for diagnosis and to ensure regression of the lesion after GFD [1]. Generally, six stages are distinguished—type 0 without lesion, type 1 (infiltrative lesion), type 2 (crypt hyperplasia), type 3 (villi atrophy: 3a: partial; 3b: subtotal; 3c: total) [51]. Furthermore, these lesions are not pathognomonic for CD, and there is a wide spectrum of diseases that can produce indistinguishable microscopic lesions.

Currently, it is considered that, in patients with high levels of antibodies, the diagnosis could be based on the combination of symptoms, antibodies determination, and genetics, omitting in this case the duodenal biopsy [11, 46], unlike what was established in the previous ESPGHAN guidelines for the diagnosis of CD. However, confirmation of CD by biopsy is considered the gold standard in the diagnosis of CD in certain types of patients.

The biopsy can be used to diagnose and monitor, but CD is a burden for patients. Therefore, less invasive and objective biomarkers are required to assess the disease. In addition, in certain patients, a challenge with gluten is necessary to make a correct diagnosis of CD. Based on this, Leonard et al. [53] investigated the ability of different biomarkers to diagnose CD after provocation. These biomarkers could, complement or replace histology in the diagnosis of CD. These authors evaluated traditional diagnostic techniques, such as biopsy, antibodies, symptomatology, as well as different biomarkers to measure the response to two levels of gluten exposure, studying interleukin-2 (IL-2), the tetramer test, and the dot enzyme-linked immunosorbent assay (Enzyme-Linked ImmunoSpot Assay, ELISpot), among others. Results showed that the measurement of IL-2 in plasma might be the first and most sensitive marker for the evaluation of gluten exposure in patients with CD. This study provides a framework for the rational design and selection of biomarkers in future gluten challenge studies with the goal of incorporating them into clinical practice.

5. Treatment of celiac disease

5.1 Diet therapy: gluten-free diet

Only effective treatment available for CD consists of following a strict GFD, excluding gluten proteins from the diet from wheat, barley, rye, and oats, as well as hybrids of these cereals such as triticale and their derivatives (starch, flour, etc.) [14]. Nevertheless, such a diet is difficult to follow due to the unintended contamination of "gluten-free" products, improper labeling, social constraints, and ubiquity of gluten proteins in raw or cooked foods and pharmaceuticals. Thus, accidental gluten encounters are likely. Most patients with CD can safely tolerate approximately 10 mg of gluten cross-contamination daily. However, there is a tremendous degree of variability within this population, and some patients may have worsening histological changes with very low daily gluten exposure [1, 10].

Strict adherence to GFD leads to remission of gastrointestinal and extra-intestinal symptoms, normalization of serological tests, and recovery of the intestinal mucosa, in most cases [14]. Initiation of strict GFD generally results in a rapid improvement of clinical symptoms, while recovery of the intestinal villi requires several years of a strict GFD (around 2 years in 34% and 5 years in 66%) [44]. Therefore, it is essential that patient with CD is aware of adherence to GFD to avoid future complications.

5.1.1 Difficulties in following a gluten-free diet: transgressions

Although adherence to GFD is the cornerstone of the treatment of patients with CD, there are conditions that prevent it from being carried out and mean that a significant percentage of patients with CD do not adhere and commit voluntary or involuntary transgressions [10]. Among the conditions that can prevent the GFD monitoring, we highlight the high economic cost of gluten-free products, which are not accessible to a large number of people with CD. Another factor to highlight that can favor its involuntary intake is the ubiquity of gluten in a high percentage of manufactured products since many of the foods that are marketed contain gluten from wheat, barley, rye, or oats, including those that intervene only as a thickener or binder. In fact, several studies carried out to determine the gluten content in natural (unprocessed) gluten-free foods or in foods labeled gluten-free reveal relatively high contamination rates, present in 9–22% of the samples analyzed [54–56]. In addition, many products contain hidden gluten, mainly due to cross-contamination with other gluten-containing foods that are processed or stored in the same place. The risk that these foods pose for patients with CD makes rigorous control of gluten content convenient [57]. Therefore, accurate detection and quantification of gluten in food are essential [10]. The Codex Alimentarius [58] has established that a food classified as "gluten-free" should not exceed 20 mg of gluten per kg of food, that is, 20 parts per million (ppm). Currently, several methods are used for the detection and quantification of gluten in foods. Enzyme-Linked ImmunoSorbent Assays (ELISAs) are the most widely used methods, as they are sensitive, rapid, and relatively easy to perform. Most commercial ELISAs use monoclonal antibodies (moAbs) such as R5 and G12 [59–64]. Other methods, such as the Polymerase Chain Reaction (PCR), developed mainly for research, are far from being able to replace ELISA, as they are not suitable for the detection of gluten in highly processed or hydrolyzed samples due to DNA degradation. Lastly, liquid chromatography/mass spectrometry methods require expensive equipment and expertise [65].

All the factors described above cause nonadherence to GFD among patients with CD. Recent studies have indicated that inadvertent gluten ingestion occurs more frequently than intentional ingestion, and gluten contamination in naturally gluten-free foods is likely to be one of the most important factors in inadvertent nonadherence [66]. Other investigations based on the study of intestinal biopsies of patients with CD on GFD for more than 2 years have suggested that transgressions are relatively frequent, detecting a lack of recovery of the intestinal villi in 36–55% of the population studied [67–69]. These inadvertent or intentional violations are the main reason for uncontrolled CD in adult patients with CD [70]. Likewise, there is a small percentage of patients with CD (approximately 0.3–10%) who do not respond to GFD and have persistent symptoms of malabsorption and intestinal villi atrophy, which is known as refractory CD (RCD) [7, 71–74].

5.1.2 Gluten-free diet monitoring methods

The existence of a reliable method that makes it possible to verify whether or not patients with CD are following a GFD is undoubtedly useful not only in monitoring the patient to avoid long-term complications, but also when diagnosing RCD [16, 44]. Among the methods to monitor adherence to GFD is the determination of specific antibodies, dietary interviews, control of symptoms, biopsies, and the detection of gluten immunogenic peptides (GIP) in stool and urine (**Table 1**) [1, 41, 47, 75].

5.1.2.1 Serological tests

Anti-tTG and anti-DGP have been used frequently to assess CD follow-up [76]. Use of these serological tests has revealed that it takes several months for the specific serology of CD to return to normal values. A significant decrease in levels during the first year suggests adherence to the diet and, therefore, patients with CD whose serology tests do not improve should be reassessed regarding their exposure to gluten [16]. However, negative serological markers do not reflect strict adherence to a GFD and are a poor predictor of dietary transgressions [17, 43, 77]. Although serology shows high accuracy for the diagnosis of CD, these tests are not as useful in follow-up, since they do not correlate with histological findings or symptoms [78]. It is important to note that a negative serology in a patient with CD on GFD does not necessarily

	Strengths	Weak points
Serological tests	• High accuracy for the diagnosis of CD	• Late positives (6–24 months to normalize) • False positives and negatives, for follow-up, no correlation with biopsy and symptoms • I need a blood draw
Dietary questionnaires, symptomatology questionnaires, and dietary interviews	• Non-invasive • Low cost	• Forgetfulness, omissions • Falsified • Tedious • Non-objective
Intestinal biopsy	• Gold standard test for the diagnosis of CD	• Invasive • Expensive, consumes hospital resources • Uncomfortable for the patient
Detection of GIP in human samples	• Simple and fast method • Non-invasive • Correlation with gluten intake	
Other bookmarks (Calprotectin)	• Simple and fast method • Non-invasive	• Non-CD specific

Table 1.
A comparison of the strengths and weaknesses of the tools used to monitor GFD in patients with CD. CD, celiac disease; GFD, gluten-free diet; GIP, gluten immunogenic peptides.

guarantee the recovery of the intestinal mucosa [14, 43]. In a recent meta-analysis, PPV of persistently positive determination of anti-tTG IgA was very low and showed a sensitivity of 38% in adults. NPV of serology in adult patients with CD on GFD for one year or more was higher, with a specificity of 80%. Therefore, the usefulness of serology in the follow-up of adult patients with CD is very limited [1].

5.1.2.2 Symptomatology

Among the most widely used methods to assess the presence of gastrointestinal symptoms in patients with CD is the Gastrointestinal Symptom Rating Scale (GSRS) questionnaire [79, 80]. This questionnaire serves to check symptoms and determine the improvement and evolution of CD. However, there are a large number of patients with CD who are asymptomatic or minimally symptomatic at the time of presentation and, in these cases, it would not be feasible to use the clinical response as an indicator of intestinal mucosal recovery and adherence to GFD [13, 70].

5.1.2.3 Intestinal biopsy

Histological lesion remains the gold standard test for the diagnosis of CD, recovery of the mucosa is the main marker of response to diet. The only method to verify this normalization of the duodenum is by performing an oral endoscopy with an intestinal biopsy, an aggressive and costly follow-up method. However, an intestinal biopsy is a method used clinically, especially in the evaluation of patients with persistent symptoms [81]. It seems advisable to perform a follow-up endoscopy in adults

1–2 years after starting GFD to ensure recovery of the mucosa [82]. In this way, it would be possible to differentiate patients who are at low risk and in whom follow-up periods can be extended, from those at high risk who may need special supervision to maintain adherence to GFD [83].

5.1.2.4 Dietary questionnaires and interviews

Adherence to GFD can be assessed through dietary interviews or questionnaires conducted by a specialist. Dietitian has an important role in providing practical advice on lifestyle and food choices [16]. Evaluation of adherence to the diet through dietary interviews has been suggested because of its low cost and because it is not invasive; however, they are difficult to standardize and are subjective.

Different questionnaires assess the frequency of food and self-reported adherence to GFD [84]. Some of the more specific questionnaires are—(a) Gluten Free Score by Biagi et al. [85], whose four items provide a score from 0 to IV and in which levels 0 and I indicate poor adherence to the diet and, (b) the Celiac Dietary Adherence Test (CDAT) developed by Leffler et al. [86], which is a brief questionnaire that allows a rapid and standardized evaluation. This last questionnaire comprises seven easy-to-apply questions with optimal psychometric characteristics that assess CD symptomatology, self-efficacy expectations, reasons for maintaining GFD, knowledge of the disease, associated risk behaviors, and the perceived degree of adherence.

Nevertheless, there is considerable controversy about the validity of dietary questionnaires in the assessment of GFD because some patients with CD do not record the actual gluten consumed intentionally in some cases. Therefore, the measurement of adherence to GFD through questionnaires appears to be subjective and imprecise and does not allow involuntary infractions to be identified [25, 84].

5.1.2.5 Detection of immunogenic gluten peptides in human samples

Recently, new noninvasive methodologies have been developed to monitor gluten exposure in patients with CD based on the detection and quantification of GIP in stool and urine samples [87–90]. These immunological methodologies (ELISA and immunochromatographic strips) based on G12 and A1 moAbs are capable of detecting GIP, which are gluten fragments resistant to gastrointestinal digestion, and mainly responsible for the immune response of patients with CD [60, 61, 91–95]. These tools make it possible to monitor adherence to GFD and detect violations cases, helping to identify the origin of clinical symptoms and avoid complications derived from gluten intake (anemia, osteoporosis, increased risk of lymphoma, etc.). These techniques have represented a revolutionary worldwide advance in the clinical practice of CD and have been introduced in the new guidelines, both European and Spanish, for monitoring the GFD of patients with CD [1, 41]. Numerous rigorous studies have evaluated the use of GIP determination in stool and/or urine to monitor adherence to GFD compared to other tools (**Table 2**). The studies included children and adults diagnosed with CD and healthy volunteers. Overall, these studies indicated that this novel technique was highly sensitive for the detection of GFD transgressions and therefore could facilitate the follow-up of patients with CD.

5.1.2.6 Other bookmarks

Other markers have been proposed for monitoring GFD, such as the permeability test [113] or fecal calprotectin [114, 115]. Determination of fecal calprotectin

	Population	Study design	References
Stool	Children	Case-control study	[87]
		Cohort study	[96]
		Prospective study	[89]
		Transversal study	[97]
		Systematic revision	[98]
		Prospective study	[99]
		Observational descriptive study	[100]
	Children and adults	Prospective study	[88]
		Transversal study	[101, 102]
	Adults	Observational prospective study	[103]
		Prospective study	[104]
		Prospective study	[105]
Urine	Children and adults	Controlled study	[106]
		Randomized controlled study	[90]
	Adults	Transversal study	[107]
		Prospective study	[108]
		Prospective study	[109]
		Prospective study	[110]
Stool and urine	Children and adults	Meta-analysis	[111]
	Adults	Prospective study	[80]
		Prospective study	[17, 77]
		Prospective study	[112]

Table 2.
Studies based on GIP determination in stool and/or urine for monitoring of gluten-free diet. GIP, gluten immunogenic peptides.

concentration has established itself in recent years as a new useful marker of gastrointestinal pathologies. Several studies show that there is an association between calprotectin levels and the degree of inflammation, so it can be used to monitor response to treatment and predict the risk of recurrence. In addition, results obtained by Oribe et al. [116] have shown that patients with positive anti-tTG IgA antibodies, that is, those in contact with gluten, showed significantly higher values of fecal calprotectin than patients undergoing GFD and non-celiac patients. These methods, by demonstrating the presence of intestinal inflammatory processes, are generally not specific for CD and, therefore, if their values are modified, it could also be due to other causes such as infectious diseases, inflammatory bowel disease (IBD), or allergic processes.

5.2 Non-dietary therapies

Since strict follow-up of GFD presents many difficulties for patients with CD, additional treatments are needed for this disease. In recent years, CD research has focused on the search for non-dietary therapies to control GFD [17, 77]. Emerging

DOI: http://dx.doi.org/10.5772/intechopen.104652

therapeutic options for CD can be broadly classified into one of the following strategies—(1) removal of toxic gluten peptides before reaching the intestinal tract, (2) regulation of the immunostimulatory effects of toxic gluten peptides, (3) modulation of intestinal permeability, (4) immune modulation and induction of gluten tolerance, and (5) restoration of imbalance in the intestinal microbiota (**Figure 1**).

	Preclinical		Phase I		Phase II		Phase III	
	Product names	Manufacturer/Research center for clinical trial management	Product names	Manufacturer/Research center for clinical trial management	Product names	Manufacturer/Research center for clinical trial management	Product names	Manufacturer/Research center for clinical trial management
Removal or reduction of toxic gluten peptides	m-TG and N-methyl-lysine	Zedira GmbH	KumaMax	PvP Biologics & Takeda Pharmaceuticals	AGY-010 antibodies	University of Alberta and IGY Inc.		
	AMY-02 exopeptidase	AMYRA Biotech AG			AN-PEP and DPP-IV	VU University Medical Centre		
					BL-7010			
					AG017	ActoBio Therapeutics, ActoBiotics		
					IMGX-003	ImmunogenX		
Regulation the gluten immunestimulatory effects	Anti-TNF Avaximab	Avaxia Biologics	tTG inhibitor	Sitari. Pharmaceuticals	PRV-015	Amgen and Provention Bio		
	BNZ-2	Bioniz Therapeutics	Hu-Mik-Beta-1	Mayo Clinic, National Cancer Institute, University of Chicago Medicine	CALY-002	Calyso Biotech SA		
	PV-267	Provid Pharmaceuticals			Anti-TNF Vedolisumab	University of Erlangen-Nuremberg, Germany		
					ZED-1227	Dr.Falk Pharma and Zedira		
Modulation of intestinal permeability							Larazotide acetate	Innovate Biopharmaceuticals
Immunemodulation and gluten tolerance	SVP Vaccine	Selecta Biosciences and Sanofi	KAN-101	Kanyos Bio, Inc., Anokion SA	Nexvax2	ImmusanT		
	Hookworm larvae	James Cook University			TAK-101	Cour Pharmaceuticals		
					Necator Americanus	Princess Alexandra Hospital, Brisbane, Australia		

Figure 1.
Clinical and preclinical trials in the development of new non-dietary therapies in CD. CD: celiac disease; PEP: prolyl endopeptidases; TNF: tumor necrosis factor; and tTG: tissue transglutaminase [128].

To date, only larazotide acetate is in phase III studies. Larazotide is an oral peptide that modulates tight junctions and prevents the passage of gluten peptides to the lamina propria by closing the intercellular junctions of enterocytes. Therefore, it could help prevent the development of the immune cascade in patients with CD, showing a reduction in symptoms as well as a reduction in anti-tTG antibody levels. In addition, some very promising therapies are PRV-015 immunotherapy, the use of oral glutenases, as well as vaccine therapies (phase II). There are many other exciting drugs that are in the early stages of research, such as tTG inhibitors, HLA blockers, and probiotics [20, 117–128]. Similarly, some therapies are being evaluated in preclinical trials and are postulated as promising treatments for the pathogenesis of CD (**Figure 1**). Thus, we are faced with many promising and emerging options for the treatment of CD.

6. Conclusions

Research on CD is changing rapidly due to a steady increase in knowledge that addresses its pathophysiology, diagnosis, follow-up, and therapeutic options.

Diagnosis of CD is based on several criteria, including positive serology, a spectrum of duodenal damage, clinical symptoms and/or risk conditions, and response to a GFD in susceptible individuals. In the absence of some of these criteria, the diagnosis of CD becomes challenging. In this regard, studies based on gluten reintroduction combined with IL-2 measurements could provide a new clinical alternative to diagnose and monitor patients who already have a GFD.

Several patients have difficulty controlling their diet they regularly consume sufficient gluten to trigger symptoms. Despite the availability of diverse traditional GFD adherence markers, such as diet tests or serology, none of them is an accurate diet evaluation method. Thus, use of GIP detection in stool and/or urine has been developed as a direct and specific test for GFD monitoring. Furthermore, non-dietary therapies have shown encouraging preliminary results in phase II and III clinical trials, such as larazotide acetate, PRV-015, IMGX-003, vaccine, and drug therapy. However, a GFD is the mainstay of CD therapy for the immediate future. For all these reasons, a health-oriented lifestyle should be promoted for better management and control of CD, responding to the growing demand of society and the empowerment of patients with CD.

Acknowledgements

This work was supported by grants from Knowledge Transfer Activities and Fundación Progreso y Salud, Consejería de Salud, Junta de Andalucía (project PI-0053-2018).

Conflict of interest

The authors declare no conflict of interest.

Appendices and nomenclature

Anti-AGA	antigliadin antibodies
Anti-DGP	antibodies deaminated gliadin peptides
Anti-EMA	anti-endomysia antibodies
Anti-tTG	anti-transglutaminase antibodies
CD	celiac disease
CDAT	celiac dietary adherence test
ELISA	enzyme-linked immunosorbent assays
GFD	gluten-free diet
GIP	gluten immunogenic peptides
IBD	inflammatory bowel disease
IELs	intraepithelial lymphocytes
IL-2	interleukin-2
NPV	negative predictive value
PPV	positive predictive value
PCR	polymerase chain reaction
RCD	refractory celiac disease
tTG	tissue transglutaminase

DOI: http://dx.doi.org/10.5772/intechopen.104652

Author details

Ángela Ruiz-Carnicer, Verónica Segura, Carolina Sousa and Isabel Comino*
Faculty of Pharmacy, Department of Microbiology and Parasitology, University of Seville, Seville, Spain

*Address all correspondence to: icomino@us.es

References

[1] Al-Toma A, Volta U, Auricchio R, Castillejo G, Sanders DS, Cellier C, et al. European society for the study of coeliac disease (ESsCD) guideline for coeliac disease and other gluten-related disorders. United European Gastroenterology Journal. 2019;**7**(5):583-613. DOI: 10.1177/2050640619844125

[2] King JA, Jeong J, Underwood FE, Quan J, Panaccione N, Windsor JW, et al. Incidence of celiac disease is increasing over time: A systematic review and meta-analysis. American Journal of Gastroenterology. 2020;**115**(4):507-525

[3] Singh P, Arora A, Strand TA, Leffler DA, Catassi C, Green PH, et al. Global prevalence of celiac disease: Systematic review and meta-analysis. Clinical Gastroenterology Hepatology. 2018;**16**(6):823-836

[4] Mardini HE, Westgate P, Grigorian AY. Racial differences in the prevalence of celiac disease in the US Population: National Health and Nutrition Examination Survey (NHANES) 2009-2012. Digestive Diseases and Sciences. 2015;**60**(6):1738-1742. DOI: 10.1007/s10620-014-3514-7

[5] Lebwohl B, Sanders DS, Green PHR. Coeliac disease. Lancet. 2018;**391**(10115):70-81. DOI: 10.1016/S0140-6736(17)31796-8

[6] Glissen JRB, Singh P. Coeliac disease. Paediatric International Child Health. 2019;**39**:23-31. DOI: 10.1080/20469047.2018.1504431

[7] Ludvigsson JF, Murray JA. Epidemiology of celiac disease. Gastroenterology Clinics of North America. 2019;**48**(1):1-18. DOI: 10.1016/j.gtc.2018.09.004

[8] Therrien A, Kelly CP, Silvester JA. Celiac disease: Extraintestinal manifestations and associated conditions. Journal of Clinical Gastroenterology. 2020;**54**(1):8-21. DOI: 10.1097/MCG.0000000000001267

[9] Sahin Y. Celiac disease in children: A review of the literature. World Journal of Clinical Pediatrics. 2021;**10**(4):53-71. DOI: 10.5409/wjcp.v10.i4.53

[10] Itzlinger A, Branchi F, Elli L, Schumann M. Gluten-free diet in celiac disease-forever and for all? Nutrients. 2018;**10**(11):1796. DOI: 10.3390/nu10111796

[11] Wolf J, Petroff D, Richter T, Auth MKH, Uhlig HH, Laass MW, et al. Validation of antibody-based strategies for diagnosis of pediatric celiac disease without biopsy. Gastroenterology. 2017;**153**(2):410-419

[12] Wieser H, Ruiz-Carnicer Á, Segura V, Comino I, Sousa C. Challenges of monitoring the gluten-free diet adherence in the management and follow-up of patients with celiac disease. Nutrients. 2021;**13**(7):2274. DOI: 10.3390/nu13072274

[13] Sharkey LM, Corbett G, Currie E, Lee J, Sweeney N, Woodward JM. Optimising delivery of care in coeliac disease—Comparison of the benefits of repeat biopsy and serological follow-up. Alimentary Pharmacology Theraphy. 2013;**38**(10):1278-1291. DOI: 10.1111/apt.12510

[14] Ludvigsson JF, Bai JC, Biagi F, Card TR, Ciacci C, Ciclitira PJ, et al. BSG Coeliac Disease Guidelines Development Group, British Society of Gastroenterology. Diagnosis and

management of adult coeliac disease: Guidelines from the British Society of Gastroenterology. Gut. 2014;**63**(8):1210-1228. DOI: 10.1136/gutjnl-2013-306578

[15] Silvester JA, Kurada S, Szwajcer A, Kelly CP, Leffler DA, Duerksen DR. Tests for serum transglutaminase and endomysial antibodies do not detect most patients with celiac disease and persistent villous atrophy on gluten-free diets: A meta-analysis. Gastroenterology. 2017;**153**(3):689-701

[16] Husby S, Bai JC. Follow-up of celiac disease. Gastroenterology Clinics of North America. 2019;**48**(1):127-136. DOI: 10.1016/j.gtc.2018.09.009

[17] Silvester JA, Comino I, Kelly CP, Sousa C, Duerksen DR, DOGGIE BAG Study Group. Most patients with celiac disease on gluten-free diets consume measurable amounts of gluten. Gastroenterology. 2020;**158**(5):1497-1499

[18] Francavilla R, Cristofori F, Stella M, Borrelli G, Naspi G, Castellaneta S. Treatment of celiac disease: From gluten-free diet to novel therapies. Minerva Pediatrica. 2014;**66**(5):501-516

[19] Leonard MM, Cureton P, Fasano A. Indications and use of the gluten contamination elimination diet for patients with nonresponsive celiac disease. Nutrients. 2017;**9**(10):1129. DOI: 10.3390/nu9101129

[20] Lindfors K, Ciacci C, Kurppa K, Lundin KEA, Makharia GK, Mearin ML, et al. Coeliac disease. Nature Reviews. 2019;**5**(1):3. DOI: 10.1038/s41572-018-0054-z

[21] Sharma N, Bhatia S, Chunduri V, Kaur S, Sharma S, Kapoor P, et al. Pathogenesis of celiac disease and other gluten related disorders in wheat and strategies for mitigating them. Frontiers in Nutrition. 2020;**7**:6. DOI: 10.3389/fnut.2020.00006

[22] Sollid LM, Khosla C. Novel therapies for coeliac disease. Journal of Internal Medicine. 2011;**269**(6):604-613

[23] Fasano A, Berti I, Gerarduzzi T, Not T, Colletti RB, Drago S, et al. Prevalence of celiac disease in at-risk and not-at-risk groups in the United States: A large multicenter study. Archives of Internal Medicine. 2003;**163**(3):286-292. DOI: 10.1001/archinte.163.3.286

[24] Kuja-Halkola R, Lebwohl B, Halfvarson J, Wijmenga C, Magnusson PK, Ludvigsson JF. Heritability of non-HLA genetics in coeliac disease: A population-based study in 107 000 twins. Gut. 2016;**65**:1793-1798. DOI: 10.1136/gutjnl-2016-311713

[25] Caio G, Volta U, Sapone A, Leffler DA, De Giorgio R, Catassi C, et al. Celiac disease: A comprehensive current review. BMC Medicine. 2019;**17**(1):142. DOI: 10.1186/s12916-019-1380-z

[26] Megiorni F, Mora B, Bonamico M, Barbato M, Nenna R, Maiella G, et al. HLA-DQ and risk gradient for celiac disease. Human Immunology. 2009;**70**(1):55-59. DOI: 10.1016/j.humimm.2008.10.018

[27] Bourgey M, Calcagno G, Tinto N, Gennarelli D, Margaritte-Jeannin P, Greco L, et al. HLA related genetic risk for coeliac disease. Gut. 2007;**56**(8):1054-1059. DOI: 10.1136/gut.2006.108530

[28] Piscaglia AC. Intestinal stem cells and celiac disease. World Journal of Stem Cells. 2014;**6**(2):213-229. DOI: 10.4252/wjsc.v6.i2.213

[29] Whyte LA, Kotecha S, Watkins WJ, Jenkins HR. Coeliac disease is more common in children with high

socio-economic status. Acta Paediatrics. 2014;**103**(3):289-294. DOI: 10.1111/apa.12494

[30] Dydensborg Sander S, Nybo Andersen AM, Murray JA, Karlstad Ø, Husby S, Størdal K. Association between antibiotics in the first year of life and celiac disease. Gastroenterology. 2019;**156**(8):2217-2229. DOI: 10.1053/j.gastro.2019.02.039

[31] Gómez-Rial J, Rivero-Calle I, Salas A, Martinón-Torres F. Rotavirus and autoimmunity. The Journal of Infection. 2020;**81**(2):183-189. DOI: 10.1016/j.jinf.2020.04.041

[32] Sánchez D, Hoffmanová I, Szczepanková A, Hábová V, Tlaskalová-Hogenová H. Contribution of infectious agents to the development of celiac disease. Microorganisms. 2021;**9**(3):547. DOI: 10.3390/microorganisms9030547

[33] Jericho H, Sansotta N, Guandalini S. Extraintestinal manifestations of celiac disease: Effectiveness of the gluten-free diet. Journal of Pediatric and Gastroenterology Nutrition. 2017;**65**(1):75-79. DOI: 10.1097/MPG.0000000000001420

[34] Jericho H, Guandalini S. Extra-intestinal manifestation of celiac disease in children. Nutrients. 2018;**10**(6):755. DOI: 10.3390/nu10060755

[35] Mearin ML, Montoro-Huguet M, Polanco I, Ribes-Koninckx C, Santolaria S. Manifestaciones clínicas de la enfermedad celíaca y criterios Diagnósticos: Diferencias entre niños, adolescentes y adultos. In: Arranz E, Fernández-Bañares F, Rosell CM, Rodrigo L, Peña AS, editors. Avances en el conocimiento de las patologías relacionadas con el gluten y evolución de los alimentos sin gluten. 2018. pp. 259-303

[36] Sansotta N, Amirikian K, Guandalini S, Jericho H. Celiac disease symptom resolution: Effectiveness of the gluten-free diet. Journal of Pediatric and Gastroenterology Nutrition. 2018;**66**(1):48-52. DOI: 10.1097/MPG.0000000000001634

[37] Hujoel IA, Reilly NR, Rubio-Tapia A. Celiac disease: Clinical features and diagnosis. Gastroenterology Clinics of North America. 2019;**48**(1):19-37. DOI: 10.1016/j.gtc.2018.09.001

[38] Lebwohl B, Rubio-Tapia A. Epidemiology, presentation, and diagnosis of celiac disease. Gastroenterology. 2021;**160**(1):63-75. DOI: 10.1053/j.gastro.2020.06.098

[39] Sallese M, Lopetuso LR, Efthymakis K, Neri M. Beyond the HLA genes in gluten-related disorders. Frontiers in Nutrition. 2020;7:575844. DOI: 10.3389/fnut.2020.575844

[40] Liu E, Lee HS, Aronsson CA, Hagopian WA, Koletzko S, Rewers MJ, et al. Risk of pediatric celiac disease according to HLA haplotype and country. The New England Journal of Medicine. 2014;**371**(1):42-49. DOI: 10.1056/NEJMoa1313977

[41] Grupo de trabajo del Protocolo para el diagnóstico precoz de la enfermedad celíaca. Protocolo para el diagnóstico precoz de la enfermedad celíaca. Ministerio de Sanidad, Servicios Sociales e Igualdad. Servicio de Evaluación del Servicio Canario de la Salud (SESCS). 2018

[42] Brown NK, Guandalini S, Semrad C, Kupfer SS. A clinician's guide to celiac disease HLA genetics. American Joural of Gastroenterology. 2019;**114**(10):1587-1592

[43] Husby S, Murray JA, Katzka DA. AGA clinical practice update on diagnosis

and monitoring of celiac disease-changing utility of serology and histologic measures: Expert review. Gastroenterology. 2019;**156**(4):885-889. DOI: 10.1053/j.gastro.2018.12.010

[44] Rubio-Tapia A, Hill ID, Kelly CP, Calderwood AH, Murray JA, American College of Gastroenterology. ACG clinical guidelines: Diagnosis and management of celiac disease. The American Journal of Gastroenterology. 2013;**108**(5):656-676

[45] Lewis NR, Scott BB. Meta-analysis: Deamidated gliadin peptide antibody and tissue transglutaminase antibody compared as screening tests for coeliac disease. Alimentary Pharmacology Theraphy. 2010;**31**(1):73-81. DOI: 10.1111/j.1365-2036.2009.04110.x

[46] Husby S, Koletzko S, Korponay-Szabó IR, Mearin ML, Phillips A, Shamir R, et al. European Society for Pediatric Gastroenterology, Hepatology, and Nutrition guidelines for the diagnosis of coeliac disease. Journal of Pediatric and Gastroenterology Nutrition. 2012;**54**(1):136-160

[47] Husby S, Koletzko S, Korponay-Szabó I, Kurppa K, Mearin ML, Ribes-Koninckx C, et al. European Society Paediatric Gastroenterology, Hepatology and Nutrition Guidelines for Diagnosing Coeliac Disease 2020. Journal of Pediatric and Gastroenterology Nutritio. 2020;**70**(1):141-156. DOI: 10.1097/MPG.0000000000002497

[48] De Leo L, Bramuzzo M, Ziberna F, Villanacci V, Martelossi S, Leo GD, et al. Diagnostic accuracy and applicability of intestinal auto-antibodies in the wide clinical spectrum of coeliac disease. eBio Medicine. 2020;**51**:102567. DOI: 10.1016/j.ebiom.2019.11.028

[49] Eiras P, Camarero C, León F, Roldán E, Asensio A, Baragaño M, et al. Linfocitos intraepiteliales en la enfermedad celíaca [Intraepitheliallymphocytes in celiacdisease]. Anales Españoles de Pediatría. 2002;**56**(3):224-232

[50] Olivencia P, Cano A, Martín MA, León F, Roy G, Redondo C. Enfermedad celíaca del adulto y linfocitos intraepiteliales. ¿Nuevas opciones para el diagnóstico? [Adultceliacdisease and intraepitheliallymphocytes. New options for diagnosis?]. Gastroenterología y Hepatología. 2008;**31**(9):555-559. DOI: 10.1157/13128293

[51] Oberhuber G, Granditsch G, Vogelsang H. The histopathology of coeliac disease: Time for a standardized report scheme for pathologists. European Journal of Gastroenterology & Hepatology. 1999;**11**(10):1185-1194. DOI: 10.1097/00042737-199910000-00019

[52] Corazza GR, Villanacci V, Zambelli C, Milione M, Luinetti O, Vindigni C, et al. Comparison of the interobserver reproducibility with different histologic criteria used in celiac disease. Clinical Gastroenterology and Hepatology. 2007;**5**(7):838-843. DOI: 10.1016/j.cgh.2007.03.019

[53] Leonard MM, Silvester JA, Leffler D, Fasano A, Kelly CP, Lewis SK, et al. Evaluating responses to gluten challenge: A randomized, double-blind, 2-dose gluten challenge trial. Gastroenterology. 2021;**160**(3):720-733

[54] Bustamante MÁ, Fernández-Gil MP, Churruca I, Miranda J, Lasa A, Navarro V, et al. Evolution of gluten content in cereal-based gluten-free products: An overview from 1998 to 2016. Nutrients. 2017;**9**(1):21. DOI: 10.3390/nu9010021

[55] Farage P, de MedeirosNóbrega YK, Pratesi R, Gandolfi L, Assunção P, Zandonadi RP. Gluten contamination

in gluten-free bakery products: A risk for coeliac disease patients. Public Health Nutrition. 2017;**20**(3):413-416. DOI: 10.1017/S1368980016002433

[56] Raju N, Joshi AKR, Vahini R, Deepika T, Bhaskarachari K, Devindra S. Gluten contamination in labelled and naturally gluten-free grain products in southern India. Food Additive Contamination. 2020;**37**(4):531-538. DOI: 10.1080/19440049.2020.1711970

[57] Mena MC, Sousa C. Analytical tools for gluten detection. Policies and regulation. In: Arranz E, Fernández-Bañares F, editors. Advances in the Understanding of Gluten Related Pathology and the Evolution of Gluten-free Foods. 2015. pp. 527-564

[58] Codex Standard 118-1979. Codex standard for foods for special dietary use for persons intolerant to gluten. Codex Alimentarius Commission; revision 2008, amendment; 2015

[59] Valdés I, García E, Llorente M, Méndez E. Innovative approach to low-level gluten determination in foods using a novel sandwich enzyme-linked immunosorbent assay protocol. European Journal of Gastroenterology & Hepatology. 2003;**15**(5):465-474. DOI: 10.1097/01.meg.0000059119.41030.df

[60] Morón B, Cebolla A, Manyani H, Alvarez-Maqueda M, Megías M, Thomas Mdel C, et al. Sensitive detection of cereal fractions that are toxic to celiac disease patients by using monoclonal antibodies to a main immunogenic wheat peptide. American Journal of Clinical Nutrition. 2008;**87**(2):405-414. DOI: 10.1093/ajcn/87.2.405

[61] Morón B, Bethune MT, Comino I, Manyani H, Ferragud M, López MC, et al. Toward the assessment of food toxicity for celiac patients: Characterization of monoclonal antibodies to a main immunogenic gluten peptide. PLoS One. 2008;**3**(5):e2294. DOI: 10.1371/journal.pone.0002294

[62] Rosell CM, Barro F, Sousa C, Mena MC. Cereals for developing gluten-free products and analytical tools for gluten detection. Journal of Cereal Science. 2014;**59**(3):354-364. DOI: 10.1016/j.jcs.2013.10.001

[63] Melini F, Melini V. Immunological methods in gluten risk analysis: A snapshot. Safety. 2018;**4**(4):56. DOI: 10.3390/safety4040056

[64] Panda R, Garber EAE. Detection and quantitation of gluten in fermented-hydrolyzed foods by antibody-based methods: Challenges, progress, and a potential path forward. Frontiers in Nutrition. 2019;**6**:97. DOI: 10.3389/fnut.2019.00097

[65] Scherf KA, Poms RE. Recent developments in analytical methods for tracing gluten. Journal of Cereal Science. 2016;**67**:112-122. DOI: 10.1016/j.jcs.2015.08.006

[66] Muhammad H, Reeves S, Jeanes YM. Identifying and improving adherence to the gluten-free diet in people with coeliac disease. Proceedings of the Nutrition Society. 2019;**78**(3):418-425. DOI: 10.1017/S002966511800277X

[67] Stoven S, Murray JA, Marietta E. Celiac disease: Advances in treatment via gluten modification. Clinical Gastroenterology and Hepatology. 2012;**10**(8):859-862. DOI: 10.1016/j.cgh.2012.06.005

[68] Hall NJ, Rubin GP, Charnock A. Intentional and inadvertent non-adherence in adult coeliac disease. A cross-sectional survey. Appetite.

2013;**68**:56-62. DOI: 10.1016/j.appet.2013.04.016

[69] Matoori S, Fuhrmann G, Leroux JC. Celiac disease: A challenging disease for pharmaceutical scientists. Pharmaceutical Research. 2013;**30**(3):619-626. DOI: 10.1007/s11095-012-0951-x

[70] Mahadev S, Murray JA, Wu TT, Chandan VS, Torbenson MS, Kelly CP, et al. Factors associated with villus atrophy in symptomatic coeliac disease patients on a gluten-free diet. Alimentary Pharmacology Theraphy. 2017;**45**(8):1084-1093. DOI: 10.1111/apt.13988

[71] Rubio-Tapia A, Murray JA. Classification and management of refractory coeliac disease. Gut. 2010;**59**(4):547-557. DOI: 10.1136/gut.2009.195131

[72] Rishi AR, Rubio-Tapia A, Murray JA. Refractory celiac disease. Expert Review of Gastroenterology & Hepatology. 2016;**10**(4):537-546. DOI: 10.1586/17474124.2016.1124759

[73] Malamut G, Cellier C. Refractory celiac disease. Gastroenterology Clinics of North America. 2019;**48**(1):137-144. DOI: 10.1016/j.gtc.2018.09.010

[74] Penny HA, Baggus EMR, Rej A, Snowden JA, Sanders DS. Non-responsive coeliac disease: A comprehensive review from the NHS England National Centre for refractory coeliac disease. Nutrients. 2020;**12**(1):216. DOI: 10.3390/nu12010216

[75] Rodrigo L, Pérez-Martinez I, Lauret-Braña E, Suárez-González A. Descriptive study of the different tools used to evaluate the adherence to a gluten-free diet in celiac disease patients. Nutrients. 2018;**10**(11):1777. DOI: 10.3390/nu10111777

[76] Ludvigsson JF, Ciacci C, Green PH, Kaukinen K, Korponay-Szabo IR, Kurppa K, et al. Outcome measures in coeliac disease trials: The Tampere recommendations. Gut. 2018;**67**(8):1410-1424. DOI: 10.1136/gutjnl-2017-314853

[77] Silvester JA, Comino I, Rigaux LN, Segura V, Green KH, Cebolla A, et al. Exposure sources, amounts and time course of gluten ingestion and excretion in patients with coeliac disease on a gluten-free diet. Alimentary Pharmacology Theraphy. 2020;**52**(9):1469-1479. DOI: 10.1111/apt.16075

[78] Silvester JA, Kurada S, Szwajcer A, Kelly CP, Leffler DA, Duerksen DR. Tests for serum transglutaminase and endomysial antibodies do not detect most patients with celiac disease and persistent villous atrophy on gluten-free diets: A meta-analysis. Gastroenterology. 2017;**153**(3):689-701

[79] Svedlund J, Sjödin I, Dotevall G. GSRS—A clinical rating scale for gastrointestinal symptoms in patients with irritable bowel syndrome and peptic ulcer disease. Digestive Diseases and Sciences. 1988;**33**(2):129-134. DOI: 10.1007/BF01535722

[80] Costa AF, Sugai E, Temprano MP, Niveloni SI, Vázquez H, Moreno ML, et al. Gluten immunogenic peptide excretion detects dietary transgressions in treated celiac disease patients. World Journal of Gastroenterology. 2019;**25**(11):1409-1420. DOI: 10.3748/wjg.v25.i11.1409

[81] Kelly CP, Bai JC, Liu E, Leffler DA. Advances in diagnosis and management of celiac disease. Gastroenterology. 2015;**148**(6):1175-1186. DOI: 10.1053/j.gastro.2015.01.044

[82] Moscoso JF, Quera PR. Enfermedad celíaca. Revisión [Update on celiac disease]. The Child. 2016;**144**(2):211-221

[83] Pekki H, Kurppa K, Mäki M, Huhtala H, Sievänen H, Laurila K, et al. Predictors and significance of incomplete mucosal recovery in celiac disease after 1 year on a gluten-free diet. The American Journal of Gastroenterology. 2015;**110**(7):1078-1085. DOI: 10.1038/ajg.2015.155

[84] Silvester JA, Weiten D, Graff LA, Walker JR, Duerksen DR. Is it gluten-free? Relationship between self-reported gluten-free diet adherence and knowledge of gluten content of foods. Nutrition. 2016;**32**(7-8):777-783. DOI: 10.1016/j.nut.2016.01.021

[85] Biagi F, Campanella J, Martucci S, Pezzimenti D, Ciclitira PJ, Ellis HJ, et al. A milligram of gluten a day keeps the mucosal recovery away: A case report. Nutrition Reviews. 2004;**62**(9):360-363. DOI: 10.1111/j.1753-4887.2004.tb00062.x

[86] Leffler DA, Dennis M, Edwards George JB, Jamma S, Magge S, Cook EF, et al. A simple validated gluten-free diet adherence survey for adults with celiac disease. Clinical Gastroenterology Hepatol. 2009;7(5):530-536

[87] Comino I, Real A, Vivas S, Síglez MÁ, Caminero A, Nistal E, et al. Monitoring of gluten-free diet compliance in celiac patients by assessment of gliadin 33-mer equivalent epitopes in feces. American Journal of Clinical Nutrition. 2012;**95**(3):670-677. DOI: 10.3945/ajcn.111.026708

[88] Comino I, Fernández-Bañares F, Esteve M, Ortigosa L, Castillejo G, Fambuena B, et al. Fecal gluten peptidesreveallimitations of serologicaltests and foodquestionnairesformonitoring gluten-free diet in celiacdiseasepatients. American Journal of Gastroentrology. 2016;**111**(10):1456-1465

[89] Comino I, Segura V, Ortigosa L, Espín B, Castillejo G, Garrote JA, et al. Prospective longitudinal study: Use of faecal gluten immunogenic peptides to monitor children diagnosed with coeliac disease during transition to a gluten-free diet. Alimentary Pharmacology Theraphy. 2019;**49**(12):1484-1492. DOI: 10.1111/apt.15277

[90] Moreno ML, Cebolla Á, Muñoz-Suano A, Carrillo-Carrion C, Comino I, Pizarro Á, et al. Detection of gluten immunogenic peptides in the urine of patients with coeliac disease reveals transgressions in the gluten-free diet and incomplete mucosal healing. Gut. 2017;**66**(2):250-257. DOI: 10.1136/gutjnl-2015-310148

[91] Comino I, Real A, de Lorenzo L, Cornell H, López-Casado MÁ, Barro F, et al. Diversity in oat potential immunogenicity: Basis for the selection of oat varieties with no toxicity in coeliac disease. Gut. 2011;**60**(7):915-922. DOI: 10.1136/gut.2010.225268

[92] Comino I, Real A, Gil-Humanes J, Pistón F, de Lorenzo L, Moreno Mde L, et al. Significant differences in coeliac immunotoxicity of barley varieties. Molecular Nutrition Food and Research. 2012;**56**(11):1697-1707. DOI: 10.1002/mnfr.201200358

[93] Comino I, Real A, Moreno Mde L, Montes R, Cebolla A, Sousa C. Immunological determination of gliadin 33-mer equivalent peptides in beers as a specific and practical analytical method to assess safety for celiac patients. Journal of the Science of Food and Agriculture. 2013;**93**(4):933-943. DOI: 10.1002/jsfa.5830

[94] Real A, Comino I, Moreno Mde L, López-Casado MÁ, Lorite P, Torres MI, et al. Identification and in vitro reactivity of celiac immunoactive peptides in an apparent gluten-free beer. PLoS One. 2014;**9**(6):e100917. DOI: 10.1371/journal.pone.0100917

[95] Moreno Mde L, Muñoz-Suano A, López-Casado MÁ, Torres MI, Sousa C, Cebolla Á. Selective capture of most celiac immunogenic peptides from hydrolyzed gluten proteins. Food Chemistry. 2016;**205**:36-42. DOI: 10.1016/j.foodchem.2016.02.066

[96] Gerasimidis K, Zafeiropoulou K, Mackinder M, Ijaz UZ, Duncan H, Buchanan E, et al. Comparison of clinical methods with the faecal gluten immunogenic peptide to assess gluten intake in coeliac disease. Journal of Pediatric and Gastroenterology Nutrition. 2018;**67**(3):356-360. DOI: 10.1097/MPG.0000000000002062

[97] Roca M, Donat E, Masip E, Crespo Escobar P, Fornes-Ferrer V, Polo B, et al. Detection and quantification of gluten immunogenic peptides in feces of infants and their relationship with diet. Revista Espanol Enfermedades Digestivas. 2019;**111**(2):106-110

[98] Myléus A, Reilly NR, Green PHR. Rate, risk factors, and outcomes of nonadherence in pediatric patients with celiac disease: A systematic review. Clinical Gastroenterology and Hepatology. 2020;**18**(3):562-573. DOI: 10.1016/j.cgh.2019.05.046

[99] Roca M, Donat E, Masip E, Crespo-Escobar P, Cañada-Martínez AJ, Polo B, et al. Analysis of gluten immunogenic peptides in feces to assess adherence to the gluten-free diet in pediatric celiac patients. European Journal of Nutrition. 2021;**60**(4):2131-2140. DOI: 10.1007/s00394-020-02404-z

[100] Fernández Miaja M, Díaz Martín JJ, Jiménez Treviño S, Suárez González M, BousoñoGarcía C. Study of adherence to the gluten-free diet in coeliac patients. Anales de Pediatría (Barcelona, Spain). 2021;**94**(6):377-384. DOI: 10.1016/j.anpedi.2020.06.017

[101] Porcelli B, Ferretti F, Biviano I, Santini A, Cinci F, Vascotto M, et al. Testing for fecal gluten immunogenic peptides: A useful tool to evaluate compliance with gluten-free diet by celiacs. Annals of Gastroenterology. 2020;**33**(6):631-637

[102] Porcelli B, Ferretti F, Cinci F, Biviano I, Santini A, Grande E, et al. Fecal gluten immunogenic peptides as indicators of dietary compliance in celiac patients. Minerva Gastroenterology Dietology. 2020;**66**(3):201-207

[103] Laserna-Mendieta EJ, Casanova MJ, Arias Á, Arias-González L, Majano P, Mate LA, et al. Poor sensitivity of fecal gluten immunogenic peptides and serum antibodies to detect duodenal mucosal damage in celiac disease monitoring. Nutrients. 2020;**13**(1):98. DOI: 10.3390/nu13010098

[104] Fernández-Bañares F, Beltrán B, Salas A, Comino I, Ballester-Clau R, Ferrer C, et al. Persistent villous atrophy in de novo adult patients with celiac disease and strict control of gluten-free diet adherence: A Multicenter Prospective Study (CADER Study). American Journal of Gastroenterology. 2021;**116**(5):1036-1043

[105] Coto L, Sousa C, Cebolla A. Individual variability in patterns and dynamics of fecal gluten immunogenic peptides excretion after low gluten intake. European Journal of Nutrition. 2022;**7**:1-17. DOI: 10.1007/s00394-021-02765-z

[106] Soler M, Estevez MC, Moreno Mde L, Cebolla A, Lechuga LM. Label-free SPR detection of gluten peptides in urine for non-invasive celiac disease follow-up. BiosensBioelectron. 2016;**79**:158-164. DOI: 10.1016/j.bios.2015.11.097

[107] Peláez EC, Estevez MC, Domínguez R, Sousa C, Cebolla A,

Lechuga LM. A compact SPR biosensor device for the rapid and efficient monitoring of gluten-free diet directly in human urine. Analytical and Bioanalytical Chemistry. 2020;**412**(24):6407-6417. DOI: 10.1007/s00216-020-02616-6

[108] Ruiz-CarnicerÁ,Garzón-BenavidesM, FombuenaB,SeguraV,García-FernándezF, Sobrino-Rodríguez S, et al. Negative predictive value of the repeated absence of gluten immunogenic peptides in the urine of treated celiac patients in predicting mucosal healing: New proposals for follow-up in celiac disease. American Journal of Clinical Nutrition. 2020;**112**(5):1240-1251

[109] Moreno ML, Sánchez-Muñoz D, Sanders D, Rodríguez-Herrera A, Sousa C. Verifying diagnosis of refractory celiac disease with urine gluten immunogenic peptides as biomarker. Frontier in Medicine (Lausanne). 2021;7:601854

[110] Coto L, Sousa C, Cebolla A. Dynamics and considerations in the determination of the excretion of gluten immunogenic peptides in urine: Individual variability at low gluten intake. Nutrients. 2021;**13**(8):2624. DOI: 10.3390/nu13082624

[111] Syage JA, Kelly CP, Dickason MA, Ramirez AC, Leon F, Dominguez R, et al. Determination of gluten consumption in celiac disease patients on a gluten-free diet. American Journal of Clinical Nutrition. 2018;**107**(2):201-207. DOI: 10.1093/ajcn/nqx049

[112] Stefanolo JP, Tálamo M, Dodds S, de la Paz TM, Costa AF, Moreno ML, et al. Real-World gluten exposure in patients with celiac disease on gluten-free diets, determined from gliadin immunogenic peptides in urine and fecal samples. Clinical Gastroenterology and Hepatology. 2021;**19**(3):484-491

[113] Duerksen DR, Wilhelm-Boyles C, Parry DM. Intestinal permeability in long-term follow-up of patients with celiac disease on a gluten-free diet. Digestive Diseases and Sciences. 2005;**50**(4):785-790. DOI: 10.1007/s10620-005-2574-0

[114] Ertekin V, Selimoğlu MA, Turgut A, Bakan N. Fecal calprotectin concentration in celiac disease. Journal of Clinical Gastroenterology. 2010;**44**(8):544-546. DOI: 10.1097/MCG.0b013e3181cadbc0

[115] Szaflarska-Popławska A, Romańczuk B, Parzęcka M. Faecal calprotectin concentration in children with coeliac disease. Przeglad Gastroenterology. 2020;**15**(1):44-47. DOI: 10.5114/pg.2020.93630

[116] Oribe J, Castro GE, Montañez SG, Pérez PC, Ruiz BB. Niveles de calprotectina fecal en pacientes con anticuerpos anti-transglutaminasa positivos y pacientes con dieta libre de gluten. Acta Bioquím Clín Latinoam. 2020;**54**(1):29-38

[117] Veeraraghavan G, Leffler DA, Kaswala DH, Mukherjee R. Celiac disease 2015 update: New therapies. Expert Review of Gastroenterology & Hepatology. 2015;**9**(7):913-927. DOI: 10.1586/17474124.2015.1033399

[118] Khaleghi S, Ju JM, Lamba A, Murray JA. The potential utility of tight junction regulation in celiac disease: Focus on larazotide acetate. Therapeutic Advances in Gastroenterology. 2016;**9**(1):37-49. DOI: 10.1177/1756283X15616576

[119] Daveson AJM, Ee HC, Andrews JM, King T, Goldstein KE, Dzuris JL, et al. Epitope-specific immunotherapy targeting cd4-positive t cells in celiac disease: Safety, pharmacokinetics,

and effects on intestinal histology and plasma cytokines with escalating dose regimens of nexvax2 in a randomized, double-blind, placebo-controlled phase 1 study. eBioMedicine. 2017;**26**:78-90. DOI: 10.1016/j.ebiom.2017.11.018

[120] Syage JA, Murray JA, Green PHR, Khosla C. Latiglutenase improves symptoms in seropositive celiac disease patients while on a gluten-free diet. Digestive Diseases and Sciences. 2017;**62**(9):2428-2432. DOI: 10.1007/s10620-017-4687-7

[121] Vaquero L, Rodríguez-Martín L, León F, Jorquera F, Vivas S. New coeliac disease treatments and their complications. Gastroenterology Hepatology. 2018;**41**(3):191-204

[122] Cellier C, Bouma G, van Gils T, Khater S, Malamut G, Crespo L, et al. Safety and efficacy of AMG 714 in patients with type 2 refractory coeliac disease: A phase 2a, randomised, double-blind, placebo-controlled, parallel-group study. The Lancet Gastroenterology & Hepatology. 2019;**4**(12):960-970. DOI: 10.1016/S2468-1253(19)30265-1

[123] Lähdeaho ML, Scheinin M, Vuotikka P, Taavela J, Popp A, Laukkarinen J, et al. Safety and efficacy of AMG 714 in adults with coeliac disease exposed to gluten challenge: A phase 2A, randomised, double-blind, placebo-controlled study. The Lancet Gastroenterology & Hepatology. 2019;**4**(12):948-959. DOI: 10.1016/S2468-1253(19)30264-X

[124] Syage JA, Green PHR, Khosla C, Adelman DC, Sealey-Voyksner JA, Murray JA. Latiglutenase Treatment for celiac disease: Symptom and quality of life improvement for seropositive patients on a gluten-free diet. GastroHep. 2019;**1**(6):293-301. DOI: 10.1002/ygh2.371

[125] Serena G, Kelly CP, Fasano A. Nondietary therapies for celiac disease. Gastroenterology Clinics of North America. 2019;**48**(1):145-163. DOI: 10.1016/j.gtc.2018.09.011

[126] Caio G, Ciccocioppo R, Zoli G, De Giorgio R, Volta U. Therapeutic options for coeliac disease: What else beyond gluten-free diet? Digestive and Liver Disease. 2020;**52**(2):130-137. DOI: 10.1016/j.dld.2019.11.010

[127] Kivelä L, Caminero A, Leffler DA, Pinto-Sanchez MI, Tye-Din JA, Lindfors K. Current and emerging therapies for coeliac disease. Nature Reviews. Gastroenterology & Hepatology. 2021;**18**(3):181-195. DOI: 10.1038/s41575-020-00378-1

[128] Segura V, Ruiz-Carnicer Á, Sousa C, Moreno ML. New insights into non-dietary treatment in celiac disease: Emerging therapeutic options. Nutrients. 2021;**13**(7):2146. DOI: 10.3390/nu13072146

Chapter 10

Host-Microbiota Interplay in IBD: The Emerging Role of Extracellular Vesicles, Perinatal Immune Priming, and Gut-Resident Immune Cells

Surbhi Mishra, Juha Saarnio and Justus Reunanen

Abstract

The human gut is populated by innumerable microorganisms which govern equilibrium and well-being. Fluctuations in the composition and function of intestinal microbiota have been shown to result in persistent ailments such as inflammatory bowel disease (IBD). Yet, conclusive cause-effect studies must be formulated in this context. This chapter features current advancements in the field of host-microbiota interactions and their association with IBD. The role of bacterial extracellular vesicles (BEVs) and modification of intestinal EV proteomes with distinctive host-microbiota interactions in IBD, perinatal immune priming in offspring from maternal IBD and the function of gut-resident immune cells in IBD have been discussed here. These compelling developments would be crucial in expanding our understanding of IBD pathogenesis, detection of novel diagnostic repertoire and therapeutic targets for this disease.

Keywords: gut microbiota, inflammatory bowel disease (IBD), host-microbiota interaction, extracellular vesicles, inflammation, immune cells

1. Introduction

A plethora of assorted microorganisms inhabits the human gastrointestinal tract. The flexibility of the hefty genome of this community allows it to adapt well within the intestinal environment and complement the host [1]. The depth of association of the microbiome with human biology is accurately demonstrated by the spectrum of tasks delegated to the microbiome including pathogen defence [2], nutrient metabolism [3], assisting immune maturation [4] and maintaining metabolic homeostasis [5]. Humans and their gut microbiota are thus known to be co-evolved in a symbiotic manner. The composition of the gut microbiota varies notably among individuals [6, 7] and determines the susceptibility of the host to several diseases including inflammatory bowel disease (IBD) [8–10]. IBD has emerged as a global health challenge in the last decade [11].

IBD is a chronic and relapsing inflammatory disorder of the intestine and has two subtypes, Crohn's disease (CD) and ulcerative colitis (UC) [12]. Although sharing some clinical features and being studied together in the past, these two diseases represent discrete pathophysiological entities. Crohn's disease is characterized by segmental inflammation with clear distinctions between affected and unaffected bowel segments. The earliest mucosal lesions appear over Peyer's patches and the terminal ileum is affected the most [13]. On the contrary, ulcerative colitis is characterized by continuous inflammation extending proximally from the rectum to the colon. Inflammation is restricted to the mucosal layer, with neutrophils permeating the lamina propria and the intestinal crypts and forming cryptic abscesses [13, 14].

Compositional and metabolic changes in the intestinal microbiota have been extensively associated with chronic inflammation; however, several aspects of our understanding of IBD pathogenesis remained unclear. This chapter highlights the significant updates in the research related to the host-microbiota interactions as well as the role of the immune system in IBD, which might provide new avenues for disease prevention and treatment.

2. Extracellular vesicles and IBD

Extracellular vesicles (EVs) have gained recognition recently as novel mediators for cell-to-cell as well as interspecies and even interkingdom interaction [15]. EVs are submicron entities found circulating in all bodily fluids and in all species, including bacteria. EVs of the eukaryotic cells emerge either from the budding of the plasma membrane or the fusion of multivesicular endosomes with the plasma membrane. EVs derived from Gram-positive and Gram-negative bacteria may disperse in extracellular space by outward budding of the prokaryotic membrane [16, 17]. EVs contain a bioactive cargo of nucleic acids (DNA, mRNA, microRNA, and other noncoding RNAs), proteins (receptors, transcription factors, enzymes, and extracellular matrix proteins), small molecular metabolites, and lipids, which can govern the functions of the recipient cell [18–20]. Based on their biogenesis and size, EVs have been categorized into microvesicles, exosomes, ectosomes, oncosomes, and outer membrane vesicles (**Table 1**) [21].

EVs produced by commensal bacteria in the gastrointestinal tract are distributed throughout the gut lumen and carry a variety of compounds with a potential role in bacterial survival and host interaction [22]. EVs have been studied in many pathological and non-pathological conditions, including colorectal cancer and IBD. The role of extracellular products from commensal bacteria in immunomodulation and maintaining the homeostasis of the intestinal tract has gained attention since 1967 [23]. A recent study of bacterial extracellular vesicles (BEVs)-host interactions by Gul *et al.* investigated the effect of BEVs derived from the gut commensal bacterium *Bacteroides thetaiotaomicron* on host immune cells. Dendritic cells, macrophages and monocytes were of particular interest as they play key roles in regulating the immune response in IBD [24].

Genes expressed in each of the immune cell-types were identified by single-cell RNA sequencing and were assumed to be all translated into functional proteins to establish the host-microbe protein-protein interaction (PPI) networks. Even though there were a large number of BEV-human PPIs, most of the bacterial proteins were hubs with the potential to interact with thousands of host proteins. It was found that a total of 48 BEV proteins comprising of hydrolases, proteases, and other catabolic enzymes without a specific cleavage site, communicate with the host immune cells (**Figure 1**). Toll-like receptor (TLR) pathway analysis revealed that targets for BEVs differ among different cells and between the same cells in healthy versus disease

Ev type	Diameter (nm)	Density (g/ml)	Origin	Morphology	Composition
Exosomes	40–150	1.13–1.19	Derived from the plasma membrane by multivesicular endosome pathway	Cup-shaped	Surrounded by a phospholipid membrane containing relatively high levels of cholesterol, sphingomyelin, and ceramide and containing detergent-resistant membrane domains
Microvesicles	100–1000	Unknown	Released from the plasma membrane during cell stress	Cup-shaped	Insufficiently known
Membrane particles	50–80, 600	1.032–1.068	The plasma membrane of epithelial cells	Cup-shaped	CD133
Apoptotic vesicles	>2000	1.16–1.28	Plasma membrane, endoplasmic reticulum	Heterogeneous	Histones, DNA, immature glycoepitopes

Table 1.
Classification of extracellular vesicles.

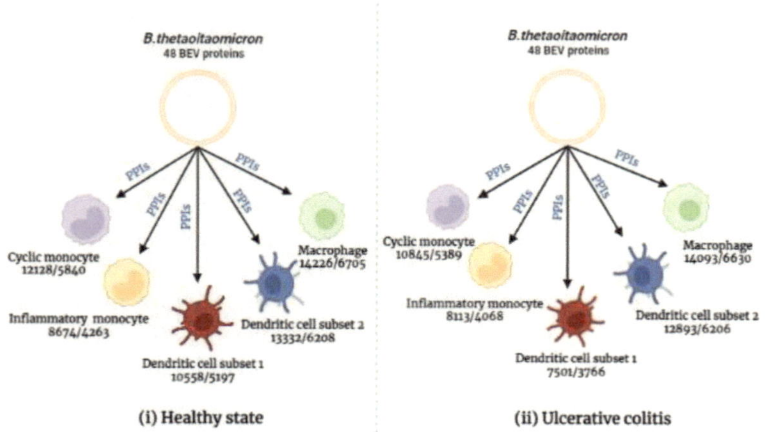

Figure 1.
Interactions of BEV proteins with immune cells in (i) Healthy state (ii) Ulcerative colitis (No. of expressed genes/No. of interacting proteins presented for each cell-type).

(ulcerative colitis) conditions [25]. These findings thus, suggest the role of cell-type as well as health status in influencing BEV-host interaction.

Zhang *et al.* [26] elucidated the association of microbiome and intestinal EV proteins in pediatric IBD. Mucosal-luminal interface samples collected from a pediatric IBD inception cohort were subjected to metaproteomic characterization for both the human and microbiota proteins. Microbial proteins related to oxidative stress responses were found to be upregulated in IBD cases compared to controls. Human proteins related to oxidative antimicrobial activities were found to be abundant in isolated free EVs and their expression was elevated in IBD cases, corresponding with the alteration of microbial functions [26]. Hence, EVs could serve as promising biomarkers with diagnostic and/or therapeutic potential in IBD.

3. Mother to child transfer of IBD

Pre- as well as post-natal bacterial colonisation plays a significant role in sculpting the immune system. Microbes transmitted from mother to infant presumably adapt to and persist in the infant gut than non-maternally acquired strains. Human trials have demonstrated the influence of maternal health status and microbiology on the development of the neonatal microbiome and immune system [27, 28]. The role of IBD in the maternal microbiome composition during pregnancy and its impact on the offspring's microbiome was investigated by Torres *et al.* by sampling pregnant women with and without IBD for their stool and saliva at each trimester, combined with their clinical and obstetric records. Post-delivery, the neonates were pursued with serial stool samples at time points of 7, 14, 30, 60, and 90 days, respectively, along with thorough health and exposure metadata. Stool samples from mother–baby pairs were then gavaged into 6–8 weeks old germ-free mice (GFM) for their immune phenotyping. 16S rRNA sequencing and microbiome analysis of the samples revealed that women with IBD maintained altered gut bacterial diversity throughout the pregnancy, with an enrichment of *Gammaproteobacteria* and a reduction in *Bacteroidetes*, compared with healthy controls. Offsprings to the IBD mothers demonstrated similarity to the bacterial diversity and composition trends of the mothers, to at least 3 months after birth compared with the offsprings to control mothers [29]. GFM inoculated with the stools from the third trimester IBD mother and 90-days infant showed a considerable reduction in the microbial diversity and fewer class-switched memory B cells and regulatory T cells in the colon, indicating the possible role of microbial factors from maternal IBD in influencing the immune system of the offspring [30].

Another study by Kim *et al.* made use of fecal calprotectin (FC) to monitor intestinal inflammation in pregnant women and their offsprings. FC is a non-glycosylated, calcium- and zinc-binding protein with antimicrobial, antiproliferative, and immunomodulatory properties, and it is used as a surrogate marker of intestinal inflammation [31]. FC levels decreased gradually in mothers with IBD during the 3 trimesters of pregnancy, contrary to the control mothers in which small gradual increase in FC levels was reported [32]. The rising levels of FC in healthy pregnancy correlated with the increase in pro-inflammatory phylum Proteobacteria and a decrease in anti-inflammatory *Faecalibacterium* [33]. Babies born to mothers with IBD presented significantly higher FC levels compared with control babies starting at 2 months of life and throughout 36 months. FC levels in both pregnant women with IBD and their babies were positively correlated with *Streptococcus* abundance and negatively correlated with that of *Alistipes* [32]. Consequently, maternal IBD has the potential to adversely affect the offspring's intestinal milieu during early life after birth, which can have significant health-related consequences later.

4. Gut-resident macrophages and microbial dysbiosis in IBD

Intestinal epithelium mononuclear phagocytes (MPs) have been designated as the 'sensors' and 'responders' to the intestinal environment by virtue of their location and function. They are represented by heterogeneous dendritic cell (DC) and macrophage subsets which are vital for the induction of immune response and regulation of inflammation [34]. Mononuclear phagocytes keep the intestinal inflammation in check either through direct regulation of microbiota or through the release of local anti-inflammatory molecules. Mononuclear phagocytes expressing the fractalkine receptor CX3CR1 and displaying a macrophage phenotype, play a key role in the uptake and sampling of bacterial and fungal antigens from the intestinal lumen [35–39].

Gut microbiota has a crucial role in maintaining tolerogenic function i.e., immunological tolerance of intestinal macrophages and bacterial dysbiosis has strongly been associated with intestinal inflammation and IBD [40–42]. Intestinal epithelium-adhering bacteria can interact with CX3CR1 MPs to regulate the immune balance in health and diseases. The enrichment of adherent-invasive *Escherichia coli* in ileal mucosa has been described in active Crohn's disease [43, 44]. This bacterium stimulates the production of IL-10 by CX3CR MPs and suppresses the Th17 immune responses [44, 45]. *Klebsiella* species derived from the oral cavity have been found to inhabit the intestine of IBD patients and induce severe colitis by the activation of Th1 proinflammatory immune response [46].

Koscsó *et al.* [47] performed extensive phenotypical, transcriptional, and functional analyses of intestinal inflammatory MPs in Salmonella colitis model. $CX3CR1^{+}$MPs were identified as the predominating inflammatory cell type and were further divided into monocyte-derived $Nos2^{+}$ $CX3CR1^{lo}$, lymph migratory $Ccr7^{+}CX3CR1^{int}$ and mucosa resident $Cxcl13^{+}CX3CR1^{hi}$ subsets. An increase in MPs in the inflamed bowel was found to be directly related to the increase in $CX3CR1^{lo}$, $CX3CR1^{int}$ and $CX3CR1^{hi}$ macrophage populations and thus, have an apparent role in the induction of pathogen-specific mucosal IgA response [34, 47]. These studies suggest that CX3CR1 MPs are crucial in maintaining immune homeostatic conditions and controlling intestinal disease development.

5. Disease-specific signatures of Crohn's disease and ulcerative colitis

Inflammatory bowel disease (IBD) involves chronic intestinal inflammation linked with critical ailment and has two subtypes- ulcerative colitis (UC), which directly affects the colon and Crohn's disease (CD) which can affect any part of the gastro-intestinal (GI) tract. Macroscopic patterns of inflammation can at times distinguish between UC and CD but an insight of mucosal and peripheral immunological as well as microbial signatures differentiating these two subtypes becomes necessary for the diagnosis, prevention of recurrence or complications, and effective treatment [48].

5.1 Microbial signatures of IBD subtypes

Gut microbiota dysbiosis has been associated with disease phenotypes in IBD and may be a causative or synergistic factor in prolonged or chronic inflammation. Microbial dysbiosis in IBD is characterized by a significant reduction in bacterial diversity and alterations in some specific taxa, including enrichment of the phyla *Proteobacteria* and *Bacteroidetes*, and a reduction in *Firmicutes* [49–52]. CD has been presented with a decrease in the proportion of Firmicutes and a slight increase in Enterobacteriaceae when compared with controls and UC patients [53]. *Bacteroides*, *Eubacterium*, *Faecalibacterium*, and *Ruminococcus* are the main bacterial genera reduced in the fecal samples of CD patients [54, 55]. A reduction in *Faecalibacterium prausnitzii* has been implicated in the etiology of CD, suggesting a critical role for the organism as an integral component of the anti-inflammatory balance in health and in CD pathogenesis. The phylum *Proteobacteria* is highly abundant in patients with active UC and decreased significantly in patients in remission, where as vice-versa for *Firmicutes*. Patients with active UC show an enrichment of *Klebsiella*, *Enterococcus*, and *Haemophilus*, while those in remission have higher numbers of *Roseburia*, *Lachnospira*, *Blautia*, and *Faecalibacterium* [56].

5.2 Immune cell signatures of IBD subtypes

An elaborated knowledge of the inflammatory landscape and immune markers of IBD in circulation and tissues become essential for the effective disease management in IBD subtypes. In this view, Mitsialis *et al.* carried out multidimensional immunophenotyping of colonic mucosa and peripheral blood of IBD (UC & CD) and non-IBD subjects to provide a deep understanding of the disease-specific immunophenotypes in UC and CD (**Figure 2**) [57]. Active ulcerative colitis (UCa) mucosa had relatively more B cells and fewer T cells and cytokine-producing effector memory (EM)-T cell subsets- $IFNG^{+}TNF^{+}$ were reduced whereas $IL17A^{++}CD161^{+}$ subsets were enriched. $CXCR3^{+}$ plasmablasts were found to be expanded in UCa. $HLA\text{-}DR^{+}CD38^{+}$ memory regulatory T cells (mTregs) were also abundant in UCa and co-expressed various chemokine receptors implying an activated memory phenotype. UCa mucosa was enriched with granulocytes expressing chemokine receptors (CXCR3, CCR6) and unconventional granulocyte markers (HLA-DR, CD38, and CD56) described to be up-regulated on granulocytes in other human diseases (**Table 2**).

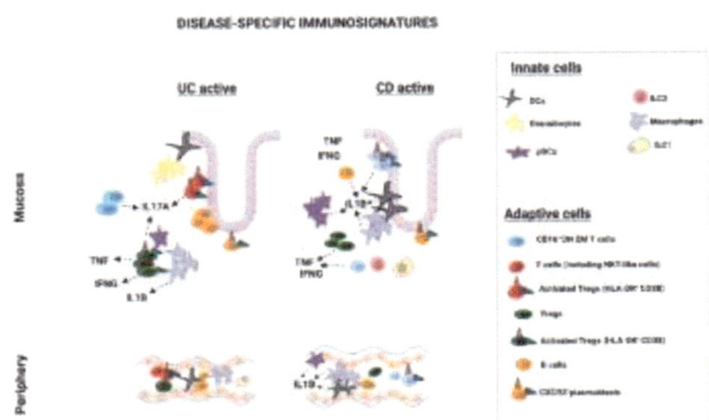

Figure 2.
Disease-specific immunosignatures of Crohn's disease (CD) and Ulcerative colitis (UC) mucosa and periphery.

Ulcerative colitis (UC) specific immunophenotypes		**Crohn's disease (CD) specific immunophenotypes**	
B:T cell ratio	+	HLA-DR+CD38+ T cells IL17A$^+$ HLA-DR$^+$ CD38$^+$ CD161$^+$ DN Effector Memory T cells IL1B$^+$ HLA-DR$^+$ CD38$^+$ T cells	- + +
Cytokine-producingeffector memory (EM)-T cell subsets: • IFNG+TNF+ • IL17A++CD161+	- +	IL1B+ IFNG+ TNF+ naïve B-cell clusters	+
CXCR3+ plasmablasts	+	CD14+ and IL1B+ macrophages/monocytes clusters	+
HLA-DR+CD38+ mTregs	+	Innate lymphoid cells (ILCs): • ILC1 and ILC1-like clusters • ILC3	+ -
Chemokine receptors CXCR3, CCR6	+		

"+"= enriched; and "-" = reduced.

Table 2.
Disease specific alterations of immune cells in IBD subtypes.

In case of active Crohn's disease mucosa (CDa), HLA-DR$^+$CD38$^+$ T cells co-expressing IFNG$^+$TNF$^+$ were diminished whereasIL17A$^+$ HLA-DR$^+$ CD38$^+$ CD161$^+$ DN EM T cells and IL1B$^+$ HLA-DR$^+$ CD38$^+$ T cells demonstrated expansion. IL1B$^+$ IFNG$^+$ TNF$^+$ naïve B-cell clusters were augmented in CDa mucosa and included CD44^{++} (marker of activated B cells), CCR7$^+$, AHR$^+$, HLA-DR$^+$, CD38$^+$ and CD11C$^+$, a marker expressed in B cells and proficient in antigen presentation linked with autoimmunity [57]. Total CD14$^+$ as well as IL1B+ macrophages/monocytes clusters were increased in peripheral CDa. Innate lymphoid cells (ILCs) signatures could differentiate Crohn's disease from ulcerative colitis. ILC1 and ILC1-like clusters were increased more in the mucosa in case of CDa than UCa whereas ILC3 were specifically reduced in UCa mucosa (**Table 2**). These findings could be explored for targeted therapeutics and possibly harnessed for personalized approaches to IBD therapy in the future.

6. Conclusion

Even though there has been a massive upsurge in the research related to host-microbiota interactions as well as the role of genetics, environmental factors, and the immune system in IBD, several facets of IBD pathogenesis remain obscure. This chapter collates the contemporary advancements in host-microbiota investigations which can be pivotal in detecting the hallmarks of IBD leading to upgraded comprehension of its pathogenesis, extension of the diagnostic repertoire and discovery of cutting-edge therapeutic targets for this disease.

EVs have emerged as prominent tools in deciphering the complex host-microbiota interactions in healthy as well as disease states. They not only regulate the gut microbiome communities, but also actively participate in the disharmony between bacteria and their hosts. EVs derived from gut commensal bacteria have been studied to play a crucial role in immunomodulation and regulating gut homeostasis in IBD [22]. The first proteomic characterization of intestinal EVs from children with new-onset IBD illustrated the presence of host defense proteins in the isolated EV samples, especially the reactive oxidant-producing enzymes responsible for increased oxidative stress in the intestine [26]. Increased oxidative stress triggers microbial defense responses and functional alterations leading to gut microbial dysbiosis and mucosal inflammation [58]. This learning is crucial for the thorough analysis of host–microbiome interactions underlying the development of IBD and the potential use of EVs as diagnostic markers and/or therapeutic agents.

Dysbiosis of microbiota in germ-free mice have been demonstrated to cause abnormal imprinting of the intestinal immune system [29]. It provides a potential link between early life exposures, microbiome and future risk of IBD, highlighting the consequences of the abnormal establishment of early life microbiome during the development of the immune system. Maternal IBD negatively impacts the development of a baby's intestinal ecosystem. Dysbiosis, in pregnant women with IBD or during early infancy can be aimed for promoting the development of a healthy microbiome in the offspring and reducing the potential risk of IBD.

Intestinal resident macrophages are acknowledged as key cellular sensors, integrating signals from the luminal microbiota to regulate intestinal homeostasis. Recent studies affirm their role in promoting anti-inflammatory environment in the healthy gut and switching to a proinflammatory state in response to any alterations in the intestinal microbiota [59]. Follow-up studies should be done to devise tools for identifying patients with compromised resident intestinal macrophages function and evaluating the clinical advantages of targeting the microbiota and immune dysfunctions within this subset of IBD patients. Intestinal macrophage subsets also exhibit peculiar activity in stimulating mucosal IgA responses [47]. This differential activity can be harnessed for designing anti-inflammatory therapies aimed at modulating macrophage function in inflammatory bowel disease.

IBD includes Crohn's disease and ulcerative colitis which are two distinct pathological conditions macroscopically, but often misinterpreted or difficult to distinguish on a deeper extent. There has been evidence of disease-specific statistical shifts in some bacterial species as well as phyla, peculiar to each subtype of IBD [56]. Single-cell analysis with CyTOF on IBD and non-IBD colonic mucosa and blood to identify disease-specific immune signatures revealed the abundance of HLA-DR$^+$CD38$^+$ T cells in both active Crohn's disease (CDa) and ulcerative colitis (UCa) mucosa [57]. CD38 has been involved in colitis in mice [60] whereas CD38$^+$ effector T cells in pediatric IBD [61], suggesting that CD38 could be targeted for IBD therapy. Various disease-specific mucosal signatures associated with differential cytokine expression were also reported. IL1B signatures particular to CD involved HLA-DR$^+$CD38$^+$ T cells, naïve B cells, and DCs. IL1B$^+$ macrophages/monocytes were augmented in both CDa and Uca mucosa, along with a specific expansion of IL1B+ monocytes to only peripheral CDa [57, 62]. Thus, exploiting IL1B can be a promising therapeutic strategy for subsets of Crohn's disease. These extrusive microbial and immunological signatures of IBD can also be of high biological and diagnostic potential. To sum up, the above-discussed studies have a robust potential of heralding state-of-art diagnostic as well as therapeutic avenues in the field of inflammatory bowel disease. Further translational work based upon these findings can lead to the upgradation of our insight and methodology towards gut disorders as critical as IBD with a prospect of personalized therapies soon.

Conflict of interest

The authors declare no conflict of interest.

Author details

Surbhi Mishra[1], Juha Saarnio[2] and Justus Reunanen[1*]

1 Biocenter Oulu, Cancer and Translational Medicine Research Unit, University of Oulu, Oulu, Finland

2 Research Unit for Surgery, Anaesthesiology and Intensive Care, University of Oulu, Oulu, Finland

*Address all correspondence to: justus.reunanen@oulu.fi

References

[1] Buffie CG et al. Precision microbiome reconstitution restores bile acid mediated resistance to clostridium difficile. Nature. 2015;**517**(7533):205-2088

[2] Cummings JH, Macfarlane GT. Role of intestinal bacteria in nutrient metabolism. Journal of Parenteral and Enteral Nutrition. 1997;**21**(6):357-365

[3] Atarashi K, Tanoue T, Oshima K, Suda W, Nagano Y, Nishikawa H, et al. Treg induction by a rationally selected mixture of Clostridia strains from the human microbiota. Nature. 2013;**500**(7461):232-236

[4] Turnbaugh PJ, Ley RE, Mahowald MA, Magrini V, Mardis ER, Gordon JI. An obesity-associated gut microbiome with increased capacity for energy harvest. Nature. 2006;**444**(7122):1027-1031

[5] Yan PG, Li JN. Advances in the understanding of the intestinal micro-environment and inflammatory bowel disease. Chinese Medical Journal. 2020;**133**(07):834-841

[6] Eckburg PB, Bik EM, Bernstein CN, Purdom E, Dethlefsen L, Sargent M, et al. Diversity of the human intestinal microbial flora. Science. 2005;**308**(5728): 1635-1638

[7] Human Microbiome Project Consortium. Structure, function and diversity of the healthy human microbiome. Nature. 2012;**486**(7402):207

[8] Hooper LV, Littman DR, Macpherson AJ. Interactions between the microbiota and the immune system. Science. 2012;**336**(6086):1268-1273

[9] Littman DR, Pamer EG. Role of the commensal microbiota in normal and pathogenic host immune responses. Cell Host & Microbe. 2011;**4**:311-323

[10] Lozupone CA, Stombaugh JI, Gordon JI, Jansson JK, Knight R. Diversity, stability and resilience of the human gut microbiota. Nature. 2012;**489**(7415):220-230

[11] Gevers D, Kugathasan S, Denson LA, Vázquez-Baeza Y, Van Treuren W, Ren B, et al. The treatment-naive microbiome in new-onset Crohn's disease. Cell Host & Microbe. 2014;**15**(3):382-392

[12] Ng SC, Shi HY, Hamidi N, Underwood FE, Tang W, Benchimol EI, et al. Worldwide incidence and prevalence of inflammatory bowel disease in the 21st century: A systematic review of population-based studies. The Lancet. 2017;**390**(10114):2769-2778

[13] Caruso R, Lo BC, Núñez G. Host–microbiota interactions in inflammatory bowel disease. Nature Reviews Immunology. 2020;**20**(7):411-426

[14] Xavier RJ, Podolsky DK. Unravelling the pathogenesis of inflammatory bowel disease. Nature. 2007;**448**(7152):427-434

[15] Buzas EI, György B, Nagy G, Falus A, Gay S. Emerging role of extracellular vesicles in inflammatory diseases. Nature Reviews Rheumatology. 2014;**10**(6):356-364

[16] Beveridge TJ. Structures of gram-negative cell walls and their derived membrane vesicles. Journal of bacteriology. 1999;**181**(16):4725-4733

[17] Avila-Calderón ED, Araiza-Villanueva MG, Cancino-Diaz JC, López-Villegas EO, Sriranganathan N, Boyle SM, et al. Roles of bacterial

membrane vesicles. Archives of Microbiology. 2015;**197**(1):10

[18] Mathieu M, Martin-Jaular L, Lavieu G, Théry C. Specificities of secretion and uptake of exosomes and other extracellular vesicles for cell-to-cell communication. Nature Cell Biology. 2019;**21**(1):9-17

[19] Fleshner M, Crane CR. Exosomes, DAMPs and miRNA: Features of stress physiology and immune homeostasis. Trends in Immunology. 2017;**38**(10):768-776

[20] Schmid M, Jensen TH. The exosome: A multipurpose RNA-decay machine. Trends in Biochemical Sciences. 2008;**33**(10):501-510

[21] Van der Pol E, Böing AN, Harrison P, Sturk A, Nieuwland R. Classification, functions, and clinical relevance of extracellular vesicles. Pharmacological Reviews. 2012;**64**(3):676-705

[22] Chang X, Wang SL, Zhao SB, Shi YH, Pan P, Gu L, et al. Extracellular vesicles with possible roles in gut intestinal tract homeostasis and IBD. Mediators of Inflammation. 2020

[23] Chatterjee SN, Das J. Electron microscopic observations on the excretion of cell-wall material by Vibrio cholerae. Microbiology. 1967;**49**(1):1

[24] Gul L, Modos D, Fonseca S, Madgwick M, Thomas JP, Sudhakar P, et al. Extracellular vesicles produced by the human commensal gut bacterium Bacteroides thetaiotaomicron affect host immune pathways in a cell-type specific manner that are altered in inflammatory bowel disease. Journal of Extracellular Vesicles. 2022;**11**(1):e12189

[25] Alberti G, Mazzola M, Gagliardo C, Pitruzzella A, Fucarini A, Giammanco M, et al. Extracellular vesicles derived from gut microbiota in inflammatory bowel disease and colorectal cancer. Biomedical Papers of the Medical Faculty of Palacky University in Olomouc. 2021;**1**:165

[26] Zhang X, Deeke SA, Ning Z, Starr AE, Butcher J, Li J, et al. Metaproteomics reveals associations between microbiome and intestinal extracellular vesicle proteins in pediatric inflammatory bowel disease. Nature Communications. 2018;**9**(1):1-4

[27] Ferretti P, Pasolli E, Tett A, Asnicar F, Gorfer V, Fedi S, et al. Mother-to-infant microbial transmission from different body sites shapes the developing infant gut microbiome. Cell Host & Microbe. 2018;**24**(1):133-145

[28] Gomez de Agüero M, Ganal-Vonarburg SC, Fuhrer T, Rupp S, Uchimura Y, Li H, et al. The maternal microbiota drives early postnatal innate immune development. Science. 2016;**351**(6279):1296-1302

[29] Torres J, Hu J, Seki A, Eisele C, Nair N, Huang R, et al. Infants born to mothers with IBD present with altered gut microbiome that transfers abnormalities of the adaptive immune system to germ-free mice. Gut. 2020;**69**(1):42-51

[30] Izcue A, Coombes JL, Powrie F. Regulatory T cells suppress systemic and mucosal immune activation to control intestinal inflammation. Immunological reviews. 2006;**212**(1):256-271

[31] Røseth AG, Schmidt PN, Fagerhol MK. Correlation between faecal excretion of indium-111-labelled granulocytes and calprotectin, a granulocyte marker protein, in patients with inflammatory bowel disease. Scandinavian Journal of Gastroenterology. 1999;**34**(1):50-54

[32] Kim ES, Tarassishin L, Eisele C, Barre A, Nair N, Rendon A, et al. Longitudinal changes in fecal calprotectin levels among pregnant women with and without inflammatory bowel disease and their babies. Gastroenterology. 2021;**160**(4):1118-1130

[33] Koren O, Goodrich JK, Cullender TC, Spor A, Laitinen K, Bäckhed HK, et al. Host remodeling of the gut microbiome and metabolic changes during pregnancy. Cell. 2012;**150**(3):470-480

[34] Gross M, Salame TM, Jung S. Guardians of the gut–murine intestinal macrophages and dendritic cells. Frontiers in Immunology. 2015;**2**(6):254

[35] Leonardi I, Li X, Semon A, Li D, Doron I, Putzel G, et al. CX3CR1+ mononuclear phagocytes control immunity to intestinal fungi. Science. 2018;**359**(6372):232-236

[36] Niess JH, Brand S, Gu X, Landsman L, Jung S, McCormick BA, et al. CX3CR1-mediated dendritic cell access to the intestinal lumen and bacterial clearance. Science. 2005;**307**(5707):254-258

[37] Hadis U, Wahl B, Schulz O, Hardtke-Wolenski M, Schippers A, Wagner N, et al. Intestinal tolerance requires gut homing and expansion of FoxP3+ regulatory T cells in the lamina propria. Immunity. 2011;**34**(2):237-246

[38] Mazzini E, Massimiliano L, Penna G, Rescigno M. Oral tolerance can be established via gap junction transfer of fed antigens from CX3CR1+ macrophages to CD103+ dendritic cells. Immunity. 2014;**40**(2):248-261

[39] Medina-Contreras O, Geem D, Laur O, Williams IR, Lira SA, Nusrat A, et al. CX3CR1 regulates intestinal macrophage homeostasis, bacterial translocation, and colitogenic Th17 responses in mice. The Journal of Clinical Investigation. 2011;**121**(12):4787-4795

[40] Ni J, Wu GD, Albenberg L, Tomov VT. Gut microbiota and IBD: Causation or correlation? Nature Reviews Gastroenterology & Hepatology. 2017;**14**(10):573-584

[41] Blander JM, Longman RS, Iliev ID, Sonnenberg GF, Artis D. Regulation of inflammation by microbiota interactions with the host. Nature Immunology. 2017;**18**(8):851-860

[42] Belkaid Y, Harrison OJ. Homeostatic immunity and the microbiota. Immunity. 2017;**46**(4):562-576

[43] Willing B, Halfvarson J, Dicksved J, Rosenquist M, Järnerot G, Engstrand L, et al. Twin studies reveal specific imbalances in the mucosa associated microbiota of patients with ileal Crohn's disease. Inflammatory Bowel Diseases. 2009;**15**(5):653-660

[44] Viladomiu M, Kivolowitz C, Abdulhamid A, Dogan B, Victorio D, Castellanos JG, et al. IgA-coated *E. coli* enriched in Crohn's disease spondyloarthritis promote TH17-dependent inflammation. Science Translational Medicine. 2017;**9**(376):eaaf9655

[45] Kim M, Galan C, Hill AA, Wu WJ, Fehlner-Peach H, Song HW, et al. Critical role for the microbiota in CX3CR1+ intestinal mononuclear phagocyte regulation of intestinal T cell responses. Immunity. 2018;**49**(1):151-163

[46] Atarashi K, Suda W, Luo C, Kawaguchi T, Motoo I, Narushima S, et al. Ectopic colonization of oral bacteria in the intestine drives TH1 cell induction and inflammation. Science. 2017;**358**(6361):359-365

[47] Koscsó B, Kurapati S, Rodrigues RR, Nedjic J, Gowda K, Shin C, et al. Gut-resident CX3CR1hi macrophages induce tertiary lymphoid structures and IgA response in situ. Science Immunology. 2020;**5**(46):eaax0062

[48] Lacroix V, Cassard A, Mas E, Barreau F. Multi-omics analysis of gut microbiota in inflammatory bowel diseases: What benefits for diagnostic, prognostic and therapeutic tools? International Journal of Molecular Sciences. 2021;**22**(20):11255

[49] Swidsinski A, Ladhoff A, Pernthaler A, Swidsinski S, Loening-Baucke V, Ortner M, et al. Mucosal flora in inflammatory bowel disease. Gastroenterology. 2002;**122**(1):44-54

[50] Ott SJ, Musfeldt M, Wenderoth DF, Hampe J, Brant O, Fölsch UR, et al. Reduction in diversity of the colonic mucosa associated bacterial microflora in patients with active inflammatory bowel disease. Gut. 2004;**53**(5):685-693

[51] Conte MP, Schippa S, Zamboni I, Penta M, Chiarini F, Seganti L, et al. Gut-associated bacterial microbiota in paediatric patients with inflammatory bowel disease. Gut. 2006;**55**(12):1760-1767

[52] Sokol H, Lay C, Seksik P, Tannock GW. Analysis of bacterial bowel communities of IBD patients: What has it revealed? Inflammatory bowel diseases. 2008;**14**(6):858-867

[53] Walker AW, Sanderson JD, Churcher C, Parkes GC, Hudspith BN, Rayment N, et al. High-throughput clone library analysis of the mucosa-associated microbiota reveals dysbiosis and differences between inflamed and non-inflamed regions of the intestine in inflammatory bowel disease. BMC Microbiology. 2011;**11**(1):1-2

[54] Takahashi K, Nishida A, Fujimoto T, Fujii M, Shioya M, Imaeda H, et al. Reduced abundance of butyrate-producing bacteria species in the fecal microbial community in Crohn's disease. Digestion. 2016;**93**(1):59-65

[55] Yu Y, Yang W, Li Y, Cong Y. Enteroendocrine cells: Sensing gut microbiota and regulating inflammatory bowel diseases. Inflammatory Bowel Diseases. 2020;**26**(1):11-20

[56] Guo X, Huang C, Xu J, Xu H, Liu L, Zhao H, et al. Gut microbiota is a potential biomarker in inflammatory bowel disease. Frontiers in Nutrition. 2021;**8**

[57] Mitsialis V, Wall S, Liu P, Ordovas-Montanes J, Parmet T, Vukovic M, et al. Single-cell analyses of colon and blood reveal distinct immune cell signatures of ulcerative colitis and Crohn's disease. Gastroenterology. 2020;**159**(2):591-608

[58] Arenas FA, Díaz WA, Leal CA, Pérez-Donoso JM, Imlay JA, Vásquez CC. The Escherichia coli btuE gene, encodes a glutathione peroxidase that is induced under oxidative stress conditions. Biochemical and Biophysical Research Communications. 2010;**398**(4):690-694

[59] Leonardi I, Li X, Iliev ID. Macrophage interactions with fungi and bacteria in IBD. Current Opinion in Gastroenterology. 2018;**34**(6):392

[60] Schneider M, Schumacher V, Lischke T, Lücke K, Meyer-Schwesinger C, Velden J, et al. CD38 is expressed on inflammatory cells of the intestine and promotes intestinal inflammation. PLoS One. 2015;**10**(5):e0126007

[61] Joosse ME, Menckeberg CL, De Ruiter LF, Raatgeep HR, van Berkel LA,

Simons-Oosterhuis Y, et al. Frequencies of circulating regulatory TIGIT+ CD38+ effector T cells correlate with the course of inflammatory bowel disease. Mucosal Immunology. 2019;**12**(1):154-163

[62] Mahida YR, Wu K, Jewell DP. Enhanced production of interleukin 1-beta by mononuclear cells isolated from mucosa with active ulcerative colitis of Crohn's disease. Gut. 1989;**30**(6):835-838

Section 4

Autoimmune Diseases of the GI Tract

Chapter 11

Molecular Impact of Dietary Fibre Metabolites on Intestinal Immunity of Host

Jitendra Kumar, Priya Sharma, Murli Dhar Mitra, Sonia Sangwan and Haribrahma Singh

Abstract

Food contains several components that are essential for health. Dietary fibres are nondigestible foods that play an important role in the maintenance of health. Nondigestible carbohydrates are an important constituent of the diet. Intestinal immunity is the bedrock of host health and holistic health maintained by nutrition and the existence of the host supported by immunity. The gastrointestinal immune barrier is exposed to the environment or food, and immunity is maintained by several factors. Dietary fibres exert molecular effects through the production of short-chain fatty acids (SCFAs) and gut microbiota. Dietary fibres and microbial communities secrete metabolites that have the potential to regulate intestinal immunity. The gastrointestinal immune barrier is a primary target for dietary fibre metabolites, and these molecules exert a signalling effect on immune cells in the intestine. In the proposed chapter, we will discuss the molecular impact of dietary fibers on intestinal immunity and how innate immune response and gut microbiota are regulated by metabolites.

Keywords: metabolites, intestinal immunity, dietary fibers, gut microbiota

1. Introduction

In addition to maintaining human growth, fertility, and health, the diet is also essential to modulating and supporting the symbiotic microbiota and the microbial communities that inhabit the digestive tract. Our gut harbors trillions of microbes that play a significant role in datary fiber metabolism. The gut microbiome modulates maturation of the immune system [1], glucose and lipid metabolism [2], and juvenile growth [3]. The microbial diversity in the gut depends on the intake of dietary fibers [4] and any alteration of dietary intake of fibres may result in dysfunction of the gut and the development of chronic inflammatory diseases like intestinal bowel disease (IBD), autoimmune diseases, colorectal cancer (CRC), and allergies. The gut microbiome is affected by diet, which in turn affects the immune system. We wondered whether the results of the high-fiber diet intervention may have coincidentally impacted participants' immune systems because the microbiota of the group differed.

 IntechOpen

Here we will discuss how dietary fiber impacts gut microbial ecology, host physiology, and health by specifically focusing on the molecular impact of dietary fiber metabolites on intestinal immunity

Here, we specifically discuss the effects of the gut microbiota on immunometabolism, and more precisely, on the intracellular metabolism of immune cells, in health and the potential consequences in diseases. In this chapter, emphasis is placed on the effects of dietary fiber metabolites as prime signaling molecules, through different signaling pathways and their link between gut microbiota and host health.

2. Dietary fibers and their metabolites in molecular function

Dietary habits, dietary patterns, and lifestyles determine the presence of different microbial species [5]. In addition to modulating the gut microbiota composition, dietary fibers directly influence biological processes and homeostasis via the metabolites that are a result of microbial fermentation of nutrients, such as short-chain unsaturated fats (SCFAs) [6]. The gut microbiota is vital for the metabolization of DFs, such as nondigestible carbohydrates (NDCs), proteins, and peptides, which have escaped digestion by host enzymes in the upper gut and have been absorbed in the lower digestive tract [7], which are known to have beneficial effects by behaving as signaling molecules via different pathways. Acetate is the most abundant SCFA produced, and it is used by many gut commensals to produce propionate and butyrate in a growth-promoting cross-feeding process. In addition, SCFA has been shown to regulate metabolic activity. Acetate affects the metabolic pathway via the G protein-coupled receptor (GPCR) and free fatty acid receptor 2 (FFAR2/GPR43), while butyrate and propionate transactivate peroxisome proliferator-activated receptors (PPAR/NR1C3) and regulate Angptl4 in colonic cells. FFAR2 regulates insulin-induced lipid accumulation in adipocytes and inflammation, while peptide tyrosine-tyrosine and glucagon-like peptide 1 control appetite. Microbiota-dependent NDCs regulate glucose homeostasis, gut integrity, and hormones via GPCR, NF-kB, and AMPK.

Dietary fibers are metabolized by the microbiota in the cecum and colon resulting in the formation of major products, such as acetate (C2), propionate (C3), and butyrate (C4) SCFAs. Acetate is a major SCFA metabolite produced from pyruvate. Propionate (C3) is created when succinate is converted to methylmalonyl-CoA through the succinate pathway. Through the classical pathway, butyrate is produced by the condensation of two molecules of acetyl-CoA and their subsequent reduction to butyryl-CoA. Butyrate is then converted to butyrate by phosphorus butyrylase and butyrate kinase [8].

3. Intestinal immunity

The SCFAs activate different G-protein-coupled receptors (GPCR) e.g. propionate (C2), which is an activator of GPR43. The expression of GPR43 has been reported in the entire gastrointestinal tract including the cells of both the nervous and immune systems. In the GI tract, GPR43 is significantly expressed in endocrine L-cells of the ileum and colon of intestinal, PYY, and GLP-1 producing cells, as well as on colonocytes and enterocytes [9], that maintains the immunity and function of the intestine [10].

4. Immunity and fiber in the diet

The share of CD4+ and CD8+ T-cells in GALT, as well as their in vitro responsiveness to mitogens, were considerably affected by the diet's fiber intake. There was a bigger proportion of CD8+ T-cells in the IEL, lamina propria, and Peyer's patches after consuming the high fermentable fiber diet, as well as a higher proportion of CD4+ T-cells in the mesenteric lymph nodes and peripheral blood except for a higher CD4:CD8 ratio [11, 12].

In the upper gastrointestinal system, prebiotic fiber is neither hydrolyzed nor absorbed, but instead assists as a selective substrate for one or a small number of beneficial colonic bacteria, modifying the gut microbiota. There is significant proof that prebiotic fibers (inulin and oligofructose) boost the percentage of good lactic acid bacteria in the human colon [13].

To yet, the mechanism(s) through which probiotics in the diet alter immune function has primarily been hypothetical. Immune activation via uninterrupted contact of the intestinal microbiota with GALT is one logical method. Small amounts of bacteria can pass through the intestinal epithelial barrier and into Peyer's patches, affecting or contributing to the activation of other immune cells [14]. The production of TNF-α and IL-6 by macrophages, as well as the production of IL-2 and IL-5 by stimulated CD4+ cells, was dramatically boosted by coculture with bifidobacteria [15].

5. Short-chain fatty acids (SCFAs) are produced by fermentation of fiber

Through the fermentation of food fibers to SCFA, the gut microbiota may influence immune cells. Increased natural killer cell activity is an outcome of SCFA. SCFA has also been presented to have anti-inflammatory properties in other investigations. In the colonic cell line HT-29, butyrate was found to decrease both constitutive and cytokine-induced production of the transcription factor NFkB [13].

Finally, SCFA generation in the colon, particularly butyrate, may lessen the glutamine need of epithelial cells, freeing it up for other cells such as immune system cells. The fact that lactulose injection can raise serum glutamine levels, and glutamine is a vital energy source for immune cells, which supports this notion [16].

The addition of fermentable fibers to the diet has been shown to boost mucin synthesis. The reduced incidence of bacterial translocation across the intestinal barrier observed in studies feeding dietary fibers could be due to increased mucin synthesis. The increased mucin formation may be due to the lower pH associated with SCFA production.

6. The effect of dietary fiber on the immune system of the gut

Carbohydrate polymers naturally occur in edible plants and are used up as vegetables, fruits, seeds, cereals, and tubers. Dietary fibers travel from the small to the large intestine, where they perform a physiological role. Fibers consist of two types soluble and insoluble. Soluble fibers undergo total fermentation in the colon whereas insoluble fiber undergoes fermentation to some extent. Dietary fiber consists of a range of organic polymers, each of which contains various monomers coupled by different glycosidic linkages, resulting in a complex and heterogeneous structure. Many methods of classifying dietary fiber, such as solubility, viscosity, and fermentability, have been formulated to aid in the correlation of physicochemical features of dietary fiber with their physiological roles. Although particular nutrients are known to play a role in the immune system's development and function, little is known about the impression of dietary fibers on immunological function. Dietary fiber is essential for good health. Higher dietary fiber consumption is linked to a lower risk of disease and mortality, according to several meta-analyses. Dietary fiber consumption is linked to a higher risk of Western diseases with immune system abnormalities, implying that dietary fiber is vital for immunological homeostasis. The preservation of the gut immune barrier is one direction through which fibers may protect against disease development[4]. The innate immune system that includes physical barriers such as the skin and mucous membranes, cell-mediated barriers such as phagocytic cells, inflammatory cells, dendritic cells, and natural killer cells, and soluble mediators such as cytokines, complement, and acute-phase proteins, delivering immunity to invading organisms without the need for prior exposure to these antigens. During the 4–5 days it takes lymphocytes to become activated, this arm of the immune system supplies the early steps of host defense that safeguard the organism. Macrophages and their precursor monocytes, as well as polymorphonuclear leukocytes (neutrophils), frame the innate immune system's core cellular component.

The human body's janitor is the gastrointestinal immunological barrier. It is made up of a mucus layer and an epithelial cell layer that keeps luminal molecules out of the immune-cell-filled lamina propria beneath. Improving intestinal barrier function by increasing dietary fiber intake could thus be a useful therapy for preventing or delaying Western immune-related illnesses. Moreover, Dietary fibers may interact directly with immunological barrier cells in the small intestine before being destroyed by microbial enzymes in the colon. The small intestine has a thin and loose mucus layer that boosts nutrient absorption while also allowing food compounds like fibers to interact directly with intestinal epithelium and immune cells. The ramifications of these interactions are that the mucus layer is strengthened, epithelial cell barrier function is improved, and intestinal immune responses are modulated as a result of these direct interactions with intestinal immune barrier cells. Dietary fibers' direct contact with the gut immune system could be one of the processes by which they improve health and prevent disease. Dietary fiber can also have an indirect positive effect on the gastrointestinal immunological barrier by stimulating the proliferation and metabolic activities of gut microbiota communities.

The large intestine holds the place of most populated in terms of microbiota and immune cells. As a result of this research, it is becoming increasingly clear that intestinal metabolism has a significant impact on human physiology. Together with a mucus layer, the vast intestinal layer of specialized epithelial cells joined by tight junction proteins serves as a barrier that separates the host's mucosa from the luminal environment. Enterocytes, which are in control of nutritional absorption, and goblet cells, which create, store, and exude mucin glycoproteins, are two key cells of the intestinal epithelium. The maximum density of goblet cells can be found here, which in turn leads to a wide range of microbiota and their subsequent conversions into products, and hence, these products lead to a large no. of consequences. SCFAs (short-chain fatty acids) are made mostly by the fermentation of non-digested carbohydrates. This fermentation produces not only the primary SCFAs of acetate, propionate, and butyrate but also lactate, a crucial intermediary in the synthesis of SCFA.

Pectin, a soluble dietary fiber with recognized modulatory effects on the gastrointestinal immunological barrier, is an essential dietary fiber. Because pectins have a positive influence on microbial communities, they may indirectly look after the intestinal barrier by increasing the growth and diversity of microbiota communities. The chemical structure of pectins has an immense impact on these actions. The well-studied prebiotics include inulin, oligofructose, and fructo-oligosaccharide (FOS). All β (2–1) linear fructans with varying degrees of polymerization are referred to as inulin. Inulin is not digestible by digestive enzymes in the small intestine due to the existence of β (2–1) bonds it enters the colon intact, where it is fermented to SCFA and gases by colonic bacteria. The well-studied prebiotics include inulin, oligofructose, and fructo-oligosaccharide (FOS). Moreover, commensal communities stimulated by dietary fiber are important taking into account, intestinal immunity. The microbial community secretes metabolites such as secondary bile acids and tryptophan, which together limit the growth of pathogens.

7. How does gut interaction affect health?

There is a substantial relationship between the intestinal barrier, the gut microbiota, and immune system cells. Increased epithelial permeability, often known as "leaky gut," permits bacteria, antigens, and toxins from the lumen to the lamina propria to enter the bloodstream, which triggers both local and systemic immune responses. The symbiotic relationship between the gut microbiota and the immune system may be disrupted by an impaired intestinal barrier function, which has been linked to the advancement of illnesses and disorders such as inflammatory bowel disease and irritability. Based on their potential effects, gut bacteria can be categorized

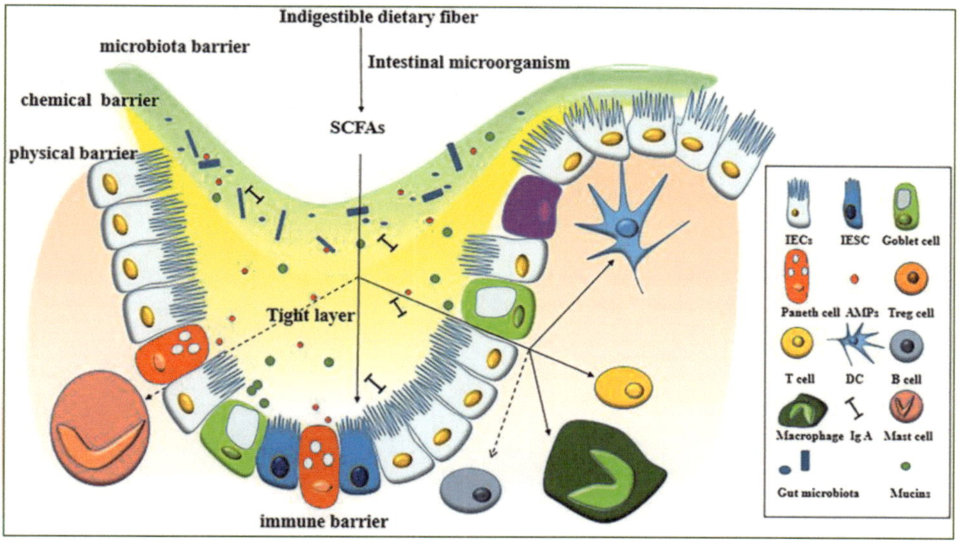

Figure 1.
Anatomical structure and composition of the gut barrier. In order of importance, there are four types of barriers: Microbiota barrier, chemical barrier, physical barrier, and immune barrier. Microorganisms, IgA, and antibacterial peptides make up the chemical barrier. It consists of IECs, goblet cells (synthesis of mucins), Paneth cells (synthesis of AMPs), and intestinal stem cells. There are T cells, B cells, macrophages, dendritic cells, and mast cells that form the immune barrier. In this image, the real arrow indicates the route by which SCFAs affect immune cells, whereas the dotted arrow indicates a possible route not described [17].

into three categories: Lactobacilli and bifidobacteria; potentially dangerous bacteria, such as some clostridia species and other commensal bacteria, such as Bacteroides, which can have both positive and negative characteristics (**Figure 1**) [18].

8. Getting enough fiber diet and how it affects immunity

The share of CD4+ and CD8+ T-cells in GALT, as well as their in vitro responsiveness to mitogens, were considerably affected by the diet's fiber intake. There was a bigger proportion of CD8+ T-cells in the IEL, lamina propria, and Peyer's patches after consuming the high fermentable fiber diet, as well as a higher proportion of CD4+ T-cells in the mesenteric lymph nodes and peripheral blood except for a higher CD4:CD8 ratio.

In the upper gastrointestinal system, prebiotic fiber is neither hydrolyzed nor absorbed, but instead assists as a selective substrate for one or a small number of beneficial colonic bacteria, modifying the gut microbiota. There is significant proof that prebiotic fibers (inulin and oligofructose) boost the percentage of good lactic acid bacteria in the human colon (**Figure 2**).

9. Dietary fiber strengthens the gut immune barrier

Pectin is a soluble dietary fiber with known modulatory effects on the gastrointestinal immunological barrier and is a noteworthy dietary fiber. Many fruits and vegetables, together with citrus fruits, apples, sugar beets, and potatoes, have had pectins separated from their primary and secondary cell walls. There are Linear 1,4-Dgalacturonan (homogalacturonan) segments and branching rhamnogalacturonan segments, which make up the majority. The degree of methyl-esterification, molecular weight, and neutral side chain topologies are all features that influence

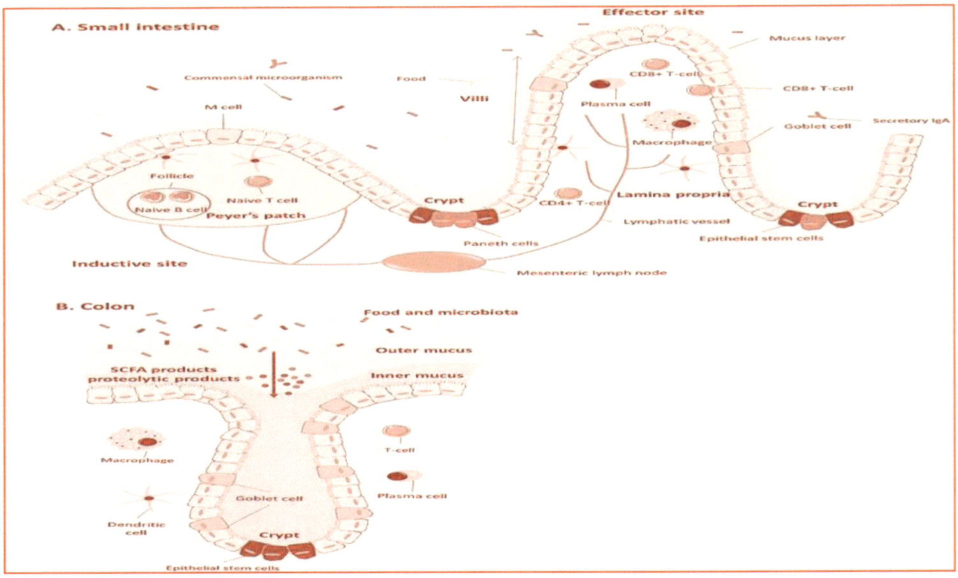

Figure 2.
The immune barrier of the small intestine consists of colonocytes and goblets [19].

pectin's functional capabilities [20]. The β-glucan and inulin metabolites also improved intestinal immunity.

10. Integrity of epithelium is preserved by SCFAs

SCFA is produced from fermented dietary fiber either through stimulation of gut microbiota which by pattern recognition receptors such as TLR2. There is no evidence that it enhances epithelial integrity under healthy conditions, but it does maintain epithelial integrity in disease states. By maintaining tight junction structures, it protects the epithelial integrity from agents that disrupt the barrier [21].

11. Conclusion

It has long been known that dietary fiber and its gut microbial metabolite like SCFAs improved metabolism of the host body. Dietary fiber via SCFA increases plasma SCFA levels and improves hepatic metabolic health. Dietary fiber intake produces SCFAs via fermentation in the gut microbiota, mainly in colon L-cells, which produce GLP-1 and PYY located mainly in the distal ileum and colon. Fiber intake suppressed the HFD-induced liver weight gain and hepatic TG accumulation along with a change in hepatic lipid metabolism, while dietary SCFA intake improved hepatic metabolic conditions by activating FFAR3. A shift in gut microbiome production of butanoate accompanied by up-regulation of microbiota and AMP-activated protein kinase (AMPK)-dependent gene expression contributes to intestinal integrity and homeostasis by influencing metabolism and transporter expression.

Conflict of interest

The authors declare no competing interest.

Notes/thanks/other declarations

We would like to acknowledge the Director, ICAR-National Dairy Research Institute, Karnal-132,001, Haryana, India.

Author details

Jitendra Kumar[1*†], Priya Sharma[1†], Murli Dhar Mitra[2], Sonia Sangwan[1] and Haribrahma Singh[3]

1 ICAR-National Dairy Research Institute, Karnal, Haryana, India

2 Department of Chemistry, Indian Institute of Technology (IIT), Banaras Hindu University (BHU), Varanasi, Uttar Pradesh, India

3 Department of Chemistry, Indian Institute of Technology (IIT), Hauz Khas South Delhi, Delhi, India

*Address all correspondence to: jitoperon@gmail.com

† These authors contributed equally.

References

[1] Rescigno M. Intestinal microbiota and its effects on the immune system. Cellular Microbiology. 2014;**16**:1004-1013

[2] Backhed F, Ding H, Wang T, Hooper LV, Koh GY, Nagy A, et al. The gut microbiota as an environmental factor that regulates fat storage. Proceedings of the National Academy of Sciences of the United States of America. 2004;**101**:15718-15723

[3] Schroeder BO, Backhed F. Signals from the gut microbiota to distant organs in physiology and disease. Nature Medicine. 2016;**22**:1079-1089

[4] Sonnenburg ED, Smits SA, Tikhonov M, Higginbottom SK, Wingreen NS, Sonnenburg JL. Diet-induced extinctions in the gut microbiota compound over generations. Nature. 2016;**529**:212-215

[5] Sender R, Fuchs S, Milo R. Are we really vastly outnumbered? Revisiting the ratio of bacterial to host cells in humans. Cell. 2016;**164**(3):337-340

[6] Gentile CL, Weir TL. The gut microbiota at the intersection of diet and human health. Science. 2018;**362**(6416):776-780

[7] Tazoe H, Otomo Y, Kaji I, Tanaka R, Karaki SI, Kuwahara AJJPP. Roles of short-chain fatty acids receptors, GPR41 and GPR43 on colonic functions. Journal of Physiology and Pharmacology. 2008;**59**(Suppl 2):251-262

[8] Louis P, Duncan SH, McCrae SI, Millar J, Jackson MS, Flint HJ. Restricted distribution of the butyrate kinase pathway among butyrate-producing bacteria from the human colon. Journal of bacteriology. 2004;**186**(7):2099-2106

[9] Kendrick SF, O'Boyle G, Mann J, Zeybel M, Palmer J, Jones DE, et al. Acetate, the key modulator of inflammatory responses in acute alcoholic hepatitis. Hepatology. 2010;**51**(6):1988-1997

[10] Kumar J, Rani K, Datt C. Molecular link between dietary fibre, gut microbiota and health. Molecular Biology Reports. 2020;**4**:1-9

[11] Schley PD, Field CJ. The immune-enhancing effects of dietary fibers and prebiotics. British Journal of Nutrition. 2002;**87**(S2):S221-S230

[12] Field CJ, McBurney MI, Massimino S, Hayek MG, Sunvold GD. The fermentable fiber content of the diet alters the function and composition of canine gut-associated lymphoid tissue. Veterinary Immunology and Immunopathology. 1999;**72**(3-4): 325-341

[13] Gibson GR, Beatty ER, Wang XIN, Cummings JH. Selective stimulation of bifidobacteria in the human colon by oligofructose and inulin. Gastroenterology. 1995;**108**(4):975-982

[14] Berg RD. Indigenous intestinal microflora and the host immune response. EOS, Journal of Immunology and Immunopharmacology. 1985;**5**(4):161-168

[15] Marin ML, Lee JH, Murtha J, Ustunol Z, Pestka JJ. Differential cytokine production in clonal macrophage and T-cell lines cultured with bifidobacteria. Journal of Dairy Science. 1997;**80**(11):2713-2720

[16] Jenkins DJ, Kendall CW, Vuksan V. Inulin, oligofructose, and intestinal

function. The Journal of Nutrition. 1999;**129**(7):1431S-1433S

[17] Liu W, Luo X, Tang J, Mo Q, Zhong H, Zhang H, et al. A bridge for short-chain fatty acids to affect inflammatory bowel disease, type 1 diabetes, and non-alcoholic fatty liver disease positively: By changing gut barrier. European Journal of Nutrition. 2021;**60**(5):2317-2330

[18] Roberfroid MB, Van Loo JA, Gibson GR. The bifidogenic nature of chicory inulin and its hydrolysis products. The Journal of Nutrition. 1998;**128**(1):11-19

[19] Cheng L, Akkerman R, Kong C, Walvoort MT, de Vos P. More than sugar in the milk: Human milk oligosaccharides as essential bioactive molecules in breast milk and current insight in beneficial effects. Critical Reviews in Food Science and Nutrition. 2021;**61**(7):1184-1200

[20] Voragen AG, Coenen GJ, Verhoef RP, Schols HA. Pectin, a versatile polysaccharide present in plant cell walls. Structural Chemistry. 2009;**20**(2):263-275

[21] Cario E, Gerken G, Podolsky DK. Toll-like receptor 2 controls mucosal inflammation by regulating epithelial barrier function. Gastroenterology. 2007;**132**(4):1359-1374

Chapter 12

Autoimmune Diseases of the GI Tract Part I: Etiology and Pathophysiology

Mahmoud Khatib A.A. Al-Ruweidi, Nada Khater, Haya Rashid Alkaabi, Maram Hasan and Huseyin C. Yalcin

Abstract

Autoimmune diseases have emerged as a pandemic in our modern societies, especially after World War II. There are currently more than 80 autoimmune diseases that compromise the lives of millions of patients around the world. There is a variety of factors that are involved in the pathogenesis of autoimmune diseases that vary from environmental factors to genetic susceptibility. The GI tract is one of the most susceptible sub-systems in human bodies for autoimmune organ-specific diseases. There are five autoimmune GI tract diseases that are most common. This review consists of two chapters. In part I, we shed the light on introducing the concept of autoimmunity, the description of the disease's pathogenesis and the diagnosis, the link between the gut and brain through what is known as the gut-brain axis, and the relationship of this axis in GI autoimmune diseases. In part II, we will shed light on the role of antibodies as markers for the prediction of the disease, artificial intelligence in GI autoimmune diseases, the nutritional role and implications in the five GI autoimmune diseases, and finally the treatment of those diseases.

Keywords: achalasia, atrophic autoimmune gastritis, celiac disease, eosinophilic esophagitis, inflammatory bowel diseases, Crohn disease, ulcerative colitis, immunological continuum, epithelial barrier dysfunction, gut-brain axis

1. Introduction

Autoimmune diseases (ADs) can be classified as the inability of the human systems to distinguish their own bodies from foreign bodies [1, 2]. There have been more than 80 autoimmune diseases reported to date [3]. The immune system remains one of the most poorly understood systems in the human body. The COVID-19 pandemic has re-shed light on the immune system once again [4, 5]. ADs can be triggered in humans due to multiple factors such as environmental factors and genetic predisposition factors. The pathogenesis of the diseases can be hugely variable but the involvement of T and B lymphocytes from the adaptive immunity remains a hallmark for this umbrella of disease [6]. The increase in the detection and classification of ADs can be owed to the development of serological tools to detect antibodies [7]. Autoimmune diseases can be classified as systematic and organ-specific diseases.

The Gastrointestinal Tract (GI tract) is a part of the digestive system in humans, and it is composed of six components: mouth, esophagus, stomach, small intestine, large intestine, and anus. The GI tract is prone to diseases, and it is affected by multiple factors in the pathogenesis of the disease and multiple manifestations of other systematic diseases appear in the GI tract. There are five common autoimmune diseases in the GI tract: (1) achalasia, (2) atrophic autoimmune gastritis (AAG), (3) celiac disease (CD), (4) eosinophilic esophagitis (EoE), and (5) inflammatory bowel diseases (IBD) which includes Crohn disease and ulcerative colitis (UC). The manifestation of other autoimmune diseases in the GI tract could be due to: (1) systematic mastocytosis, (2) systematic sclerosis and CREST syndrome, (3) autoimmune enteropathy, (4) autoimmune hepatitis, (5) autoimmune pancretitis, (6) mixed connective tissues disease, (7) primary sclerosing cholangitis and autoimmune sclerosing cholangitis, and (8) systemic lupus erythematosus (SLE). The review of the manifestation in the GI tract is beyond the scope of this chapter. This chapter covers the definition and etiology of autoimmune diseases, the relationship between autoinflammation and autoimmunity, an overview of the five diseases, the common antibodies that are used as a predictor factor for the disease, the role of gut-brain axis, and the psychological link in the GI tract autoimmune diseases, the role of nutrition in GI autoimmune diseases, and the treatment available for the diseases.

2. Definition, pathophysiology, and etiology of autoimmune diseases

ADs are a cluster of heterogenous pathological events with an increasing number of registered cases worldwide and a prevalence of around 10% in the western populations, according to thorough epidemiological studies [8]. The hallmark of AD is the tissue injury and consequent malfunction resulting from a system or organ-specific inflammatory reaction due to the failure of self-antigen tolerance [9, 10]. Shaping the classification of AD advanced and these diseases were re-defined over the years: autoimmune inflammatory diseases used to be typically divided into autoimmune diseases and autoinflammatory diseases [11]; this separation is based on the involvement of either the innate or the adaptive immune systems and the detection of increased titer of autoantibodies [11]. So according to this definition set, ADs are distinguished by the involvement of the adaptive immune system represented by T and B lymphocytes and the presence of autoantibodies [11]. Over a decade ago, McGonagle and McDermott suggested the "continuum model" in immunology, in which a spectrum of diseases is established including the autoinflammatory and AD in both extremities of the created spectrum [12]. Moreover, the authors suggested criteria to set the boundaries for diseases to be considered clinically autoinflammatory or autoimmune, based on the genetic mutations that occur in each type of disease [12]. Thus, margins of ADs are set by mutations linked to monogenic autoimmune diseases which exhibit predisposed to the adaptive immune system and the existence of autoantibodies. On the contrary, the margins are set for autoinflammatory diseases by mutations in elements that take place in the innate immune system mutations specifically in tissues that are prone to pathological events onset, where no evidence of the involvement of autoimmune mechanisms [12–14]. In the last few years, the concept of autoinflammation and autoimmune diseases kept being refined as several monogenic and polygenic common and novel disorders have been recognized, feeding into the updating of knowledge of the pathophysiology of autoinflammatory and ADs [15, 16]. Thorough investigations associated with convincing pathophysiological hypotheses in model diseases from any extremity of the continuum reveal the intimate relationship between the mechanisms of innate and adaptive systems [17]. Advanced modern approaches including molecular imaging technologies, genome-wide association studies, and the characterization of tissue-associated factors in some diseases supported the idea of the interplay between innate and adaptive immune mechanisms in specific ADs [9, 13]. Eventually, these techniques also helped in the verification of the continuum model that was suggested by McGonagle and co-workers.

3. Inflammation to autoinflammation and epithelial barrier dysfunction: a brief look into the developmental stage

The most challenging aspect in immunological diseases such as autoinflammatory and autoimmunity diseases is to identify the early events that trigger immune dysregulation [18]. Autoinflammatory and autoimmunity are closely correlated and are sometimes confused by mistake. Although there are similarities and perhaps a continuum between them, but they nevertheless do not refer to the same thing. The biggest distinction line that can be drawn between autoinflammatory and autoimmune disease is that the autoinflammatory diseases is that autoinflammatory was referred to the dysregulations related to adaptive immunity while autoimmunity was defined due to the dysregulation in innate immunity [9, 19]. This definition is not entirely accurate and a little bit outdated as we will preview in this section how the innate and adaptive immunity are involved in both autoinflammation and autoimmunity and what are the links between them. The pathological nature of the two process that are self-destructive and systematic that include monogenic and polygenic diseases [20]. A chronic activation of the immune system happens in both processes that lead to tissue inflammation or damage.

3.1 The role of innate and adaptive immunity

Immunologic defenses in vertebrates consist of two immunologic subsystems—innate and adaptive immunities. Innate is the natural immunity by birth while the adaptive immunity is the acquired immunity [21].

3.1.1 Innate immunity

The innate immune system constitutes the first line of defense in an individual's immune system. As a result, it detects pathogens as well as other harmful triggers that may cause inflammation and trigger adaptive immunity [22]. Among the effector cells of innate immunity are macrophages, dendritic cells, and antigen-presenting cells (APCs) [23]. Innate immunity identifies and recognizes the molecular patterns expressed by pathogens (pathogen associated molecular patterns, PAMPs) or by damaged cells (damage associated molecular patterns, DAMPs) [24]. There are three types of pattern recognition receptors (PRR): TLRs (Toll-like receptors), RLRs (RIG-I-like receptors), and NLRs (nucleotide-binding oligomerization domain-like receptors) [25]. It is thought that upon recognition of foreign molecules, intracellular signal transduction pathways are induced, which induces the expression of interferon alpha (IFN) sequences, IFN-α sequences, TNF sequences, and interleukin 1 (IL-1) sequences. Both autoinflammatory and autoimmune diseases can be caused by the dysregulation of these receptors, which mostly involves excessive or prolonged activation [11, 26]. The activation of the inflammasome is crucial to host defense against pathogens. Several inflammasomes are implicated in the immunological process of diseases, including NALP1 and NALP3, or cryopyrin inflammasomes. By activating pro-caspase-1, the inflammasome mediates the conversion of pro-IL-1ß and IL-18 into the active forms. Genetic mutations in either pyrin, cryopyrin, or TNF receptor super-family genes have been associated with autoinflammatory diseases [27]. It is unclear how inflammasomes contribute to autoimmunity. Nevertheless, its role is still yet to be discovered as NLRI and IL-1ß as a primary suspect to look at. The upregulation of IL-2 receptor that leads to B cell proliferation and enhanced antibody production is caused by the crucial role of IL-1ß, which affects both B and T cells, thereby prolonging T cell survival [28]. Furthermore, they drive differentiation of the Th17 cells as well. Therefore, IL-1ß stimulates T and B cells and may play a crucial role in linking the NLR activation with adaptive immunity response [20].

3.1.2 Adaptive immunity

For adaptive immunity to mature, it requires between three and five days. B cells, T cells, and cytotoxic T cells are involved in adaptive immunity [19]. Antigens are recognized by specific antigen receptors, primarily B and T cell receptors (BCR and TCR), which are highly specific. As such, innate immunity provides a first line of defense against damage and infection. Adaptive immunity, however, provides a more effective but slower resistance.

A significant role is played by adaptive immunity in the development and maintenance of autoimmune diseases. Despite this, different mechanisms contribute to the disease by the innate immune system. The autoimmune process is divided into two phases: During the initiation phase (phase 1), self-nucleic acids released by apoptotic cells are recognized and internalized by dendritic cells (DC) through TLR, causing IFN-α, production. The IFN- α stimulates the maturation of dendritic cells, the presentation of autoantigens, the recruitment of B and T cells, and the production of autoantibodies. After entering a second phase, (self-sustaining amplification) plasmocytoid dendritic cells internalize autoantibodies and nucleic acids through Fcγ-receptors (FcγRs). DC and T cells are stimulated and activated by IFN- α, resulting in self-perpetuation of antibody production and inflammation [6, 29].

3.2 The immunological disease continuum

IL-1 1β and type I interferon (IFN) are also polarized cytokines that are related to innate immunity, with IFN being more associated with autoimmune diseases, specifically Systemic Lupus Erythematosus (SLE), while IL-1β is associated with pure innate immunity. The importance of recognizing Type I IFN dysregulation driving autoimmunity, as well as NLR dysfunction driving classical autoinflammatory diseases without autoantibody formation, has led to a polarization in the classification of immune diseases. From the original recognition of autoinflammatory diseases being linked to NLR cytoplasmic resident innate immune receptors (NLRP3 in particular). There have been several reports linking innate immune-mediated pathologies to inflammasomes, including NLRC4. It is remarkable that NLR family members show consistent association with both monogenic and polygenic autoinflammatory disease, whereas TLRs do not, possibly attributable to functional redundancy in TLRs.

3.3 The epithelial barrier dysfunction: leaky gut as a third element of pathogenesis

The epithelium of the gastrointestinal tract or commonly known as the epithelial barrier is the largest mucosal lining that forms an interface between a mammalian host and the external environment [30]. Protecting the body from pathogens and foreign substances is the primary function of the epithelial barrier [31]. It is through the anatomical structure of the GI that processes such as digestion, absorption, and neuroendocrine network, as well as immune function balance, take place. Trillions of microbial inhabitants inhabit the lumen of the gut. These microbes play an important role in digestion and modulate the immune system [32]. The regulation of molecular trafficking between the intestinal lumen and the submucosa via the paracellular space maintains the capability of the intestinal permeability which interacts continuously with various bodies such as foodborne pathogens and antigens. Paracellular space is estimated to measure between 10 and 15 Å. Physiologically, solutes with a molecular radius of over 15 Å (~3.5 kDa) are not susceptible to uptake through this pathway [33]. The intestinal permeability is the property that allows solutes and fluids to pass between the lumen and tissues. Additionally, intestinal barrier function is determined by how well mucus and other extracellular components, such as mucus, prevent this exchange [32].

Transfers of macromolecules are largely affected by the paracellular permeability of epithelium, which is influenced by the intercellular tight junctions (TJs) [34]. The TJs are highly dynamic structures that serve a variety of functions both physiologically and pathologically in the intestinal epithelium [35]. The Zonulin protein appears to modulate intercellular TJs, and it has been shown that Zonulin expression is elevated in conditions associated with dysfunction of TJs, such as celiac disease [36–40]. An impaired epithelial barrier is associated with a wide range of chronic diseases, including allergies, autoimmune diseases, and metabolic disorders [41]. Changes in the permeability of the GI tract's epithelial lining facilitate a passage for commensal bacteria and their products from the lumen into the bloodstream creating what is known as a "leaky gut". There has been a growing interest over the past decade in the role of leaky gut's association with autoimmune diseases. There have been some suggestions that the leakage of pathogens into the body system results in autoimmunity making the leaky gut a third source of pathogenesis besides environmental triggers and genetic predisposition [42]. A dysbiosis which is a perturbation of the structural dynamics of the microbial community in the intestinal tract causes leaky gut condition and it is closely entangled with autoimmune diseases. As discussed here, the microbiota and particularly the intestinal microbes are important in the immune system and their disturbance can be associated with autoimmune diseases.

There are various immune cells such as T and B cells as well as macrophages and dendritic cells which are found beneath the layer of lamina propria of the intestinal epithelium. These cells are crucial for the maintenance of hemostasis in the intestinal epithelium. Epithelial cells suppress inflammation by generating regulatory dendritic cells, regulatory T and B cells, as well as anti-inflammatory cytokines [43]. In the event of a leaky gut and damage to the epithelial barrier, some pathogens such as *Staphylococcus aureus* may colonize areas such as leaky barrier areas [41]. In turn, dysbiotic microbiota moves to the interepithelial and subepithelial spaces, activating a local or systemic inflammatory response suspected to contribute to many immune-mediated diseases. There then follows a series of events that lead to chronic periepithelial inflammation with leaky epithelial barriers. It is not understood that the autoimmune response occurs before the epithelial barrier insult or post the insult. The causes of the epithelial barrier's insults could be variable and include but are not limited to genetic predisposition such as filaggrins and TJ polymorphisms, environmental factors such as microplastics and food emulsifiers [44, 45], allergens such as house dust [46], microbiota's flora, surfactants, and dietary factors [47]. For a detailed review, we refer the reader to [41].

The intestinal commensal is exposed to the host's immune system in various organs due to epithelial intestinal barrier leakage and autoimmune diseases. It has been observed that few of the GI intestinal epithelial cells (IECs) are essential for maintaining intestinal homeostasis and in the function of the intestinal epithelium, as well as participating in IBD pathogenesis [48]. There is collective evidence about the role of the epithelial barrier in EoE. It is reported that EoE–linked calpain 14 is an IL-13–induced protease that mediates esophageal epithelial barrier impairment [49]. There is also a reported role of TGF-β1 in the alterial esophageal epithelial barrier function by attenuation of claudin-7 in EoE [50]. The role of epithelial barrier dysfunction is well established in EoE and we refer the reader for in-depth scope review [51]. In ADs, there is an association between leaky gut and the development of AD. For instance, in CD, an increase in the number of apoptotic IEC in the peritoneal mucosa is reported as well as impaired epithelial barrier function [52]. It is reported that epithelial barrier is dysfunctional through TJs defects [52]. This is a growing area of research and by shedding the light more on the relationship between epithelial barrier dysfunction and what is known as leaky gut syndrome, the association between it and GI autoimmune diseases if confirmed can provide a therapeutic route in the treatment and prevention of GI autoimmune diseases.

4. The 5 common GI autoimmune diseases

4.1 Achalasia

Achalasia is a rare autoimmune motility disorder that is caused by the degeneration of the myenteric neuronal esophageal plexus that consequently results in an aperistalsis and impaired incomplete relaxation of the lower esophageal sphincter (LES) and ineffective contractions in the esophageal body [53]. In the distal esophagus and the lower esophageal sphincter, achalasia is characterized by a functional loss of myenteric plexus ganglion cells or chronic ganglionitis [54]. Since there is no known cause for the initial loss of inhibitory neurons in individuals suffering from achalasia, it could be considered an idiopathic disorder [55]. Nevertheless, the onset of neuronal degeneration may be caused by an indolent viral infection such as herpes simplex virus 1 (HSV-1), measles, and human papillomavirus have been proposed as potential antigens. Evidence indicates that HSV-1 DNA has been detected in esophageal tissue, and that isolated T cells from achalasia are monoclonal in nature and that they proliferate and release cytokines upon exposure to HSV-1 antigens [56, 57]. It is possible that this is since HSV-1 is a neurotropic virus with a predilection for squamous epithelium, which causes selection loss of enteric neurons in the esophagus. Nevertheless, this theory is not entirely accurate, as HSV-1 DNA was also frequently detected in control individuals' esophagus [58]. Thus, it might be argued that HSV-1 only triggers persistent immune activation and subsequent loss of enteric neurons in individuals with genetically suspected hosts [59]. In patients who have an immunogenetic variation, viral infection may trigger a disordered immune reaction. Achalasia may also be caused by muscular eosinophilia in some cases. It has been demonstrated that such inflammatory processes decrease, gradually destroy, or eventually eliminate the esophageal myenteric plexus (MP) [60]. It has been found that achalasia is associated with several genes and immunological markers including Interleukin-10 promoter polymorphism [61] and Interleukin 23 receptor [62], HLA class II gene polymorphisms [63], KIT (KIT proto-oncogene, receptor tyrosine kinase) [64, 65], and vasoactive intestinal peptide receptor 1 [66, 67], among others.

Achalasia is reported to have an annual incidence of 1 per 100,000 individuals worldwide [68]. There is an equal frequency of achalasia in men and women when they are adults [69] and among different ethnicities [70]. Other autoimmune diseases are prevalent in achalasia patients such as diabetes. Progressive dysphagia to both solids and liquids is the hallmark symptom associated with a diagnosis of achalasia [71]. In addition, regurgitation of undigested food, respiratory symptoms such as nocturnal coughs, recurrent respiratory infections, pneumonia, chest pains, and loss of weight may occur [58, 72, 73]. According to conventional manometry, the characteristics of achalasia are as follows: (1) absence of peristalsis, sometimes with increased intra-esophageal pressure associated with the stasis of food and saliva, (2) The LOS remains partially relaxed on deglutition (residual pressure > 10 mm Hg), and (3) the LOS often exhibits a raised resting tone.

4.2 Atrophic autoimmune gastritis (AAG)

Atrophic autoimmune gastritis (AAG) is an immune-mediated disorder characterized by nonspecific symptoms [74–76]. A diagnosis of AAG is confirmed by the presence of circulating antibodies against the adenosine triphosphate enzyme H/K (parietal cell antibodies, PCA); the same antibodies are also found against anti intrinsic factor (anti-IF) [77]. In AAG, the native gastric glands within the mucosa gradually disappear or shrink over time [78]. Consequently, mucosal atrophy occurs sparing the antrum and extensive pseudopyloric or intestinal metaplasia occurs [79]. There may be involvement of both the antrum and corpus, but the corpus only has apparent functional and clinical consequences [80]. Multiple modifications may precede atrophy, including focal atrophy, lymphoplasmacytic infiltrate in the lamina propria, parietal cells pseudohypertrophy, and enterochromaffin-like (ECL) cell hyperplasia. It has long been recognized that AAG, as well as other autoimmune disorders, tend to cluster in families, which could reinforce the genetic component of disease. Through using mouse models, it has been possible to discover AAG susceptibility genes (Gasa 1, 2, 3, and 4) on chromosomes 4 and 6 and H2 region, three of which are located on the same locus as non-obese diabetic mouse diabetes mellitus susceptibility genes [12, 13]. The prevalence of autoimmune atrophic gastritis is relatively low. It may be attributed at least in part to the underdiagnosis of *Helicobacter pylori*-induced gastritis in many cases, and the absence of clinical manifestations in the early stages of the disease [81]. The incidence of AAG is three times higher in women than in men [82]. There is an age-dependent increase in the prevalence of AAG of 2% [83]. AAG occurs in 25 out of every 100,000 people each year. Patients with AAG have 3–5 higher risks of developing other autoimmune diseases, such as oral erosive lichen [84], myasthenia gravis [85], vitiligo [86], diabetes mellitus (DM) [87], autoimmune thyroid disease [88], and Addison's disease [89]. Patients are usually diagnosed in advanced stages when the disease is irreversible or threatening symptoms have occurred, including abnormalities such as pernicious anemia, and neurological or gastric oncological complications [90–92]. The symptoms of AAG appear slowly and may remain asymptomatic for a long period of time. Symptoms of the disease range from mild weakness to severe psychological manifestations such as paranoia (megaloblastic madness). Pernicious anemia is the main clinical manifestation of AAG. A common symptom of iron deficiency is fatigue, restless legs, brittle nails, hair loss, impaired immune function, and poor wound healing. Iron deficiency is independent of and precedes anemia. Shortness of breath, dizziness, tachycardia, and lightheadedness are some of the symptoms of anemia (regardless of the cause) [93]. The presence of AAG can be asymptomatic or cause symptoms, depending on the level of atrophy that affects the absorption of vitamin B12 or other substances, such as folate and iron. Deficiency in vitamin D can develop over a long period of time, and patients may not show symptoms until reserves are exhausted. The diagnosis of AAG can be done through serological tests, endoscopy, and histopathology biopsy. Antibodies that are used for serological tests such as APCA, anti-ID antibodies, and anti-*H. pylori* antibodies (anti-HP-IgM and anti-HP-IgG). AAG patients who have oxyntic gland atrophy often have elevated levels of gastrin (including Gastrin-17) and it is measured in many cases to confirm the diagnosis. Endoscopy has been often used in the diagnosis, although it has many limitations such as low sensitivity and specificity. There is, however, a golden rule when it comes to diagnosing AAG through endoscopy, which is the absence of normal capillaries resembling honeycombs and collecting venules in regular shape and appearance. Biopsy histology is the most reliable method. Before oxyntic mucosa is lost completely, AAG appears as a series of features: (1) infiltrated lymphocytes and plasma cells in lamina propria, (2) focal atrophy of oxyntic mucosa along with SPEM or IM, (3) pseudohypertrophy of parietal cells and (4) hyperplasia of the ECL [93].

4.3 Celiac disease

Celiac disease (CD) is a multisystem disorder characterized by enteropathy [94]. Genetically predisposed individuals develop CD when the immune system reacts inappropriately to a T cell-mediated immune response [95]. Almost any organ system can be affected by celiac disease, approximately half to two-thirds of patients suffer from extra-intestinal symptoms; some studies claim that they may be more common than gastrointestinal symptoms [96]. CD patients can be classified into two categories symptomatic and asymptomatic. Asymptomatic CD patients are those who at the time of their initial diagnosis of CD do not exhibit any symptoms even if they are directly questioned about their condition. The term symptomatic CD refers to those individuals who demonstrate clinically visible gastrointestinal and/or extraintestinal symptoms related to gluten consumption [97, 98]. Symptomatic celiac disease can be further divided into classical and nonclassical celiac disease. Some genes have been involved in CD. It is often considered that CD can be viewed as a polygenic disorder that involves both major histocompatibility complex MHC (human leukocyte antigen [HLA]) and non-MHC genes [99]. Currently, it is well-established that six MHC and 39 non-MHC loci, as well as several independent genetic variants, contribute to disease risk. The genetic variants are responsible for roughly 31% of CD heritability, and the MHC is responsible for 25% [100]. In CD, HLA-DQ2 and -DQ8 are key genetic markers, and an autoantigen is involved (tissue transglutaminase 2: tTG2). Approximately 25–35% of the general population has HLA-DQ2/DQ8 with only 3% of these individuals developing CD [101]. Globally, CD affects between 0.6% and 1% of the population [102]. CD affects both children and adults. The mean age at the diagnosis is 38, but 20% of the patients are diagnosed over the age of 60 [103]. Women however are diagnosed at an earlier age and present more often with constipation, bloating, and anemia of iron deficiency than men [104, 105]. Gluten is the main etiology of CD. Gluten is a mixture of proteins found in grains of wheat (including gliadins and glutenins). CD can be caused by the presence of proteins from barley (hordeins) and rye (secalins). Among these, the gliadin peptides are the most immunogenic for CD [106]. Any case with malabsorption is defined as a classical disease and all other cases as nonclassical. Neoclassic CD manifests with largely extraintestinal symptoms, often monosymptomatic (e.g. iron deficiency anemia, premature metabolic bone disease, infertility, elevated transaminase levels) in the absence of clinical malabsorption. Over time, diarrhea has become less common at presentation, but it remains the most common gastrointestinal symptom [104]. Potential CD is a clinical term to describe suspected CD patients. Potential CD is characterized by normal small intestinal mucosa with positive CD serologic findings [107]. The diagnosis of CD remains challenging as it is estimated that currently only 20% of patients who have CD have been diagnosed [108]. CD cannot be diagnosed with one tool only. There is always a need for a combination of clinical features, serology, and histology are needed together to confirm the diagnosis [109]. In serological tests, patients should be on gluten-containing diets. Positivity in tests for Serum immunoglobulin A (IgA) anti-tissue transglutaminase antibody (anti-tTG-IgA) is widely accepted for the diagnosis but has low specificity. Serum immunoglobulin A (IgA) anti-tissue transglutaminase antibody (anti-tTG-IgA) are 100% specific but less sensitive [110–116]. Deamidated gliadin peptide (DGP) antibodies of the IgG class are advantageous for younger children [117]. All patients with suspected CD should undergo a duodenal biopsy. Regardless of CD serology results, duodenal biopsies should be performed in high-risk symptomatic patients [118]. There is a four out of five rule that is common in the diagnosis of CD. According to this rule, four of the following criteria are sufficient to establish CD diagnosis: (1) apparent and typical signs and symptoms of diarrhea and malabsorption, (2) positive serological tests of antibodies, (3) a patient with HLA-DQ2 or HLA-DQ8 positivity, (4) damage to the intestines, such as villous atrophy and lesions and (5) the response of the patient to GFD. This rule is important in the diagnosis of the diseases as many CD subtypes can be classified naming the non-classical CD which has no malabsorption or diarrhea, seronegative CD patients who do not show responses to serological antibodies, and a potential CD which has no damage to the intestines, and non-responsive CD who show no responses to GFD [109].

4.4 Eosinophilic esophagitis

Eosinophilic esophagitis (EoE) is an immune-mediated condition in which eosinophils infiltrate into the esophageal mucosa and lead to symptoms of esophageal dysfunction [119]. In the absence of secondary causes, the disease is considered to belong to the spectrum of eosinophilic gastrointestinal disorders [120]. In the absence of treatment, EoE can lead to esophageal fibrosis, the formation of strictures, and esophageal narrowing leading to esophageal dysfunction [119, 121]. Throughout the world, the health care systems are burdened by EoE, a major factor in upper gastrointestinal morbidity [122, 123]. The US healthcare system is estimated to spend $350 to $947 million burden annually on EoE [122]. It has been found that the EoE disease prevalence has been associated with Single Nucleotide Polymorphisms (SNP) in the Thymic stromal lymphopoietin (TSLP) and TSLP-Rwhich is correlated with increase in the TSLP levels [124]. There are several environmental allergens implicated. One of these allergens is food. Food allergens trigger EoE and the disease can be put into remission by removal of specific foods, either via elimination diets or hypoallergenic elemental formulas [125–127]. It is commonly accepted that EoE, is due to a Th2 inflammation driven by TSLP secreted by esophageal epithelial cells and is under the influence of genetic predisposition [124, 128–130]. EoE Th2 inflammation with a non-IgE-mediated trigger has been found to be triggered by certain foods [131]. It was reported that food that causes vomiting and abdominal pain is soy, wheat, egg, and milk [132, 133]. An elimination diet known as the six-food elimination diet (SFED) refers to the removal from the diet of EoE patients of wheat, milk, eggs, nuts, soy, fish, and shellfish that are considered to be allergens [134]. Th2 cytokines result in an increased Th2 response from T cells, basophils, Invariant natural killer T iNKTs, and mast cells in EoE. Th2 cytokines also enhance eosinophil survival and activation, thus resulting in fibrotic modification [135]. It has been reported that IL-4 enhances eotaxin-3 secretion by epithelial cells, which is responsible for the increased migration of eosinophils. IL-4 also causes fibroblasts to release periostin, collagen, and B-actin, promoting local fibrosis [136]. Eosinophils are mainly differentiated, recruited, and survived by cytokine IL-5 [137]. The cytokine TSLP is primarily produced by epithelial cells at barrier surfaces such as skin, gut, and lungs because of danger signals, infectious agents, cytokines produced by atopic cytokines (IL-4, IL-13, TNFα), and environmental allergens [138]. The Th2 inflammation observed in EoE is most likely caused by TSLP. EoE prevalence estimates vary with location. The highest incidence occurs in western countries where EoE is more easily diagnosed and has an estimated prevalence of 56 per 100,000 people [139] in some statistics. However, several estimates place the prevalence of EoE at between 0.5 and 1 case per 1000 individuals, yet the disease is detected in between 2.4% and 6.6% of patients undergoing endoscopy for any reason [140–144]. The primary symptoms of EoE in adolescents and adults is dysphagia, which affects 60–100% of patients, food impaction can affect more than 25%, and 30–60% of patients report heartburn and 44% report noncardiac chest pain [145–150]. Diagnostic criteria must include both clinical and histological features: symptoms of esophageal dysfunction, the presence of at least 15 eosinophils in a high-power field, and exclusion of alternative causes of eosinophilia in the esophagus [119, 151].

4.5 Inflammatory bowel diseases

Inflammatory bowel diseases (IBDs) refer to both ulcerative colitis (UC) and Crohn's disease, as well as other non-infectious inflammations of the bowel that are symptomatic of relapsing chronic disorders of the bowel [152]. There has been an increasing incidence of inflammatory bowel disease (IBD) globally. It has been commonly agreed that genetic susceptibility, external environment, microbial flora of the intestine, and immune responses are all components of IBD pathogenesis [153, 154]. Globally, IBD affects 4.2 million people, including 1.5 million Americans, and 2.2 million Europeans [154, 155]. First-degree relatives are five times more likely to develop IBD than those without IBD. There is a possibility that some genes are shared by both diseases since Crohn's Disease and UC can occur within the same family. In both diseases, environmental factor leads to triggering events [156]. In recent studies, 163 IBD-associated gene loci have been identified, of which 110 are associated with both diseases, 30 with Crohn and 23 with UC. Genetic analyses have revealed the essential role of autophagy in immune responses to IBD and identified two autophagy-related genes, ATG16L1 and IRGM [157–159]. Autophagy plays an important role in intracellular homeostasis, working to degrade cytosolic contents and organelles and resist infection, and eliminate microbes inside the cell. The coding mutation T300A is associated with an increased risk of Crohn since ATG16L1 is essential for all forms of autophagy [160]. Recent studies have demonstrated a link between IBD and IL23R (a coder for pro-inflammatory IL23 cytokine) [161]. IL23 is involved in Th17 cells [162]. IBD is well established to be caused by the Th17 and IL-23 pathway, with susceptibility loci for UC and Crohn, identified in IL23R, IL12B, JAK2, and STAT3 [163, 164]. The incidence and prevalence of IBD are increasing worldwide but are highest in westernized areas. In Europe, the highest prevalence values have been reported (ulcerative colitis 505 per 100,000 in Norway; Crohn's disease 322 per 100,000 in Germany). Several countries in Europe and North America have a prevalence of inflammatory bowel disease that exceeds 0.3%. Caucasians are prone to IBD more than Africans and Asians. Crohn is slightly more likely than UC to have a family history of the disorder, although both disorders are polygenic [165, 166]. The peak incidence for UC and Crohn is in the second to fourth decade, and no significance influence on prevalence by gender [167]. IBD patients compared to that in healthy control have unstable gut microbiota and it is often considered one of the triggering factors of IBD [168, 169]. There is a reduction in the diversity of microbiota in IBD, possibly making the host more vulnerable to pathogens or pathobionts colonizing it [170]. A high level of NO2 and SO2 correlate with an increased risk of IBD with elevated levels of air pollution. This is related to an increase in circulating polymorphonuclear leukocytes and plasma cytokines in IBD [171–173]. Environmental priming with triggering events is involved in manifestation of both diseases. Several triggers exist, including geography, social stress, a fast-paced lifestyle, smoking, diet, and drugs [174]. It has also been correlated that low vitamin is associated with IBD [175]. Stress has been commonly associated with IBD patients and it has been labeled as a trigger cause through multiple immunopathogenic pathways [176]. Persistent diarrhea with blood and mucous is a common symptom of IBD patients. The results of laboratory tests could be valuable in diagnosing IBD. Among the initial tests to be performed are complete blood count (CBC), renal function tests, liver enzyme tests, stools cultures, and C difficile toxin [156, 177].

4.5.1 Crohn disease

Crohn's disease is an inflammatory bowel disease characterized by skip lesions and transmural inflammation, leading to inflammation throughout the entire gastrointestinal tract, from the mouth to the anus [178] There are three major locations of Crohn's disease, involving the terminal ileum, the colon, and the small bowel in about 55% of patients, and the colon in about 20% of cases. In 25% of patients, fissures and fistulas may develop, as well as upper gastrointestinal disease or extraintestinal manifestations. In 10% of patients, isolated perianal complaints may develop [179]. Although Crohn's etiology is not completely understood, genetics, immunology, and environment contribute to the onset and progression of this disease [180]. The annual incidence of Crohn is approximately 3 to 20 cases per 100,000 people [154]. Incidence of Crohn is highest among patients younger than 30 years of age, although it is increasing in older individuals [181]. North America and Western Europe are the most common places where individuals experience Crohn's disease, but Asia and Latin America are experiencing it in an increasingly more often manner as well [174, 182]. In general, the gender ratio of CD is almost similar with slightly higher prevalence in women. In western countries, there is no sex difference is apparent in incidence, whereas in Asian populations, Crohn is slightly more prevalent in men than in women. It was identified that more than 200 loci were associated with Crohn's risk [183]. The coding variation in the intracellular pattern recognition receptor gene NOD2 (also known as CARD15), is selectively associated with Crohn's risk. There is considerable variance at a few loci that are associated with aggregate heritable risk, including IL23R, the IL-2 receptor gene, and NOD2 [184, 185]. Being homozygous at NOD2 increases the risk of developing Crohn's by 20–40 times while being heterozygous increases it by 2–4 times [186]. Crohn's pathogenesis is linked to NOD2 c.3019-3020insC and ATG16L1p.Thr300Ala, respectively as has been shown by novel immunopathogenesis study [187]. It has been shown that patients with early onset of Crohn have mutations in IL-10 receptor genes [188]. Crohn can be divided into three phenotypic subtypes: inflammatory, structuring, and fistulizing. The inflammatory Crohn phenotype is characterized by inflammation of the gastrointestinal tract with no evidence of stricturing or fistulizing. Over time, this inflammation may result in fibrosis and luminal narrowing, resulting in stricturing disease. The fibrosis is reversible and there would be a need for surgical intervention. Transmural inflammation can also result in the development of a fistulous tract or sinus in patients with fistulizing Crohn. The bowel can develop a fistula with any adjacent organ (such as the bladder, vagina, or other parts of the bowel) [189]. Crohn typically manifests as weight loss, diarrhea with blood, iron deficiency, chronic and postprandial abdominal pain, fever, lack of rectal urgency, and nighttime awakenings [190]. C-reactive protein levels, sedimentation rates, or other acute phase reactants (e.g. ferritin and platelets) are commonly elevated in Crohn patients. Low B12 levels are also common. Family histories of IBD also significantly influence Crohn patients [178]. Just like other GI autoimmune diseases, some tests are used together to confirm the diagnosis such as serological tests, endoscopy, and histological tests for biopsies. Crohn is diagnosed by autoantibodies, such as anti- Saccharomyces cerevisiae antibodies (ASCAs), anti-outer membrane porin C antibodies, anti-Pseudomonas fluorescens-associated sequence I2 antibodies, and anti-CBir1 antibodies. Additionally, perinuclear antineutrophil cytoplasmic antibodies (pANCAs), antimannobioside carbohydrate antibodies, anti-laminaribioside carbohydrate antibodies, anti-chitobioside carbohydrate antibodies, as well as anti-laminarin antibodies [191–193]. In endoscopy findings for a diagnosis of Crohn are often characterized by a patchy distribution of inflammation and skip lesions. It might be apparent in endoscopy the presence of aphthous erosions or longitudinal ulcers.

4.5.2 Ulcerative colitis

UC is a chronic inflammatory disease of the colon characterized by a continuous mucosal inflammation extending from the rectum to the proximal colon with a variation in the degree of extent [194]. Colitis of the colon and rectum is characterized by continuous areas of inflammation and ulceration, without any segments of normal tissue present. It typically affects only the innermost lining [195]. Tenesmus and bloody diarrhea are hallmark symptoms of UC. The patients also report mild tenderness, lower abdominal cramping as well as fatigue due to the blood loss. Even though the etiology of UC remains unclear, increasing evidence suggests it may be an autoimmune condition [196]. The disease can develop at any age but is most commonly diagnosed before the age of 30 [197]. It has been reported that the prevalence of UC varies globally between 2.42 and 298.5/100,000, with the highest incidence occurring in North America and Northern Europe [154]. UC affects both sexes equally, and it affects all ethnicity with 3–6 more prevalence in Jewish people [198]. A "Western diet," left-handedness, and depression may increase risk for UC [199–203]. As part of the diagnosis of UC, it was recommended to test for CBC to check for intestinal blood loss and anemia [195]. Additionally, erythrocyte sedimentation rate (ESR), C-reactive protein (CRP), fecal lactoferrin, and fecal calprotectin levels are used to assess inflammation [198]. Nutritional status and deficiencies are assessed using serum albumin, iron studies, and vitamin B12 levels. UC is usually confirmed by sigmoidoscopy or colonoscopy if it has not been ruled out. It has a hallmark of the presence of continuous colonic inflammation characterized by erythema, loss of normal vascular pattern, granularity, erosions, friability, bleeding, and ulcerations, with a clear distinct demarcation between inflamed and non-inflamed bowel (**Table 1**) [198].

5. The role of psychological association with GI tract autoimmunity

Stress occurs when a demand of the environment exceeds an individual's ability to adapt [204]. Described by Selye in 1936, stress is defined as an organism attempting to maintain homoeostasis that faces either an actual threat (physical) or a perceived threat (psychological) for which it must adapt its behavior to survive [205]. Stress is not necessarily a negative effect as sometimes it can be a positive aspect for people to motivate them and enhance their performance in life in general and in this case, it's

	Disease description	Etiology	Prevalence (%)	Clinical manifestation
Achalasia	Damage at the lower end of the esophagus prevents food from entering the stomach	Nerve damage	0.001	Dysphagia, regurgitation of undigested food, nocturnal coughs, pneumonia
AAG	The parietal cells of the stomach's corpus and fundus are destroyed	Pernicious anemia is caused by a lack of vitamin B12	0.025	Variable symptoms depend on level of atrophy but mainly iron deficiency symptoms
CD	An enteropathy-associated multisystem disorder	Gluten intolerance	0.6–1	Asymptomatic; symptomatic includes constipation, bloating, diarrhea, malabsorption, and iron deficiency anemia
EoE	An immune-mediated condition that causes esophageal dysfunction when eosinophils invade the esophageal mucosa	Environmental allergens	0.05–1	Dysphagia, food impaction, heartburn, chest pain
Crohn disease	The entire GI tract exhibits skip lesions and transmural inflammation	Genetics, immunological, and environmental factors	0.02	Weight loss, diarrhea with blood, iron deficiency, abdominal pain, lack of rectal urgency, nighttime awakening
UC	A continuous mucosal inflammation from the rectum to the proximal colon with varying degrees of extent	Genetics, immunological, and environmental factors	0.002–0.3	Tenesmus and bloody diarrhea, abdominal cramps, and fatigue

Table 1.
Summary of the 5 GI autoimmune diseases.

called eustress [206]. Trauma and significant life stressors, such as a loss of a loved one or a natural disaster, occur to almost all people at some point in their lives [207]. A decrease in an individual's ability to adapt to environmental factors can lead to events that are negative and cause distress [208]. There are different kinds of stressors that can cause a disturbance: acute (lasting for minutes), brief (for a short duration), or chronic (lasting for a long time) [209]. As a result of their exposure to these stresses, a considerable number of people will develop serious psychiatric reactions, including posttraumatic stress disorder (PTSD). In addition, the acute stress reaction may be best defined as a reaction triggered by a life event that causes great stress or a change in life that triggers an acute stress reaction. Hence, it is apparent that stress and psychiatric disorder can cause physiologic changes. It is possible that the hypothalamic-pituitary-adrenal axis and the autonomic nervous system are disrupted, impairing immune function, and making people more susceptible to physical diseases. Many autoimmune diseases have unknown etiology. It has been speculated that there is a huge influence on the psychiatric reactions to life stressors and their influence in relation to autoimmune diseases. Many animal studies have suggested a close link between them [210]. Nevertheless, human studies are limited [211, 212].

5.1 Pathways for stress and related disorders

Stress causes activation of endocrine processes that provides a key pathway for further effects on health. Two main endocrine systems are involved: hypothalamic-pituitary-adrenocortical (HPA) and sympathetic-adrenal-medullary (SAM) axes. Cortisol is the main culprit controlling several physiological processes due to HPA activation like anti-inflammatory processes, metabolism of fats, carbohydrates, proteins, and gluconeogenesis. In a similar way, catecholamine released because of SAM activation regulates several functions like cardiovascular, pulmonary, hepatic, skeletal muscles, and immune system in collaboration with autonomic nervous system. Exposure to stressors causes activation of these endocrine systems. Prolonged activation of HPA and SAM causes an impaired control of physiological systems responsible for physical and psychiatric illness.

5.2 Stress as a trigger for autoimmune diseases

Stress was shown to lead glandular disturbance including autoimmune endocrine disorders [213]. In autoimmune diseases, there is a close link between stress and major stress hormones as an etiological factor [214]. Immune dysregulation could lead to atopic autoimmune diseases due to the infiltration of cytokine production and increased host defense. The repetition and the duration of stress could lead to an acute phase response that results in a chronic inflammatory process [215]. The inflammatory response is contained within the stress response, implying that stress can affect the innate immune system and causing an inflammatory response [216]. There is an association between some sort of psychological stress especially PTSD with elevated T cells that can lead to hyperreactive immune responses, higher igM, and lower dehydroepiandrosterone levels which is found in many cardiovascular and autoimmune diseases [217].

5.3 Phycological associations with the GI autoimmune diseases

Many of the patients who have GI autoimmune diseases suffer from psychiatric comorbidities such as anxiety, stress, and depression. The link between them could be bi-directional as anxiety, stress, and depression could be an etiology as well as comorbidities due to autoimmune diseases.

5.3.1 Psychological association with achalasia

It was noted that achalasia can occur after a long episode of chronic stress. Since achalasia is a rare disorder, not so many studies try to research the relationship between stress and anxiety and achalasia. In 2020, Kalantari et al. conducted a study that maps the experience of achalasia patients from initial symptoms to management of symptoms. In their findings, they found that people who had achalasia before the diagnosis had anxiety due to the uncertainty about their diagnosis [218]. According to a study from Germany, after the diagnosis of achalasia, patients were more likely to develop depression at significantly higher rates than those without the condition. Regardless of other comorbidities and the clinical characteristics of the patients, achalasia is associated with an increased incidence of depression according to their study [219]. The question whether stress, anxiety, or depression are a contributor or trigger for achalasia, or they are a secondary outcome of achalasia yet needs to be further studied.

5.3.2 AAG

It was also reported that acute stress can be a cause for AAG [220]. There is more evidence that AAG may lead to vitamin B12 deficiency, which may manifest as neuropsychiatric disorders, such as emotional instability, cognitive deficits, depression, and personality change [221]. In 2015, Tenca et al. found that the psychopathological profile has a role in symptoms occurrence in AAG [222]. It was also reported that those with AG have a significantly higher risk of experiencing psychological distress, with younger females (<50 years) displaying the highest risk, regardless of whether they have an infection with *H. pylori* (HP) [223]. Zhao et al. found that chronic atrophic gastritis patients were 54.5% likely to experience depression, as the regression analysis indicated that interpersonal sensitivity correlated positively with depression [224].

5.3.3 Celiac disease

In CD, there is a clear relationship that associates celiac diseases with stress, anxiety, and depression. Just like the other autoimmune GI diseases, the debate is not yet settled. However, it is suggested that CD has a role in these manifestations [225]. CD presents in many clinical presentations that are poorly understood such as changes in behavior are evident in cases of anxiety, depression, short-term memory loss, sleep disturbances, cognitive impairment, psychosis, and attention deficit disorder [226]. In CD, many patients have reported the symptoms of CD after stressful life events [227]. Addolorato et al. reported in a longitudinal study that 71% of people with celiac disease suffered from high levels of anxiety, the levels of anxiety were high in 24% of the control subjects, and 26% of the newly diagnosed celiac patients demonstrated anxiety [228]. In a Swedish study that evaluated patients with CD between 1973 and 2016, they concluded that children with CD have an increased risk of developing psychiatric disorders in adulthood [229]. According to Wahab et al., CD is associated with anxiety and oppositional defiant behavior when it is combined with HLA-DQ2 or HLA-DQ8 risk alleles [230] as a conclusion for their study on CD Autoimmunity and Emotional and Behavioral Problems in Childhood. Depression has been reported in association with CD since 1951 [231]. Several studies have shown that people with CD are more likely to suffer from depression than people without CD [232–239].

Butwicka et al. found that children with CD had a 1.4-fold greater risk of developing mental disorders compared with the general population. Childhood CD was identified as a risk factor for mood disorders, anxiety disorders, eating disorders, behavioral disorders, ADHD, ASD, and intellectual disabilities in their study. Moreover, mood, eating, or behavioral disorders were more common before celiac disease diagnosis [240]. Individuals with CD have an increased risk of anxiety disorders, according to several studies [228, 241, 242]. These come in agreement with Clappison's systematic review and meta-analysis on the psychiatric disorders association with CD [243]. Psychological symptoms before diagnosis could be caused by the general health of the patient, or by hypoperfusion of the brain in certain regions, a result of vitamin deficiency due to malabsorption. Also, Hyperphomocysteiemia can damage the blood-brain barrier, exposing the neuronal tissue to neuro-irritative substances [226]. Additionally, they may be associated with Ads such as thyroid disease, a risk factor for depression, panic disorder, and type 1 diabetes. There is speculation that one of the possible explanations could be due to the cytokines that are produced by the immune reactions, which can affect the brain circuits that control mood [226].

5.4 EoE

There is some evidence that EoE and its treatments can significantly reduce psychological functioning, resulting in increased anxiety and depression [244]. Mental distress is a common problem among adult EoE patients, with an increased risk of significant anxiety among those younger than 35 years of age [245]. Mechanistically speaking, the protein Eotaxin-1/CCL11 which is involved in Eosinophil Recruitment could be the reason for the pathopsychological involvement in EoE patients. There has been evidence that eotaxin affects the central nervous system, and it was noted that eotaxin-1/CCL11 crosses the blood-brain barrier unaltered [246]. Eotaxin-1/CCL11 inhibits neural progenitor cell proliferation in isolated neurons and neurons derived from neurospheres, as well as in hippocampal slices without affecting their ability to form neurons or astrocytes in vitro [247]. Neurons were not directly affected by eotaxin-1/CCL11. However, related chemokines were able to promote microglial migration and activation, producing reactive oxygen species, which exacerbated glutamate-induced neurodegeneration [248]. The serum levels of 22 cytokines/chemokines, including eotaxin-1 and CCL11, were assessed in 49 patients with major depression, and 49 matched controls reported increased levels of the molecule in an inflammatory context [249].

5.5 Inflammatory bowel disease: celiac disease and UC

Stressors (i.e. environmental events) can affect the expression of symptoms in people with Crohn. It was suggested by Crohn himself, in his book Regional ileitisi in 1949 [250]. It has been reported that stressful life events cause the disease to manifest since 1960 [251–255]. UC has been shown to be psychosomatic disease since 1969 [256]. Psychological stress has been shown to promote systemic and mucosal proinflammatory responses, which could contribute to the exacerbation of UC in everyday life [257]. The UC patients exhibit hostility, somatization, anxiety, and depression even during remission, which is not surprising since the disease was entirely reversible [258]. In general, IBD exhibit more psychological disorder. IBD patients suffer from high rates of psychological distress and comorbid conditions, including depression, anxiety disorders, and bipolar disorder according to a cohort study [259]. An analysis of 1078 patients with IBD, including 303 patients with Crohn's disease and 775 patients with ulcerative colitis, found that 75% of patients believed that psychologic stress caused an exacerbation of their symptoms [260].

5.5.1 The bidirectionality in IBD

According to a study that assessed perceptions of stress over time (2 years) in three subgroups, those with chronically active symptoms had the greatest perceptions of stress over time [261]. Over time, those with chronically inactive symptoms displayed the lowest levels of perceived stress [261]. Perceived stress scores were intermediate between those whose symptoms fluctuated from inactive to active over the 2-year period. In these studies, the directionality of the association between adverse mental health and active symptoms of disease could not be established [262]. They do indicate, however, that adverse mental health is a problem for those whose disease is symptomatic. Psychological comorbidity is three times more prevalent in people with IBD than in the general population [262, 263]. More than 25% of patients with IBD may suffer from depression and more than 30% from anxiety during their lifetime [262, 264, 265]. Study of the Manitoba IBD Cohort Study population that underwent the CIDI and comparison with the Canadian Community Health Survey population that did the CIDI revealed that people with IBD were twice as likely to have a lifetime history of mood disorders than controls both within 12 months of diagnosis and within a year following diagnosis [262, 264]. Nearly 80% of those with IBD and an anxiety disorder had their anxiety disorder diagnosed more than two years before their IBD diagnosis. It is estimated that more than 50% of those with mood disorders were diagnosed before they were diagnosed with IBD. Therefore, it seems that not just chronic disease symptoms can lead to an increased level of anxiety and depression, but the presence of these psychological diseases could also predispose a person to develop IBD [262].

6. Microbiome and autoimmune diseases: the gut-brain axis in GI autoimmune diseases

Mammalian microbiota consists of a variety of microorganisms, such as bacteria, archaea, fungi, and viruses. A symbiotic relationship exists between humans and bacteria, most of which are present in the gastrointestinal system [266]. An essential component of the host's health and well-being is the gut microbiota, the collection of intestinal microorganisms throughout the GI tract [267]. Assemblage of the gut microbiome begins during birth, primarily from the mother's vaginal and fecal microbiomes if naturally delivered, or from the skin and environmental microbes if delivered via cesarean section [268–270]. There are more than 100 trillion microorganisms living within the GI tract, which together form the microbiota, a complex biosystem. Microbiota are organisms belonging to all domains of life, including Eukaryotes, Bacteria, and Archaea. The main components that comprise this micro-universe belong to the bacterial group and are divided into four phyla: Actinobacteria, Proteobacteria, Bacteroidetes, and Firmicutes [271]. In terms of host health, microbes found in the gut are involved in nondigestible carbohydrates metabolism, immune system development, and drug metabolism. Human diseases linked to gut microbiota include IBD, metabolic diseases, allergic diseases, and neurodevelopmental diseases [272, 273]. The microbiome of a newborn infant is affected by nutrition, physio-chemicals, and biological properties of the body, as well as life events [274]. In this period of life, while breast milk is the primary source of nutrients, there are big shifts in bacterial taxa and much more variation between infants than between adults. Different immune responses to the microbes colonizing the host or other lifestyle factors could account for the large functional and phylogenetic variability [275]. The gut and brain developed from the same tissue, the neural crest, during embryogenesis and influence each other tightly [276].

As both are parts of the immune system, the gut, the brain, and multiple organs can all be affected by disturbances in this system. Communication between the gut and brain is known as gut-brain axis [277]. Communication between the brain and gut involves neural pathways, such as the enteric nervous system (ENS), vagus, sympathetic, and spinal nerves, as well as humoral pathways involving cytokines, hormones, and neuropeptides as signaling molecules [278]. The ENS controls the functions of the gut and includes blood flow absorption, motility, and secretion, and these four compromise the main function of the gut-brain axis [267]. The alteration of the gut microbiota by any factor can lead to signaling to ENS resulting in an alteration in the hormone secretion. Chemical signals from the intestinal epithelium, enteric endocrine system, and immune system are highly receptive to this area, and it provides input to sensory pathways that signal the emotional and cognitive centers of the brain. ENS also receives efferent information from the brain via autonomic neural connections (sympathetic and parasympathetic) and hormonal pathways that modulate digestive functions [279]. Food intake regulation, glucose metabolism, and modulation of the GI-associated immune system include digestive processes, GI tract synchronization of physical and emotional states are all part of brain-gut axis interactions [267]. The relationship between gut-brain axis and stress, depression, and anxiety is well established. These psychological conditions have biological mechanisms and manifestations. Allostasis is process in which the body's ability to restore homeostasis can result these psychological conditions. In allostasis, the hypothalamic-pituitary-adrenal (HPA) axis regulates the body's stress response systems, including neuroendocrine signaling and the glucocorticoids it produces, and BDNF it regulates, help with memory and learning [280]. Glucocorticoids is released from adrenal glands during stressful events, and it controls the homeostatic conditions. However, it can result anti-inflammatory responses [281]. It was shown that the gut microbiota helps to regulate the stress response as its absence results in an overproduction of Glucocorticoids after stressful events, particularly through Lactobacillus spp. in stress. In addition, it was shown that *Lactobacillus rhamnosus* reduces anxiety and Bifidobacterium spp. improves stress.

There is bidirectionality between gut and the microbiota in stress management. Through the release of cytokines and neurotransmitters, inflammation of the GI tract stresses the microbiome [282]. In conjunction with the increase in intestinal permeability, these molecules then travel systemically. Rogue molecules from the permeable gut (leaky gut) are amplified when blood levels of TNF-a and MCP (monocyte chemoattractant protein) are elevated [283, 284]. Anxiety, depression, and memory loss result from their release [285]. It was reported that there is a relationship between elevation of IL-5 and TNF-a with depression and anxiety that suggest that these pro-inflammatory cytokines are involved in the development of anxiety and depression which is also manifested in chronic inflammation and altered immune cells in the peripheral blood [284]. The hypothalamicpituitary- adrenal (HPA) axis can be simulated with pro-inflammatory cytokines. The hypothalamus can release corticotropin releasing factors simulating the adenohypophysis to release adrenocorticotropic hormone (ACTH). The ACTH can induce the release of cortisol which is a stress hormone from the adrenal gland which acts as negative feedback in the pro-inflammatory signal transduction. Hyperactivity of the HPA axis is a major cause of psychological responses such as stress, anxiety, and depression [286].

The research focus on gut-brain axis is recent and it's not proceeding with the pace that it was expected to be. Nevertheless, it's a complicated area of research due to variety of factors that are involved in the process and the multiple pathways that could play a vital role in the processes. In addition, the components of the gut microbiota are huge. Several neural, hormonal, metabolic, immunological, and microbial signals drive gut-brain communication [287]. In autoimmune diseases, many patients have reported psychological comorbidities and it's not confirmed whether these comorbidities are due to quality of life with the disease, or they are one of the inducers that are involved in the pathogenesis of the disease. It could be bidirectional, nevertheless. There is an autoimmune component to major psychiatric disorders. In psychiatric disorders, disequilibrium of cellular processes in the GI tract is likely to contribute to immune dysfunction [288]. Symptoms of gastrointestinal diseases worsen psychological complaints and vice versa, suggesting a significant role for an imbalance in the gut-brain axis in both conditions. The gut is strongly implicated in a variety of neurological diseases via direct and indirect mechanisms, according to growing evidence. Intestinal microbes and their products (e.g., metabolites) as well as immune education in the mucosal immune system, including the release of proinflammatory cytokines, are key components. The intestinal epithelium regulates these processes by translating signals from bacteria and inflammation to the immune system and secreting hormones and peptides which are involved in the metabolic processing of dietary nutrients [289]. Some of GI autoimmune diseases mechanisms of gut-brain axis role and the clear direct relationship of stress, anxiety and depression are well established such as IBD [290, 291] and CD [292]. Since this is a growing area of the research, more investigations need to be done to cover the relationship between the involved components.

7. Conclusions

To conclude, GI autoimmune diseases can be compromising to the patients' life, and they can be due to multiple factors. Over the past few decades, the number of GI autoimmune diseases have increased exponentially. GI autoimmune diseases although they are organ specific. Nevertheless, there is a need for multidisciplinary approaches for diagnosing and understanding the pathogenesis of these diseases. Antibodies provide a current excellent predictor for those diseases. Nevertheless, the investment in biotechnology to develop more specific and sensitive tools for is needed. The understanding of the interwinding between the brain and gut as well as other etiological factors can provide better approach in preventive medicine in dealing with this disease as well as increase in the quality of the life of the patients beside the current available pharmacological and surgical available options.

Acknowledgements

This section of your manuscript may also include funding information. The publication of this chapter is funded by Qatar National Library.

Conflict of interest

The authors declare no conflict of interest.

Author details

Mahmoud Khatib A.A. Al-Ruweidi[1,2], Nada Khater[3], Haya Rashid Alkaabi[4], Maram Hasan[1] and Huseyin C. Yalcin[1]*

1 Biomedical Research Center, Qatar University, Doha, Qatar

2 Pharmaceutical Sciences Department, College of Pharmacy, QU Health, Qatar University, Doha, Qatar

3 Department of Sport Science, College of Arts and Sciences, Qatar University, Doha, Qatar

4 Weill Cornell Medical College Qatar, Doha, Qatar

*Address all correspondence to: hyalcin@qu.edu.qa

References

[1] Singh K, Chang C, Gershwin ME. IgA deficiency and autoimmunity. Autoimmunity Reviews. 2014;**13**(2):163-177. DOI: 10.1016/j.autrev.2013.10.005

[2] Sinha AA, Lopez MT, McDevitt HO. Autoimmune diseases: The failure of self tolerance. Science. 1990; **248**(4961):1380-1388. DOI: 10.1126/science.1972595

[3] Agmon-Levin N, Lian Z, Shoenfeld Y. Explosion of autoimmune diseases and the mosaic of old and novel factors. Cellular & Molecular Immunology. 2011;**8**(3):189-192. DOI: 10.1038/cmi.2010.70

[4] Ehrenfeld M, Tincani A, Andreoli L, Cattalini M, Greenbaum A, Kanduc D, et al. Covid-19 and autoimmunity. Autoimmunity Reviews. 2020;**19**(8):102597. Available from: https://linkinghub.elsevier.com/retrieve/pii/S1568997220301610

[5] Galeotti C, Bayry J. Autoimmune and inflammatory diseases following COVID-19. Nature Reviews Rheumatology. 2020;**16**(8):413-414. DOI: 10.1038/s41584-020-0448-7

[6] Navegantes KC, de Souza GR, Pereira PAT, Czaikoski PG, Azevedo CHM, Monteiro MC. Immune modulation of some autoimmune diseases: The critical role of macrophages and neutrophils in the innate and adaptive immunity. Journal of Translational Medicine. 2017;**15**(1):36. DOI: 10.1186/s12967-017-1141-8

[7] Eggert M, Zettl UK, Neeck G. Autoantibodies in autoimmune diseases. Current Pharmaceutical Design. 2010;**16**(14):1634-1643. DOI: 10.2174/138161210791164144

[8] Alexander T, Sharrack B, Rovira M, et al. Autoimmune Disease. In: Carreras E, Dufour C, Mohty M, et al., editors. The EBMT Handbook: Hematopoietic Stem Cell Transplantation and Cellular Therapies [Internet]. 7th edition. Cham (CH), Heidelberg, Germany: Springer; 2019. Chapter 91. Available from: https://www.ncbi.nlm.nih.gov/books/NBK553982/ DOI: 10.1007/978-3-030-02278-5_91

[9] Masters SL, Simon A, Aksentijevich I, Kastner DL. Horror autoinflammaticus: The molecular pathophysiology of autoinflammatory disease (*). Annual Review of Immunology. 2009;**27**:621-668. DOI: 10.1146/annurev.immunol.25.022106.141627

[10] Delves PJ, Roitt IM. Encyclopedia of Immunology. Michigan, USA: Academic Press; 1998. Available from: https://agris.fao.org/agris-search/search.do?recordID=US201300022761

[11] Hedrich CM. Shaping the spectrum—From autoinflammation to autoimmunity. Clinical Immunology. 2016;**165**:21-28. Available from: https://www.sciencedirect.com/science/article/pii/S1521661616300298

[12] McGonagle D, McDermott MF. A proposed classification of the immunological diseases. PLoS Medicine. 2006;**3**(8):e297. DOI: 10.1371/journal.pmed.0030297

[13] Surace AEA, Hedrich CM. The role of epigenetics in autoimmune/inflammatory disease. Frontiers in Immunology. 2019;**10**:1525. DOI: 10.3389/fimmu.2019.01525

[14] Wang L, Wang F-S, Gershwin ME. Human autoimmune diseases: A

comprehensive update. Journal of Internal Medicine. 2015;**278**(4):369-395. DOI: 10.1111/joim.12395

[15] Aksentijevich I. Update on genetics and pathogenesis of autoinflammatory diseases: The last 2 years. Seminars in Immunopathology. 2015;**37**(4):395-401. DOI: 10.1007/s00281-015-0478-4

[16] Canna SW, Goldbach-Mansky R. New monogenic autoinflammatory diseases—A clinical overview. Seminars in Immunopathology. 2015;**37**(4):387-394. Available from: http://link.springer.com/10.1007/s00281-015-0493-5

[17] Masters SL. Broadening the definition of autoinflammation. Seminars in Immunopathology. 2015;**37**(4):311-312. DOI: 10.1007/s00281-015-0497-1

[18] Pollard KM, Hultman P, Kono DH. Toxicology of autoimmune diseases. Chemical Research in Toxicology. 2010;**23**(3):455-466. DOI: 10.1021/tx9003787

[19] Caso F, Costa L, Nucera V, Barilaro G, Masala IF, Talotta R, et al. From autoinflammation to autoimmunity: Old and recent findings. Clinical Rheumatology. 2018;**37**(9):2305-2321. DOI: 10.1007/s10067-018-4209-9

[20] Doria A, Zen M, Bettio S, Gatto M, Bassi N, Nalotto L, et al. Autoinflammation and autoimmunity: Bridging the divide. Autoimmunity Reviews. 2012;**12**(1):22-30. DOI: 10.1016/j.autrev.2012.07.018

[21] Clark R, Kupper T. Old meets new: The interaction between innate and adaptive immunity. The Journal of Investigative Dermatology. 2005;**125**(4):629-637. DOI: 10.1111/j.0022-202X.2005.23856.x

[22] Iwasaki A, Medzhitov R. Regulation of adaptive immunity by the innate immune system. Science. 2010;**327**(5963):291-295. DOI: 10.1126/science.1183021

[23] Gaudino SJ, Kumar P. Cross-talk between antigen presenting cells and T cells impacts intestinal homeostasis, bacterial infections, and tumorigenesis. Frontiers in Immunology. 2019;**10**:360. DOI: 10.3389/fimmu.2019.00360

[24] Roh JS, Sohn DH. Damage-associated molecular patterns in inflammatory diseases. Immune Network. 2018;**18**(4):e27. DOI: 10.4110/in.2018.18.e27

[25] Kawai T, Akira S. The roles of TLRs, RLRs and NLRs in pathogen recognition. International Immunology. 2009;**21**(4):317-337. DOI: 10.1093/intimm/dxp017

[26] Kany S, Vollrath JT, Relja B. Cytokines in inflammatory disease. International Journal of Molecular Sciences. 2019;**20**(23):6008. DOI: 10.3390/ijms20236008

[27] McVey MJ, Steinberg BE, Goldenberg NM. Inflammasome activation in acute lung injury. American Journal of Physiology. Lung Cellular and Molecular Physiology. 2021;**320**(2):L165-L178. DOI: 10.1152/ajplung.00303.2020

[28] Liao W, Lin J-X, Leonard WJ. Interleukin-2 at the crossroads of effector responses, tolerance, and immunotherapy. Immunity. 2013;**38**(1):13-25. DOI: 10.1016/j.immuni.2013.01.004

[29] Limmer A, Sacher T, Alferink J, Nichterlein T, Arnold B, Hämmerling GJ. A two-step model for the induction of organ-specific autoimmunity.

In: Novartis Foundation Symposia. Chichester, UK: John Wiley & Sons, Ltd; 2007. pp. 159-171. (Novartis Foundation symposium). Available from: https://onlinelibrary.wiley.com/doi/10.1002/9780470515525.ch12

[30] Montgomery RK, Mulberg AE, Grand RJ. Development of the human gastrointestinal tract: Twenty years of progress. Gastroenterology. 1999;**116**(3):702-731. DOI: 10.1016/s0016-5085(99)70193-9

[31] Celebi Sözener Z, Cevhertas L, Nadeau K, Akdis M, Akdis CA. Environmental factors in epithelial barrier dysfunction. The Journal of Allergy and Clinical Immunology. 2020;**145**(6):1517-1528. Available from: https://www.sciencedirect.com/science/article/pii/S009167492030573X

[32] Lee SH. Intestinal permeability regulation by tight junction: Implication on inflammatory bowel diseases. Intestinal Research. 2015;**13**(1):11-18. DOI: 10.5217/ir.2015.13.1.11

[33] Christen U, von Herrath MG. Induction, acceleration or prevention of autoimmunity by molecular mimicry. Molecular Immunology. 2004;**40**(14-15):1113-1120. DOI: 10.1016/j.molimm.2003.11.014

[34] Tang VW, Goodenough DA. Paracellular ion channel at the tight junction. Biophysical Journal. 2003;**84**(3):1660-1673. DOI: 10.1016/S0006-3495(03)74975-3

[35] Fasano A, Shea-Donohue T. Mechanisms of disease: The role of intestinal barrier function in the pathogenesis of gastrointestinal autoimmune diseases. Nature Clinical Practice. Gastroenterology & Hepatology. 2005;**2**(9):416-422. Available from: https://www.nature.com/articles/ncpgasthep0259

[36] Wang W, Uzzau S, Goldblum SE, Fasano A. Human zonulin, a potential modulator of intestinal tight junctions. Journal of Cell Science. 2000;**113**(Pt 24):4435-4440. DOI: 10.1242/jcs.113.24.4435

[37] Fasano A, Not T, Wang W, Uzzau S, Berti I, Tommasini A, et al. Zonulin, a newly discovered modulator of intestinal permeability, and its expression in coeliac disease. Lancet. 2000;**355**(9214):1518-1519. DOI: 10.1016/S0140-6736(00)02169-3

[38] Fasano A. Regulation of intercellular tight junctions by zonula occludens toxin and its eukaryotic analogue zonulin. Annals of the New York Academy of Sciences. 2000;**915**:214-222. DOI: 10.1111/j.1749-6632.2000.tb05244.x

[39] Fasano A. All disease begins in the (leaky) gut: Role of zonulin-mediated gut permeability in the pathogenesis of some chronic inflammatory diseases. F1000Research. 2020;**9**:69. Available from: https://f1000research.com/articles/9-69/v1

[40] Martinez EE, Mehta NM, Fasano A. The zonulin pathway as a potential mediator of gastrointestinal dysfunction in critical illness. Pediatric Critical Care Medicine. 2022. DOI: 10.1097/PCC.0000000000002985

[41] Yazici D, Ogulur I, Kucukkase O, Li M, Rinaldi AO, Pat Y, et al. Epithelial barrier hypothesis and the development of allergic and autoimmune diseases. Allergo Journal International. 2022;**31**(4):91-102. DOI: 10.1007/s40629-022-00211-y

[42] Fasano A. Leaky gut and autoimmune diseases. Clinical Reviews in

Allergy and Immunology. 2012;**42**(1):71-78. DOI: 10.1007/s12016-011-8291-x

[43] Jutel M, Akdis M, Budak F, Aebischer-Casaulta C, Wrzyszcz M, Blaser K, et al. IL-10 and TGF-beta cooperate in the regulatory T cell response to mucosal allergens in normal immunity and specific immunotherapy. European Journal of Immunology. 2003;**33**(5):1205-1214. Available from: https://onlinelibrary.wiley.com/doi/10.1002/eji.200322919

[44] Mitamura Y, Ogulur I, Pat Y, Rinaldi AO, Ardicli O, Cevhertas L, et al. Dysregulation of the epithelial barrier by environmental and other exogenous factors. Contact Dermatitis. 2021;**85**(6):615-626. DOI: 10.1111/cod.13959

[45] Akdis CA. The epithelial barrier hypothesis proposes a comprehensive understanding of the origins of allergic and other chronic noncommunicable diseases. The Journal of Allergy and Clinical Immunology. 2022;**149**(1):41-44. DOI: 10.1016/j.jaci.2021.11.010

[46] Akbarshahi H, Menzel M, Ramu S, Mahmutovic Persson I, Bjermer L, Uller L. House dust mite impairs antiviral response in asthma exacerbation models through its effects on TLR3. Allergy. 2018;**73**(5):1053-1063. DOI: 10.1111/all.13378

[47] Zhu Y-T, Yuan Y-Z, Feng Q-P, Hu M-Y, Li W-J, Wu X, et al. Food emulsifier polysorbate 80 promotes the intestinal absorption of mono-2-ethylhexyl phthalate by disturbing intestinal barrier. Toxicology and Applied Pharmacology. 2021;**414**:115411. DOI: 10.1016/j.taap.2021.115411

[48] Coskun M. Intestinal epithelium in inflammatory bowel disease. Frontiers in Medicine. 2014;**1**:24. DOI: 10.3389/fmed.2014.00024

[49] Davis BP, Stucke EM, Khorki ME, Litosh VA, Rymer JK, Rochman M, et al. Eosinophilic esophagitis-linked calpain 14 is an IL-13-induced protease that mediates esophageal epithelial barrier impairment. JCI Insight. 2016;**1**(4):e86355. DOI: 10.1172/jci.insight.86355

[50] Nguyen N, Fernando SD, Biette KA, Hammer JA, Capocelli KE, Kitzenberg DA, et al. TGF-β1 alters esophageal epithelial barrier function by attenuation of claudin-7 in eosinophilic esophagitis. Mucosal Immunology. 2018;**11**(2):415-426. DOI: 10.1038/mi.2017.72

[51] Rochman M, Azouz NP, Rothenberg ME. Epithelial origin of eosinophilic esophagitis. The Journal of Allergy and Clinical Immunology. 2018;**142**(1):10-23. DOI: 10.1016/j.jaci.2018.05.008

[52] Schumann M, Siegmund B, Schulzke JD, Fromm M. Celiac disease: Role of the epithelial barrier. Cellular and Molecular Gastroenterology and Hepatology. 2017;**3**(2):150-162. DOI: 10.1016/j.jcmgh.2016.12.006

[53] Rieder E, Fernandez-Becker NQ, Sarosiek J, Guillaume A, Azagury DE, Clarke JO. Achalasia: Physiology and diagnosis. Annals of the New York Academy of Sciences. 2020;**1482**(1):85-94. DOI: 10.1111/nyas.14510

[54] Goldblum JR, Whyte RI, Orringer MB, Appelman HD. Achalasia. A morphologic study of 42 resected specimens. The American Journal of Surgical Pathology. 1994;**18**(4):327-337. Available from: https://www.ncbi.nlm.nih.gov/pubmed/8141427

[55] Pandolfino JE, Gawron AJ. Achalasia: A systematic review. Journal of the American Medical Association. 2015;**313**(18):1841-1852. DOI: 10.1001/jama.2015.2996

[56] Castagliuolo I, Brun P, Costantini M, Rizzetto C, Palù G, Costantino M, et al. Esophageal achalasia: Is the herpes simplex virus really innocent? Journal of Gastrointestinal Surgery. 2004;**8**(1):24-30; discussion 30. DOI: 10.1016/j.gassur.2003.10.004

[57] Facco M, Brun P, Baesso I, Costantini M, Rizzetto C, Berto A, et al. T cells in the myenteric plexus of achalasia patients show a skewed TCR repertoire and react to HSV-1 antigens. The American Journal of Gastroenterology. 2008;**103**(7):1598-1609. DOI: 10.1111/j.1572-0241.2008.01956.x

[58] Boeckxstaens GE, Zaninotto G, Richter JE. Achalasia. Lancet. 2014;**383**(9911):83-93. DOI: 10.1016/S0140-6736(13)60651-0

[59] Boeckxstaens GE. Achalasia: Virus-induced euthanasia of neurons? The American Journal of Gastroenterology. 2008;**103**(7):1610-1612. DOI: 10.1111/j.1572-0241.2008.01967.x

[60] Spechler SJ, Konda V, Souza R. Can eosinophilic esophagitis cause achalasia and other esophageal motility disorders? The American Journal of Gastroenterology. 2018;**113**(11):1594-1599. DOI: 10.1038/s41395-018-0240-3

[61] Nuñez C, García-González MA, Santiago JL, Benito MS, Mearín F, de la Concha EG, et al. Association of IL10 promoter polymorphisms with idiopathic achalasia. Human Immunology. 2011;**72**(9):749-752. DOI: 10.1016/j.humimm.2011.05.017

[62] de León AR, de la Serna JP, Santiago JL, Sevilla C, Fernández-Arquero M, de la Concha EG, et al. Association between idiopathic achalasia and IL23R gene. Neurogastroenterology and Motility. 2010;**22**(7):734-738, e218. DOI: 10.1111/j.1365-2982.2010.01497.x

[63] Pressman A, Behar J. Etiology and pathogenesis of idiopathic achalasia. Journal of Clinical Gastroenterology. 2017;**51**(3):195-202. DOI: 10.1097/MCG.0000000000000780

[64] Palmieri O, Mazza T, Merla A, Fusilli C, Cuttitta A, Martino G, et al. Gene expression of muscular and neuronal pathways is cooperatively dysregulated in patients with idiopathic achalasia. Scientific Reports. 2016;**6**:31549. DOI: 10.1038/srep31549

[65] Alahdab YO, Eren F, Giral A, Gunduz F, Kedrah AE, Atug O, et al. Preliminary evidence of an association between the functional c-kit rs6554199 polymorphism and achalasia in a Turkish population. Neurogastroenterology and Motility. 2012;**24**(1):27-30. Available from: https://onlinelibrary.wiley.com/doi/10.1111/j.1365-2982.2011.01793.x

[66] Aggestrup S, Uddman R, Sundler F, Fahrenkrug J, Håkanson R, Sørensen HR, et al. Lack of vasoactive intestinal polypeptide nerves in esophageal achalasia. Gastroenterology. 1983;**84**(5 Pt 1):924-927. Available from: https://www.ncbi.nlm.nih.gov/pubmed/6832568

[67] Paladini F, Cocco E, Cascino I, Belfiore F, Badiali D, Piretta L, et al. Age-dependent association of idiopathic achalasia with vasoactive intestinal peptide receptor 1 gene. Neurogastroenterology and Motility. 2009;**21**(6):597-602. Available from: https://onlinelibrary.wiley.com/doi/10.1111/j.1365-2982.2009.01284.x

[68] Samo S, Carlson DA, Gregory DL, Gawel SH, Pandolfino JE, Kahrilas PJ. Incidence and prevalence of achalasia in Central Chicago, 2004-2014, since the widespread use of high-resolution manometry. Clinical Gastroenterology and Hepatology. 2017;**15**(3):366-373. DOI: 10.1016/j.cgh.2016.08.030

[69] Sadowski DC, Ackah F, Jiang B, Svenson LW. Achalasia: Incidence, prevalence and survival. A population-based study. Neurogastroenterology and Motility. 2010;**22**(9):e256-e261. Available from: https://onlinelibrary.wiley.com/doi/10.1111/j.1365-2982.2010.01511.x

[70] Sonnenberg A. Hospitalization for achalasia in the United States 1997-2006. Digestive Diseases and Sciences. 2009;**54**(8):1680-1685. DOI: 10.1007/s10620-009-0863-8

[71] Ng KY, Li KF, Lok KH, Lai L, Ng CH, Li KK, et al. Ten-year review of epidemiology, clinical features, and treatment outcome of achalasia in a regional hospital in Hong Kong. Hong Kong Medical Journal. 2010;**16**(5):362-366. Available from: https://www.ncbi.nlm.nih.gov/pubmed/20890000

[72] Vantrappen G, Hellemans J, Deloof W, Valembois P, Vandenbroucke J. Treatment of achalasia with pneumatic dilatations. Gut. 1971;**12**(4):268-275. DOI: 10.1136/gut.12.4.268

[73] Richter JE, Boeckxstaens GE. Management of achalasia: Surgery or pneumatic dilation. Gut. 2011;**60**(6): 869-876. DOI: 10.1136/gut.2010.212423

[74] Di Sabatino A, Lenti MV, Giuffrida P, Vanoli A, Corazza GR. New insights into immune mechanisms underlying autoimmune diseases of the gastrointestinal tract. Autoimmunity Reviews. 2015;**14**(12):1161-1169. DOI: 10.1016/j.autrev.2015.08.004

[75] Massironi S, Zilli A, Elvevi A, Invernizzi P. The changing face of chronic autoimmune atrophic gastritis: An updated comprehensive perspective. Autoimmunity Reviews. 2019;**18**(3):215-222. DOI: 10.1016/j.autrev.2018.08.011

[76] Miceli E, Vanoli A, Lenti MV, Klersy C, Di Stefano M, Luinetti O, et al. Natural history of autoimmune atrophic gastritis: A prospective, single Centre, long-term experience. Alimentary Pharmacology & Therapeutics. 2019;**50**(11-12):1172-1180. Available from: https://onlinelibrary.wiley.com/doi/10.1111/apt.15540

[77] Centanni M, Marignani M, Gargano L, Corleto VD, Casini A, Delle Fave G, et al. Atrophic body gastritis in patients with autoimmune thyroid disease: An underdiagnosed association. Archives of Internal Medicine. 1999;**159**(15):1726-1730. DOI: 10.1001/archinte.159.15.1726

[78] Rugge M, Correa P, Dixon MF, Fiocca R, Hattori T, Lechago J, et al. Gastric mucosal atrophy: Interobserver consistency using new criteria for classification and grading. Alimentary Pharmacology & Therapeutics. 2002;**16**(7):1249-1259. DOI: 10.1046/j.1365-2036.2002.01301.x

[79] Dixon MF, Genta RM, Yardley JH, Correa P. The participants. Classification and grading of gastritis: The updated Sydney system. The American Journal of Surgical Pathology. 1996;**20**(10):1161. Available from: https://journals.lww.com/ajsp/fulltext/1996/10000/classification_and_grading_of_gastritis__the.1.aspx

[80] Neumann WL, Coss E, Rugge M, Genta RM. Autoimmune atrophic

gastritis—Pathogenesis, pathology and management. Nature Reviews. Gastroenterology & Hepatology. 2013;**10**(9):529-541. Available from: https://www.nature.com/articles/nrgastro.2013.101

[81] Rodriguez-Castro KI, Franceschi M, Miraglia C, Russo M, Nouvenne A, Leandro G, et al. Autoimmune diseases in autoimmune atrophic gastritis. Acta Bio-Medica. 2018;**89**(8-S):100-103. Available from: https://www.mattioli1885journals.com/index.php/actabiomedica/article/view/7919

[82] Lenti MV, Miceli E, Cococcia S, Klersy C, Staiani M, Guglielmi F, et al. Determinants of diagnostic delay in autoimmune atrophic gastritis. Alimentary Pharmacology & Therapeutics. 2019;**50**(2):167-175. DOI: 10.1111/apt.15317

[83] Jacobson DL, Gange SJ, Rose NR, Graham NM. Epidemiology and estimated population burden of selected autoimmune diseases in the United States. Clinical Immunology and Immunopathology. 1997;**84**(3):223-243. DOI: 10.1006/clin.1997.4412

[84] López-Jornet P, Parra-Perez F, Pons-Fuster A. Association of autoimmune diseases with oral lichen planus: A cross-sectional, clinical study. Journal of the European Academy of Dermatology and Venereology. 2014;**28**(7):895-899. Available from: https://onlinelibrary.wiley.com/doi/10.1111/jdv.12202

[85] Kanazawa M, Shimohata T, Tanaka K, Nishizawa M. Clinical features of patients with myasthenia gravis associated with autoimmune diseases. European Journal of Neurology. 2007;**14**(12):1403-1404. DOI: 10.1111/j.1468-1331.2007.01978.x

[86] Amerio P, Tracanna M, De Remigis P, Betterle C, Vianale L, Marra ME, et al. Vitiligo associated with other autoimmune diseases: Polyglandular autoimmune syndrome types 3B+C and 4. Clinical and Experimental Dermatology. 2006;**31**(5):746-749. DOI: 10.1111/j.1365-2230.2006.02171.x

[87] De Block CEM, De Leeuw IH, Van Gaal LF. Autoimmune gastritis in type 1 diabetes: A clinically oriented review. The Journal of Clinical Endocrinology and Metabolism. 2008;**93**(2):363-371. DOI: 10.1210/jc.2007-2134

[88] Fallahi P, Ferrari SM, Ruffilli I, Elia G, Biricotti M, Vita R, et al. The association of other autoimmune diseases in patients with autoimmune thyroiditis: Review of the literature and report of a large series of patients. Autoimmunity Reviews. 2016;**15**(12):1125-1128. DOI: 10.1016/j.autrev.2016.09.009

[89] Zelissen PM, Bast EJ, Croughs RJ. Associated autoimmunity in Addison's disease. Journal of Autoimmunity. 1995;**8**(1):121-130. DOI: 10.1006/jaut.1995.0009

[90] Toh BH, van Driel IR, Gleeson PA. Pernicious anemia. The New England Journal of Medicine. 1997;**337**(20):1441-1448. DOI: 10.1056/NEJM199711133372007

[91] Vannella L, Lahner E, Osborn J, Annibale B. Systematic review: Gastric cancer incidence in pernicious anaemia. Alimentary Pharmacology & Therapeutics. 2013;**37**(4):375-382. DOI: 10.1111/apt.12177

[92] La Rosa S, Vanoli A. Gastric neuroendocrine neoplasms and related precursor lesions. Journal of Clinical Pathology. 2014;**67**(11):938-948. DOI: 10.1136/jclinpath-2014-202515

[93] Minalyan A, Benhammou JN, Artashesyan A, Lewis MS, Pisegna JR. Autoimmune atrophic gastritis: Current perspectives. Clinical and Experimental Gastroenterology. 2017;**10**:19-27. DOI: 10.2147/CEG.S109123

[94] Oxentenko AS, Rubio-Tapia A. Celiac disease. Mayo Clinic Proceedings. 2019;**94**(12):2556-2571. DOI: 10.1016/j.mayocp.2019.02.019

[95] Ludvigsson JF, Leffler DA, Bai JC, Biagi F, Fasano A, Green PHR, et al. The Oslo definitions for coeliac disease and related terms. Gut. 2013;**62**(1):43-52. Available from: https://gut.bmj.com/content/62/1/43?62/1/43r62/1/43

[96] Hujoel IA, Reilly NR, Rubio-Tapia A. Celiac disease: Clinical features and diagnosis. Gastroenterology Clinics of North America. 2019;**48**(1):19-37. DOI: 10.1016/j.gtc.2018.09.001

[97] Green PHR, Rostami K, Marsh MN. Diagnosis of coeliac disease. Best Practice & Research. Clinical Gastroenterology. 2005;**19**(3):389-400. DOI: 10.1016/j.bpg.2005.02.006

[98] Rubio-Tapia A, Hill ID, Kelly CP, Calderwood AH, Murray JA. American College of Gastroenterology. ACG clinical guidelines: Diagnosis and management of celiac disease. The American Journal of Gastroenterology. 2013;**108**(5):656-676; quiz 677. Available from: https://journals.lww.com/00000434-201305000-00007

[99] Sollid LM, Lie BA. Celiac disease genetics: Current concepts and practical applications. Clinical Gastroenterology and Hepatology. 2005;**3**(9):843-851. DOI: 10.1016/s1542-3565(05)00532-x

[100] Dieli-Crimi R, Cénit MC, Núñez C. The genetics of celiac disease: A comprehensive review of clinical implications. Journal of Autoimmunity. 2015;**64**:26-41. DOI: 10.1016/j.jaut.2015.07.003

[101] Mazzilli MC, Ferrante P, Mariani P, Martone E, Petronzelli F, Triglione P, et al. A study of Italian pediatric celiac disease patients confirms that the primary HLA association is to the DQ (α1∗ 0501, β1∗ 0201) heterodimer. Human Immunology. 1992;**33**(2):133-139. Available from: https://www.sciencedirect.com/science/article/pii/019888599290064T

[102] Fasano A, Catassi C. Clinical practice. Celiac disease. New England Journal of Medicine. 2012;**367**(25):2419-2426. DOI: 10.1056/NEJMcp1113994

[103] Ludvigsson JF, Rubio-Tapia A, van Dyke CT, Melton LJ 3rd, Zinsmeister AR, Lahr BD, et al. Increasing incidence of celiac disease in a north American population. The American Journal of Gastroenterology. 2013;**108**(5):818-824. DOI: 10.1038/ajg.2013.60

[104] Dominguez Castro P, Harkin G, Hussey M, Christopher B, Kiat C, Liong Chin J, et al. Changes in presentation of celiac disease in Ireland from the 1960s to 2015. Clinical Gastroenterology and Hepatology. 2017;**15**(6):864-871.e3. Available from: https://www.sciencedirect.com/science/article/pii/S1542356516311168

[105] Schøsler L, Christensen LA, Hvas CL. Symptoms and findings in adult-onset celiac disease in a historical Danish patient cohort. Scandinavian Journal of Gastroenterology. 2016;**51**(3):288-294. DOI: 10.3109/00365521.2015.1092576

[106] Scanlon SA, Murray JA. Update on celiac disease—Etiology, differential diagnosis, drug targets, and management advances. Clinical and Experimental

Gastroenterology. 2011;**4**:297-311. DOI: 10.2147/CEG.S8315

[107] Biagi F, Trotta L, Alfano C, Balduzzi D, Staffieri V, Bianchi PI, et al. Prevalence and natural history of potential celiac disease in adult patients. Scandinavian Journal of Gastroenterology. 2013;**48**(5):537-542. DOI: 10.3109/00365521.2013.777470

[108] Rubio-Tapia A, Ludvigsson JF, Brantner TL, Murray JA, Everhart JE. The prevalence of celiac disease in the United States. American Journal of Gastroenterology. 2012;**107**(10):1538-1544. quiz 1537, 1545. DOI: 10.1038/ajg.2012.219

[109] Raiteri A, Granito A, Giamperoli A, Catenaro T, Negrini G, Tovoli F. Current guidelines for the management of celiac disease: A systematic review with comparative analysis. World Journal of Gastroenterology. 2022;**28**(1):154-175. DOI: 10.3748/wjg.v28.i1.154

[110] Stern M, Working Group on Serologic Screening for Celiac Disease. Comparative evaluation of serologic tests for celiac disease: A European initiative toward standardization. Journal of Pediatric Gastroenterology and Nutrition. 2000;**31**(5):513-519. DOI: 10.1097/00005176-200011000-00012

[111] Carroccio A, Vitale G, Di Prima L, Chifari N, Napoli S, La Russa C, et al. Comparison of anti-transglutaminase ELISAs and an anti-endomysial antibody assay in the diagnosis of celiac disease: A prospective study. Clinical Chemistry. 2002;**48**(9):1546-1550. Available from: https://www.ncbi.nlm.nih.gov/pubmed/12194932

[112] Korponay-Szabó IR, Dahlbom I, Laurila K, Koskinen S, Woolley N, Partanen J, et al. Elevation of IgG antibodies against tissue transglutaminase as a diagnostic tool for coeliac disease in selective IgA deficiency. Gut. 2003;**52**(11):1567-1571. DOI: 10.1136/gut.52.11.1567

[113] Granito A, Muratori L, Muratori P, Petrolini N, Bianchi FB, Volta U. Antitransglutaminase antibodies and giardiasis. American Journal of Gastroenterology. 2004;**99**(12):2505-2506. DOI: 10.1111/j.1572-0241.2004.41389_9.x

[114] Reeves GEM, Squance ML, Duggan AE, Murugasu RR, Wilson RJ, Wong RC, et al. Diagnostic accuracy of coeliac serological tests: A prospective study. European Journal of Gastroenterology & Hepatology. 2006;**18**(5):493-501. DOI: 10.1097/00042737-200605000-00006

[115] Volta U, Fabbri A, Parisi C, Piscaglia M, Caio G, Tovoli F, et al. Old and new serological tests for celiac disease screening. Expert Review of Gastroenterology & Hepatology. 2010;**4**(1):31-35. DOI: 10.1586/egh.09.66

[116] Sood A, Khurana MS, Mahajan R, Midha V, Puri S, Kaur A, et al. Prevalence and clinical significance of IgA anti-tissue transglutaminase antibodies in patients with chronic liver disease. Journal of Gastroenterology and Hepatology. 2017;**32**(2):446-450. Available from: https://onlinelibrary.wiley.com/doi/10.1111/jgh.13474

[117] Mozo L, Gómez J, Escanlar E, Bousoño C, Gutiérrez C. Diagnostic value of anti-deamidated gliadin peptide IgG antibodies for celiac disease in children and IgA-deficient patients. Journal of Pediatric Gastroenterology and Nutrition. 2012;**55**(1):50-55. DOI: 10.1097/MPG.0b013e31824703c7

[118] Ludvigsson JF, Bai JC, Biagi F, Card TR, Ciacci C, Ciclitira PJ, et al. Diagnosis and management of adult coeliac disease: Guidelines from the British Society of Gastroenterology. Gut. 2014;**63**(8):1210-1228. DOI: 10.1136/gutjnl-2013-306578

[119] Liacouras CA, Furuta GT, Hirano I, Atkins D, Attwood SE, Bonis PA, et al. Eosinophilic esophagitis: Updated consensus recommendations for children and adults. Journal of Allergy and Clinical Immunology. 2011;**128**(1):3-20.e6. quiz 21-2. DOI: 10.1016/j.jaci.2011.02.040

[120] Reed CC, Dellon ES. Eosinophilic esophagitis. The Medical Clinics of North America. 2019;**103**(1):29-42. DOI: 10.1016/j.mcna.2018.08.009

[121] Lucendo AJ, Molina-Infante J, Arias Á, von Arnim U, Bredenoord AJ, Bussmann C, et al. Guidelines on eosinophilic esophagitis: Evidence-based statements and recommendations for diagnosis and management in children and adults. United European Gastroenterology Journal. 2017;**5**(3):335-358. DOI: 10.1177/2050640616689525

[122] Jensen ET, Kappelman MD, Martin CF, Dellon ES. Health-care utilization, costs, and the burden of disease related to eosinophilic esophagitis in the United States. The American Journal of Gastroenterology. 2015;**110**(5):626-632. DOI: 10.1038/ajg.2014.316

[123] Dellon ES, Hirano I. Epidemiology and natural history of eosinophilic esophagitis. Gastroenterology. 2018;**154**(2):319-332.e3. DOI: 10.1053/j.gastro.2017.06.067

[124] Sherrill JD, Gao P-S, Stucke EM, Blanchard C, Collins MH, Putnam PE, et al. Variants of thymic stromal lymphopoietin and its receptor associate with eosinophilic esophagitis. The Journal of Allergy and Clinical Immunology. 2010;**126**(1):160-5.e3. DOI: 10.1016/j.jaci.2010.04.037

[125] Kelly KJ, Lazenby AJ, Rowe PC, Yardley JH, Perman JA, Sampson HA. Eosinophilic esophagitis attributed to gastroesophageal reflux: Improvement with an amino acid-based formula. Gastroenterology. 1995;**109**(5):1503-1512. DOI: 10.1016/0016-5085(95)90637-1

[126] Liacouras CA, Spergel JM, Ruchelli E, Verma R, Mascarenhas M, Semeao E, et al. Eosinophilic esophagitis: A 10-year experience in 381 children. Clinical Gastroenterology and Hepatology. 2005;**3**(12):1198-1206. DOI: 10.1016/s1542-3565(05)00885-2

[127] Arias Á, González-Cervera J, Tenias JM, Lucendo AJ. Efficacy of dietary interventions for inducing histologic remission in patients with eosinophilic esophagitis: A systematic review and meta-analysis. Gastroenterology. 2014;**146**(7):1639-1648. Available from: https://www.sciencedirect.com/science/article/pii/S0016508514002170

[128] Blanchard C, Wang N, Stringer KF, Mishra A, Fulkerson PC, Abonia JP, et al. Eotaxin-3 and a uniquely conserved gene-expression profile in eosinophilic esophagitis. The Journal of Clinical Investigation. 2006;**116**(2):536-547. DOI: 10.1172/JCI26679

[129] Spergel JM, Brown-Whitehorn TF, Beausoleil JL, Franciosi J, Shuker M, Verma R, et al. 14 years of eosinophilic esophagitis: Clinical features and prognosis. Journal of Pediatric Gastroenterology and Nutrition. 2009;**48**(1):30-36. DOI: 10.1097/MPG.0b013e3181788282

[130] Rothenberg ME, Spergel JM, Sherrill JD, Annaiah K, Martin LJ, Cianferoni A, et al. Common variants at 5q22 associate with pediatric eosinophilic esophagitis. Nature Genetics. 2010;**42**(4):289-291. DOI: 10.1038/ng.547

[131] Simon D, Cianferoni A, Spergel JM, Aceves S, Holbreich M, Venter C, et al. Eosinophilic esophagitis is characterized by a non-IgE-mediated food hypersensitivity. Allergy. 2016;**71**(5):611-620. Available from: https://onlinelibrary.wiley.com/doi/10.1111/all.12846

[132] Fogg MI, Ruchelli E, Spergel JM. Pollen and eosinophilic esophagitis. The Journal of Allergy and Clinical Immunology. 2003;**112**(4):796-797. DOI: 10.1016/s0091-6749(03)01715-9

[133] Spergel JM, Beausoleil JL, Mascarenhas M, Liacouras CA. The use of skin prick tests and patch tests to identify causative foods in eosinophilic esophagitis. The Journal of Allergy and Clinical Immunology. 2002;**109**(2):363-368. DOI: 10.1067/mai.2002.121458

[134] Cotton CC, Eluri S, Wolf WA, Dellon ES. Six-food elimination diet and topical steroids are effective for eosinophilic esophagitis: A meta-regression. Digestive Diseases and Sciences. 2017;**62**(9):2408-2420. DOI: 10.1007/s10620-017-4642-7

[135] Blanchard C, Mishra A, Saito-Akei H, Monk P, Anderson I, Rothenberg ME. Inhibition of human interleukin-13-induced respiratory and oesophageal inflammation by anti-human-interleukin-13 antibody (CAT-354). Clinical and Experimental Allergy. 2005;**35**(8):1096-1103. Available from: https://onlinelibrary.wiley.com/doi/10.1111/j.1365-2222.2005.02299.x

[136] Muir AB, Lim DM, Benitez AJ, Modayur Chandramouleeswaran P, Lee AJ, Ruchelli ED, et al. Esophageal epithelial and mesenchymal cross-talk leads to features of epithelial to mesenchymal transition in vitro. Experimental Cell Research. 2013;**319**(6):850-859. DOI: 10.1016/j.yexcr.2012.12.002

[137] Wenzel SE, Schwartz LB, Langmack EL, Halliday JL, Trudeau JB, Gibbs RL, et al. Evidence that severe asthma can be divided pathologically into two inflammatory subtypes with distinct physiologic and clinical characteristics. American Journal of Respiratory and Critical Care Medicine. 1999;**160**(3):1001-1008. DOI: 10.1164/ajrccm.160.3.9812110

[138] Lyles J, Rothenberg M. Role of genetics, environment, and their interactions in the pathogenesis of eosinophilic esophagitis. Current Opinion in Immunology. 2019;**60**:46-53. DOI: 10.1016/j.coi.2019.04.004

[139] Dellon ES. Eosinophilic esophagitis. Gastroenterology Clinics of North America. 2013;**42**(1):133-153. DOI: 10.1016/j.gtc.2012.11.008

[140] Aceves SS, Newbury RO, Dohil R, Schwimmer J, Bastian JF. Distinguishing eosinophilic esophagitis in pediatric patients: Clinical, endoscopic, and histologic features of an emerging disorder. Journal of Clinical Gastroenterology. 2007;**41**(3):252-256. DOI: 10.1097/01.mcg.0000212639.52359.f1

[141] Veerappan GR, Perry JL, Duncan TJ, Baker TP, Maydonovitch C, Lake JM, et al. Prevalence of eosinophilic esophagitis in an adult population undergoing upper endoscopy: A prospective study. Clinical Gastroenterology and Hepatology. 2009;7(4):420-426.e2. DOI: 10.1016/j.cgh.2008.10.009

[142] Sealock RJ, Kramer JR, Verstovsek G, Richardson P, Rugge M, Parente P, et al. The prevalence of oesophageal eosinophilia and eosinophilic oesophagitis: A prospective study in unselected patients presenting to endoscopy. Alimentary Pharmacology & Therapeutics. 2013;**37**(8):825-832. DOI: 10.1111/apt.12268

[143] Joo MK, Park J-J, Kim S-H, Kim KH, Jung W, Yun J-W, et al. Prevalence and endoscopic features of eosinophilic esophagitis in patients with esophageal or upper gastrointestinal symptoms. Journal of Digestive Diseases. 2012;**13**(6):296-303. Available from: https://onlinelibrary.wiley.com/doi/10.1111/j.1751-2980.2012.00589.x

[144] Assiri AM, Saeed A. Incidence and diagnostic features of eosinophilic esophagitis in a group of children with dysphagia and gastroesophageal reflux disease. Saudi Medical Journal. 2014;**35**(3):292-297. Available from: https://www.ncbi.nlm.nih.gov/pubmed/24623210

[145] Gonsalves N, Policarpio-Nicolas M, Zhang Q, Rao MS, Hirano I. Histopathologic variability and endoscopic correlates in adults with eosinophilic esophagitis. Gastrointestinal Endoscopy. 2006;**64**(3):313-319. DOI: 10.1016/j.gie.2006.04.037

[146] Dellon ES, Gibbs WB, Fritchie KJ, Rubinas TC, Wilson LA, Woosley JT, et al. Clinical, endoscopic, and histologic findings distinguish eosinophilic esophagitis from gastroesophageal reflux disease. Clinical Gastroenterology and Hepatology. 2009;7(12):1305-1313. quiz 1261. DOI: 10.1016/j.cgh.2009.08.030

[147] Sperry SLW, Crockett SD, Miller CB, Shaheen NJ, Dellon ES. Esophageal foreign-body impactions: Epidemiology, time trends, and the impact of the increasing prevalence of eosinophilic esophagitis. Gastrointestinal Endoscopy. 2011;**74**(5):985-991. DOI: 10.1016/j.gie.2011.06.029

[148] Dellon ES, Kim HP, Sperry SLW, Rybnicek DA, Woosley JT, Shaheen NJ. A phenotypic analysis shows that eosinophilic esophagitis is a progressive fibrostenotic disease. Gastrointestinal Endoscopy. 2014;**79**(4):577-85.e4. DOI: 10.1016/j.gie.2013.10.027

[149] Dellon ES, Liacouras CA. Advances in clinical management of eosinophilic esophagitis. Gastroenterology. 2014;**147**(6):1238-1254. DOI: 10.1053/j.gastro.2014.07.055

[150] Hiremath GS, Hameed F, Pacheco A, Olive A, Davis CM, Shulman RJ. Esophageal food impaction and eosinophilic esophagitis: A retrospective study, systematic review, and meta-analysis. Digestive Diseases and Sciences. 2015;**60**(11):3181-3193. DOI: 10.1007/s10620-015-3723-8

[151] Dellon ES, Gonsalves N, Hirano I, Furuta GT, Liacouras CA, Katzka DA, et al. ACG clinical guideline: Evidenced based approach to the diagnosis and management of esophageal eosinophilia and eosinophilic esophagitis (EoE). The American Journal of Gastroenterology. 2013;**108**(5):679-692. quiz 693. DOI: 10.1038/ajg.2013.71

[152] Strober W, Fuss I, Mannon P. The fundamental basis of inflammatory bowel disease. The Journal of Clinical Investigation. 2007;**117**(3):514-521. DOI: 10.1172/JCI30587

[153] Hou JK, El-Serag H, Thirumurthi S. Distribution and manifestations of inflammatory bowel disease in Asians, Hispanics, and African Americans: A systematic review. The American Journal of Gastroenterology.

2009;**104**(8):2100-2109. DOI: 10.1038/ajg.2009.190

[154] Molodecky NA, Soon IS, Rabi DM, Ghali WA, Ferris M, Chernoff G, et al. Increasing incidence and prevalence of the inflammatory bowel diseases with time, based on systematic review. Gastroenterology. 2012;**142**(1):46-54.e42. quiz e30. DOI: 10.1053/j.gastro.2011.10.001

[155] Cosnes J, Gower-Rousseau C, Seksik P, Cortot A. Epidemiology and natural history of inflammatory bowel diseases. Gastroenterology. 2011;**140**(6):1785-1794. DOI: 10.1053/j.gastro.2011.01.055

[156] Sairenji T, Collins KL, Evans DV. An update on inflammatory bowel disease. Primary Care. 2017;**44**(4):673-692. DOI: 10.1016/j.pop.2017.07.010

[157] Hampe J, Franke A, Rosenstiel P, Till A, Teuber M, Huse K, et al. A genome-wide association scan of nonsynonymous SNPs identifies a susceptibility variant for Crohn disease in ATG16L1. Nature Genetics. 2007;**39**(2):207-211. DOI: 10.1038/ng1954

[158] Rioux JD, Xavier RJ, Taylor KD, Silverberg MS, Goyette P, Huett A, et al. Genome-wide association study identifies new susceptibility loci for Crohn disease and implicates autophagy in disease pathogenesis. Nature Genetics. 2007;**39**(5):596-604. Available from: https://www.nature.com/articles/ng2032

[159] McCarroll SA, Huett A, Kuballa P, Chilewski SD, Landry A, Goyette P, et al. Deletion polymorphism upstream of IRGM associated with altered IRGM expression and Crohn's disease. Nature Genetics. 2008;**40**(9):1107-1112. Available from: https://www.nature.com/articles/ng.215

[160] Liu T-C, Stappenbeck TS. Genetics and pathogenesis of inflammatory bowel disease. Annual Review of Pathology. 2016;**11**:127-148. DOI: 10.1146/annurev-pathol-012615-044152

[161] Duerr RH, Taylor KD, Brant SR, Rioux JD, Silverberg MS, Daly MJ, et al. A genome-wide association study identifies IL23R as an inflammatory bowel disease gene. Science. 2006;**314**(5804):1461-1463. DOI: 10.1126/science.1135245

[162] Liu Z-J, Yadav P-K, Su J-L, Wang J-S, Fei K. Potential role of Th17 cells in the pathogenesis of inflammatory bowel disease. World Journal of Gastroenterology. 2009;**15**(46):5784-5788. Available from: http://www.wjgnet.com/1007-9327/full/v15/i46/5784.htm

[163] Lees CW, Satsangi J. Genetics of inflammatory bowel disease: Implications for disease pathogenesis and natural history. Expert Review of Gastroenterology & Hepatology. 2009;**3**(5):513-534. DOI: 10.1586/egh.09.45

[164] Polgar N, Csongei V, Szabo M, Zambo V, Melegh BI, Sumegi K, et al. Investigation of JAK2, STAT3 and CCR6 polymorphisms and their gene-gene interactions in inflammatory bowel disease. International Journal of Immunogenetics. 2012;**39**(3):247-252. Available from: https://onlinelibrary.wiley.com/doi/10.1111/j.1744-313X.2012.01084.x

[165] Khor B, Gardet A, Xavier RJ. Genetics and pathogenesis of inflammatory bowel disease. Nature. 2011;**474**(7351):307-317. DOI: 10.1038/nature10209

[166] Ahmad T, Satsangi J, McGovern D, Bunce M, Jewell DP. Review article: The genetics of inflammatory bowel disease. Alimentary Pharmacology

& Therapeutics. 2001;**15**(6):731-748. Available from: http://doi.wiley.com/10.1046/j.1365-2036.2001.00981.x

[167] Ponder A, Long MD. A clinical review of recent findings in the epidemiology of inflammatory bowel disease. Clinical Epidemiology. 2013;**5**:237-247. DOI: 10.2147/CLEP.S33961

[168] Manichanh C, Borruel N, Casellas F, Guarner F. The gut microbiota in IBD. Nature Reviews. Gastroenterology & Hepatology. 2012;**9**(10):599-608. DOI: 10.1038/nrgastro.2012.152

[169] Ni J, Wu GD, Albenberg L, Tomov VT. Gut microbiota and IBD: Causation or correlation? Nature Reviews. Gastroenterology & Hepatology. 2017;**14**(10):573-584. Available from: https://www.nature.com/articles/nrgastro.2017.88

[170] Ananthakrishnan AN, Bernstein CN, Iliopoulos D, Macpherson A, Neurath MF, Ali RAR, et al. Environmental triggers in IBD: A review of progress and evidence. Nature Reviews. Gastroenterology & Hepatology. 2018;**15**(1):39-49. DOI: 10.1038/nrgastro.2017.136

[171] van Eeden SF, Tan WC, Suwa T, Mukae H, Terashima T, Fujii T, et al. Cytokines involved in the systemic inflammatory response induced by exposure to particulate matter air pollutants (PM10). American Journal of Respiratory and Critical Care Medicine. 2001;**164**(5):826-830. DOI: 10.1164/ajrccm.164.5.2010160

[172] Thia KT, Loftus EV Jr, Sandborn WJ, Yang S-K. An update on the epidemiology of inflammatory bowel disease in Asia. The American Journal of Gastroenterology. 2008;**103**(12):3167-3182. DOI: 10.1111/j.1572-0241.2008.02158.x

[173] Kaplan GG, Hubbard J, Korzenik J, Sands BE, Panaccione R, Ghosh S, et al. The inflammatory bowel diseases and ambient air pollution: A novel association. The American Journal of Gastroenterology. 2010;**105**(11):2412-2419. Available from: https://journals.lww.com/00000434-201011000-00017

[174] Loftus EV Jr. Clinical epidemiology of inflammatory bowel disease: Incidence, prevalence, and environmental influences. Gastroenterology. 2004;**126**(6):1504-1517. DOI: 10.1053/j.gastro.2004.01.063

[175] Fletcher J, Cooper SC, Ghosh S, Hewison M. The role of vitamin D in inflammatory bowel disease: Mechanism to management. Nutrients. 2019;**11**(5):10.3390/nu11051019

[176] Sgambato D, Miranda A, Ranaldo R, Federico A, Romano M. The role of stress in inflammatory bowel diseases. Current Pharmaceutical Design. 2017;**23**(27):3997-4002. DOI: 10.2174/1381612823666170228123357

[177] Travis SPL, Stange EF, Lémann M, Oresland T, Chowers Y, Forbes A, et al. European evidence based consensus on the diagnosis and management of Crohn's disease: Current management. Gut. 2006;**55**(Suppl. 1):i16-i35. DOI: 10.1136/gut.2005.081950b

[178] Tripathi K, Feuerstein JD. New developments in ulcerative colitis: Latest evidence on management, treatment, and maintenance. Drugs Context. 2019;**8**:212572. Available from: https://www.drugsincontext.com/new-developments-in-ulcerative-colitis:-latest-evidence-on-management,-treatment,-and-maintenance

[179] Cheifetz AS. Management of active Crohn disease. Journal of the American Medical Association. 2013;**309**(20):2150-2158. DOI: 10.1001/jama.2013.4466

[180] Torres J, Mehandru S, Colombel J-F, Peyrin-Biroulet L. Crohn's disease. Lancet. 2017;**389**(10080):1741-1755. DOI: 10.1016/S0140-6736(16)31711-1

[181] Kappelman MD, Rifas-Shiman SL, Kleinman K, Ollendorf D, Bousvaros A, Grand RJ, et al. The prevalence and geographic distribution of Crohn's disease and ulcerative colitis in the United States. Clinical Gastroenterology and Hepatology. 2007;**5**(12):1424-1429. DOI: 10.1016/j.cgh.2007.07.012

[182] Ng SC, Bernstein CN, Vatn MH, Lakatos PL, Loftus EV Jr, Tysk C, et al. Geographical variability and environmental risk factors in inflammatory bowel disease. Gut. 2013;**62**(4):630-649. DOI: 10.1136/gutjnl-2012-303661

[183] Huang H, Fang M, Jostins L, Umićević Mirkov M, Boucher G, Anderson CA, et al. Fine-mapping inflammatory bowel disease loci to single-variant resolution. Nature. 2017;**547**(7662):173-178. Available from: https://www.nature.com/articles/nature22969

[184] Jostins L, Ripke S, Weersma RK, Duerr RH, McGovern DP, Hui KY, et al. Host–microbe interactions have shaped the genetic architecture of inflammatory bowel disease. Nature. 2012;**491**(7422):119-124. Available from: https://www.nature.com/articles/nature11582

[185] Ellinghaus D, Jostins L, Spain SL, Cortes A, Bethune J, Han B, et al. Analysis of five chronic inflammatory diseases identifies 27 new associations and highlights disease-specific patterns at shared loci. Nature Genetics. 2016;**48**(5):510-518. Available from: https://www.nature.com/articles/ng.3528

[186] Hugot JP, Chamaillard M, Zouali H, Lesage S, Cézard JP, Belaiche J, et al. Association of NOD2 leucine-rich repeat variants with susceptibility to Crohn's disease. Nature. 2001;**411**(6837):599-603. DOI: 10.1038/35079107

[187] Roda G, Chien Ng S, Kotze PG, Argollo M, Panaccione R, Spinelli A, et al. Crohn's disease. Nature Reviews Disease Primers. 2020;**6**(1):22. DOI: 10.1038/s41572-020-0156-2

[188] Zhu L, Shi T, Zhong C, Wang Y, Chang M, Liu X. IL-10 and IL-10 receptor mutations in very early onset inflammatory bowel disease. Gastroenterology Research and Practice. 2017;**10**(2):65-69. DOI: 10.14740/gr740w

[189] Feuerstein JD, Cheifetz AS. Crohn disease: Epidemiology, diagnosis, and management. Mayo Clinic Proceedings. 2017;**92**(7):1088-1103. DOI: 10.1016/j.mayocp.2017.04.010

[190] Danese S, Fiorino G, Mary J-Y, Lakatos PL, D'Haens G, Moja L, et al. Development of red flags index for early referral of adults with symptoms and signs suggestive of Crohn's disease: An IOIBD initiative. Journal of Crohn's & Colitis. 2015;**9**(8):601-606. DOI: 10.1093/ecco-jcc/jjv067

[191] Plevy S, Silverberg MS, Lockton S, Stockfisch T, Croner L, Stachelski J, et al. Combined serological, genetic, and inflammatory markers differentiate non-IBD, Crohn's disease, and ulcerative colitis patients. Inflammatory Bowel Diseases. 2013;**19**(6):1139-1148. DOI: 10.1097/MIB.0b013e318280b19e

[192] Mitsuyama K, Niwa M, Takedatsu H, Yamasaki H, Kuwaki K,

Yoshioka S, et al. Antibody markers in the diagnosis of inflammatory bowel disease. World Journal of Gastroenterology. 2016;**22**(3):1304-1310. DOI: 10.3748/wjg.v22.i3.1304

[193] Gu P, Kapur A, Li D, Haritunians T, Vasiliauskas E, Shih DQ, et al. Serological, genetic and clinical associations with increased health-care resource utilization in inflammatory bowel disease. Journal of Digestive Diseases. 2018;**19**(1):15-23. Available from: https://onlinelibrary.wiley.com/doi/10.1111/1751-2980.12566

[194] Silverberg MS, Satsangi J, Ahmad T, Arnott IDR, Bernstein CN, Brant SR, et al. Toward an integrated clinical, molecular and serological classification of inflammatory bowel disease: Report of a working party of the 2005 Montreal world congress of gastroenterology. Canadian Journal of Gastroenterology. 2005;**19**(Suppl. A):5A-36A. DOI: 10.1155/2005/269076

[195] Colitis-Pathophysiology U. Inflammatory bowel disease part I: Ulcerative colitis-pathophysiology and conventional and alternative treatment options. Alternative Medicine Review. 2003;**8**(3):247-283. Available from: http://www.anaturalhealingcenter.com/documents/Thorne/articles/bowel_disease8-3.pdf

[196] Das KM, Biancone L. Is IBD an autoimmune disorder? Inflammatory Bowel Diseases. 2008;**14**(Suppl. 2):S97-S101. DOI: 10.1002/ibd.20723

[197] Wei S-C, Sollano J, Hui YT, Yu W, Santos Estrella PV, Llamado LJQ, et al. Epidemiology, burden of disease, and unmet needs in the treatment of ulcerative colitis in Asia. Expert Review of Gastroenterology & Hepatology. 2021;**15**(3):275-289. DOI: 10.1080/17474124.2021.1840976

[198] Gajendran M, Loganathan P, Jimenez G, Catinella AP, Ng N, Umapathy C, et al. A comprehensive review and update on ulcerative colitis. Disease-a-Month. 2019;**65**(12):100851. DOI: 10.1016/j.disamonth.2019.02.004

[199] Persson PG, Ahlbom A, Hellers G. Diet and inflammatory bowel disease: A case-control study. Epidemiology. 1992;**3**(1):47-52. DOI: 10.1097/00001648-199201000-00009

[200] Tragnone A, Valpiani D, Miglio F, Elmi G, Bazzocchi G, Pipitone E, et al. Dietary habits as risk factors for inflammatory bowel disease. European Journal of Gastroenterology & Hepatology. 1995;**7**(1):47-51. Available from: https://www.ncbi.nlm.nih.gov/pubmed/7866810

[201] Morris DL, Montgomery SM, Galloway ML, Pounder RE, Wakefield AJ. Inflammatory bowel disease and laterality: Is left handedness a risk? Gut. 2001;**49**(2):199-202. DOI: 10.1136/gut.49.2.199

[202] Frolkis AD, Vallerand IA, Shaheen A-A, Lowerison MW, Swain MG, Barnabe C, et al. Depression increases the risk of inflammatory bowel disease, which may be mitigated by the use of antidepressants in the treatment of depression. Gut. 2019;**68**(9):1606-1612. Available from: https://gut.bmj.com/content/68/9/1606.abstract

[203] Nahon S, Lahmek P, Durance C, Olympie A, Lesgourgues B, Colombel J-F, et al. Risk factors of anxiety and depression in inflammatory bowel disease. Inflammatory Bowel Diseases. 2012;**18**(11):2086-2091. DOI: 10.1002/ibd.22888

[204] Edwards JR, Caplan RD, Van Harrison R. Person-environment fit theory. Theories of Organizational

Stress. 1998;**28**(1):67-94. Available from: https://public.kenan-flagler.unc.edu/faculty/edwardsj/edwardsetal1998.pdf

[205] Selye H. A syndrome produced by diverse nocuous agents. Nature. 1936;**138**(3479):32-32. Available from: https://www.nature.com/articles/138032a0

[206] Mark LF, Matheny J, Kolt GS. Eustress, distress, and interpretation in occupational stress. Journal of Managerial Psychology. 2003;**18**(7):726-744. DOI: 10.1108/02683940310502412

[207] Bhargava D, Trivedi H, Others. A study of causes of stress and stress management among youth. IRA-International Journal of Management & Social Sciences. 2018;**11**(03):108-117. Available from: https://www.academia.edu/download/71278322/1ee3ebf6bf36184a12247dc2d7b17dd80371.pdf

[208] Dean J, Keshavan M. The neurobiology of depression: An integrated view. Asian Journal of Psychiatry. 2017;**27**:101-111. DOI: 10.1016/j.ajp.2017.01.025

[209] Wheaton B. The nature of stressors. In: Horwitz AV, editor. A Handbook for the Study of Mental Health: Social Contexts, Theories, and Systems. Vol. xvii. New York, NY, US: Cambridge University Press; 1999. pp. 176-197. Available from: https://psycnet.apa.org/fulltext/1999-02431-009.pdf

[210] Harbuz MS, Richards LJ, Chover-Gonzalez AJ, Marti-Sistac O, Jessop DS. Stress in autoimmune disease models. Annals of the New York Academy of Sciences. 2006;**1069**:51-61. DOI: 10.1196/annals.1351.005

[211] Glaser R, Kiecolt-Glaser JK. Stress-induced immune dysfunction: Implications for health. Nature Reviews. Immunology. 2005;**5**(3):243-251. DOI: 10.1038/nri1571

[212] Song H, Fang F, Tomasson G, Arnberg FK, Mataix-Cols D, Fernández de la Cruz L, et al. Association of stress-related disorders with subsequent autoimmune disease. Journal of the American Medical Association. 2018;**319**(23):2388-2400. DOI: 10.1001/jama.2018.7028

[213] Selye H. The stress of life. repositorio.cenpat-conicet.gob.ar; 1956. Available from: http://www.repositorio.cenpat-conicet.gob.ar/bitstream/handle/123456789/415/theStressOfLife.pdf?sequence=1

[214] Frieri M. Neuroimmunology and inflammation: Implications for therapy of allergic and autoimmune diseases. Annals of Allergy, Asthma & Immunology. 2003;**90**(6 Suppl. 3):34-40. DOI: 10.1016/s1081-1206(10)61658-4

[215] Stojanovich L, Marisavljevich D. Stress as a trigger of autoimmune disease. Autoimmunity Reviews. 2008;**7**(3):209-213. DOI: 10.1016/j.autrev.2007.11.007

[216] Chida Y, Sudo N, Kubo C. Social isolation stress exacerbates autoimmune disease in MRL/lpr mice. Journal of Neuroimmunology. 2005;**158**(1-2):138-144. DOI: 10.1016/j.jneuroim.2004.09.002

[217] Stam R. PTSD and stress sensitisation: A tale of brain and body part 1: Human studies. Neuroscience and Biobehavioral Reviews. 2007;**31**(4):530-557. DOI: 10.1016/j.neubiorev.2006.11.010

[218] Kalantari M, Hollywood A, Lim R, Hashemi M. Mapping the experiences of people with achalasia from initial symptoms to long-term management. Health Expectations. 2021;**24**(1):131-139.

Available from: https://onlinelibrary.wiley.com/doi/10.1111/hex.13160

[219] Loosen SH, Kandler J, Luedde T, Kostev K, Roderburg C. Achalasia is associated with a higher incidence of depression in outpatients in Germany. PLoS ONE. 2021;**16**(4):e0250503. DOI: 10.1371/journal.pone.0250503

[220] Elseweidy MM. Brief review on the causes, diagnosis and therapeutic treatment of gastritis disease. Alternative and Integrative Medicine. 2017;**06**(01):1-6. Available from: https://www.esciencecentral.org/journals/brief-review-on-the-causes-diagnosis-and-therapeutic-treatment-of-gastritisdisease-2327-5162-1000231.php?aid=86738

[221] Miceli E, Brondino N, Lenti MV, Di Stefano M, Staiani M, Zugnoni F, et al. Impaired quality of life in patients with autoimmune atrophic gastritis. Digestive Diseases and Sciences. 2021;**66**(10):3322-3329. DOI: 10.1007/s10620-020-06656-x

[222] Tenca A, Massironi S, Pugliese D, Consonni D, Mauro A, Cavalcoli F, et al. Gastro-esophageal reflux and antisecretory drugs use among patients with chronic autoimmune atrophic gastritis: A study with pH-impedance monitoring. Neurogastroenterology and Motility. 2016;**28**(2):274-280. DOI: 10.1111/nmo.12723

[223] Takeoka A, Tayama J, Kobayashi M, Sagara I, Ogawa S, Saigo T, et al. Psychological effects of Helicobacter pylori-associated atrophic gastritis in patients under 50 years: A cross-sectional study. Helicobacter. 2017;**22**(6):e12445. Available from: https://onlinelibrary.wiley.com/doi/10.1111/hel.12445

[224] Zhao X, Wu M, Zhang D, Sun Y, Yang Y, Xie H, et al. The relationship of interpersonal sensitivity and depression among patients with chronic atrophic gastritis: The mediating role of coping styles. Journal of Clinical Nursing. 2018;**27**(5-6):e984-e991. DOI: 10.1111/jocn.14114

[225] Smith LB, Lynch KF, Kurppa K, Koletzko S, Krischer J, Liu E, et al. Psychological manifestations of celiac disease autoimmunity in young children. Pediatrics. 2017;**139**(3):10.1542/peds.2016-2848

[226] Zingone F, Swift GL, Card TR, Sanders DS, Ludvigsson JF, Bai JC. Psychological morbidity of celiac disease: A review of the literature. United European Gastroenterology Journal. 2015;**3**(2):136-145. DOI: 10.1177/2050640614560786

[227] Mårild K, Frostell AS, Ludvigsson JF. Psychological stress and coeliac disease in childhood: A cohort study. BMC Gastroenterology. 2010;**10**(106):1-6. DOI: 10.1186/1471-230X-10-106

[228] Addolorato G, Capristo E, Ghittoni G, Valeri C, Mascianà R, Ancona C, et al. Anxiety but not depression decreases in coeliac patients after one-year gluten-free diet: A longitudinal study. Scandinavian Journal of Gastroenterology. 2001;**36**(5):502-506. DOI: 10.1080/00365520119754

[229] Lebwohl B, Haggård L, Emilsson L, Söderling J, Roelstraete B, Butwicka A, et al. Psychiatric disorders in patients with a diagnosis of celiac disease during childhood from 1973 to 2016. Clinical Gastroenterology and Hepatology. 2021;**19**(10):2093-2101.e13. DOI: 10.1016/j.cgh.2020.08.018

[230] Wahab RJ, Beth SA, Derks IPM, Jansen PW, Moll HA, Kiefte-de Jong JC. Celiac disease autoimmunity and emotional and behavioral problems in childhood. Pediatrics.

2019;**144**(4):e20183933. DOI: 10.1542/peds.2018-3933

[231] Prugh DG. A preliminary report on the role of emotional factors in idiopathic celiac disease. Psychosomatic Medicine. 1951;**13**(4):220-241. DOI: 10.1097/00006842-195107000-00002

[232] Addolorato G, Stefanini GF, Capristo E, Caputo F, Gasbarrini A, Gasbarrini G. Anxiety and depression in adult untreated celiac subjects and in patients affected by inflammatory bowel disease: A personality "trait" or a reactive illness? Hepato-Gastroenterology. 1996;**43**(12):1513-1517. Available from: https://www.ncbi.nlm.nih.gov/pubmed/8975957

[233] Carta MG, Hardoy MC, Boi MF, Mariotti S, Carpiniello B, Usai P. Association between panic disorder, major depressive disorder and celiac disease: A possible role of thyroid autoimmunity. Journal of Psychosomatic Research. 2002;**53**(3):789-793. DOI: 10.1016/s0022-3999(02)00328-8

[234] Fera T, Cascio B, Angelini G, Martini S, Guidetti CS. Affective disorders and quality of life in adult coeliac disease patients on a gluten-free diet. European Journal of Gastroenterology & Hepatology. 2003;**15**(12):1287-1292. DOI: 10.1097/00042737-200312000-00006

[235] Cicarelli G, Della Rocca G, Amboni M, Ciacci C, Mazzacca G, Filla A, et al. Clinical and neurological abnormalities in adult celiac disease. Neurological Sciences. 2003;**24**(5):311-317. DOI: 10.1007/s10072-003-0181-4

[236] Pynnönen PA, Isometsä ET, Aronen ET, Verkasalo MA, Savilahti E, Aalberg VA. Mental disorders in adolescents with celiac disease. Psychosomatics. 2004;**45**(4):325-335. DOI: 10.1176/appi.psy.45.4.325

[237] Siniscalchi M, Iovino P, Tortora R, Forestiero S, Somma A, Capuano L, et al. Fatigue in adult coeliac disease. Alimentary Pharmacology & Therapeutics. 2005;**22**(5):489-494. DOI: 10.1111/j.1365-2036.2005.02619.x

[238] Ludvigsson JF, Reutfors J, Ösby U, Ekbom A, Montgomery SM. Coeliac disease and risk of mood disorders—A general population-based cohort study. Journal of Affective Disorders. 2007;**99**(1-3):117-126. Available from: https://www.sciencedirect.com/science/article/pii/S0165032706003636?casa_token=M7hqiBpFN-oAAAAA:vXR_Qk5XdsAIQyOSpDnwsxNjDw8qYSKJ_w03LMBFadsbH7kdvTWpRPJ4fO2w8A-13dPKBKiCmrI

[239] Mazzone L, Reale L, Spina M, Guarnera M, Lionetti E, Martorana S, et al. Compliant gluten-free children with celiac disease: An evaluation of psychological distress. BMC Pediatrics. 2011;**11**:46. DOI: 10.1186/1471-2431-11-46

[240] Butwicka A, Lichtenstein P, Frisén L, Almqvist C, Larsson H, Ludvigsson JF. Celiac disease is associated with childhood psychiatric disorders: A population-based study. The Journal of Pediatrics. 2017;**184**:87-93.e1. DOI: 10.1016/j.jpeds.2017.01.043

[241] Carta MG, Conti A, Lecca F, Sancassiani F, Cossu G, Carruxi R, et al. The burden of depressive and bipolar disorders in celiac disease. Clinical Practice and Epidemiology in Mental Health. 2015;**11**(1):180-185. Available from: https://clinical-practice-and-epidemiology-in-mental-health.com/VOLUME/11/PAGE/180/

[242] Addolorato G, Mirijello A, D'Angelo C, Leggio L, Ferrulli A,

Vonghia L, et al. Social phobia in coeliac disease. Scandinavian Journal of Gastroenterology. 2008;**43**(4):410-415. DOI: 10.1080/00365520701768802

[243] Clappison E, Hadjivassiliou M, Zis P. Psychiatric manifestations of coeliac disease, a systematic review and meta-analysis. Nutrients. 2020;**12**(1):142-158. DOI: 10.3390/nu12010142

[244] Taft TH, Guadagnoli L, Edlynn E. Anxiety and depression in eosinophilic esophagitis: A scoping review and recommendations for future research. Journal of Asthma and Allergy. 2019;**12**:389-399. DOI: 10.2147/JAA.S193045

[245] de Rooij WE, Bennebroek Evertsz' F, Lei A, Bredenoord AJ. Mental distress among adult patients with eosinophilic esophagitis. Neurogastroenterology and Motility. 2021;**33**(7):e14069. DOI: 10.1111/nmo.14069

[246] Erickson MA, Morofuji Y, Owen JB, Banks WA. Rapid transport of CCL11 across the blood-brain barrier: Regional variation and importance of blood cells. The Journal of Pharmacology and Experimental Therapeutics. 2014;**349**(3):497-507. DOI: 10.1124/jpet.114.213074

[247] Krathwohl MD, Kaiser JL. Chemokines promote quiescence and survival of human neural progenitor cells. Stem Cells. 2004;**22**(1):109-118. DOI: 10.1634/stemcells.22-1-109

[248] Parajuli B, Horiuchi H, Mizuno T, Takeuchi H, Suzumura A. CCL11 enhances excitotoxic neuronal death by producing reactive oxygen species in microglia. Glia. 2015;**63**(12):2274-2284. DOI: 10.1002/glia.22892

[249] Simon NM, McNamara K, Chow CW, Maser RS, Papakostas GI, Pollack MH, et al. A detailed examination of cytokine abnormalities in major depressive disorder. European Neuropsychopharmacology. 2008;**18**(3):230-233. DOI: 10.1016/j.euroneuro.2007.06.004

[250] Crohn BB, Burrill B, Ginzburg, Leon, Oppenheimer, Gordon D. Regional Ileitis: A pathologic and clinical entity. JAMA. 1984;**251**(1):73-79. Available from: https://jamanetwork.com/journals/jama/article-abstract/390889

[251] Parfitt HL. Psychiatric aspects of regional enteritis. Canadian Medical Association Journal. 1967;**97**(13):807-811. Available from: https://www.ncbi.nlm.nih.gov/pubmed/6050900

[252] Ford CV, Glober GA, Castelnuovo-Tedesco P. A psychiatric study of patients with regional enteritis. Journal of the American Medical Association. 1969;**208**(2):311-315. Available from: https://www.ncbi.nlm.nih.gov/pubmed/4388299

[253] Cohn EM, Lederman II, Shore E. Regional enteritis and its relation to emotional disorders. The American Journal of Gastroenterology. 1970;**54**(4):378-387. Available from: https://www.ncbi.nlm.nih.gov/pubmed/5478850

[254] Schwartz RA, Schwartz IK. Psychiatric disorders associated with Crohn's disease. International Journal of Psychiatry in Medicine. 1982;**12**(1):67-73. DOI: 10.2190/fl8y-1wu6-1n5l-1fqv

[255] Helzer JE, Chammas S, Norland CC, Stillings WA, Alpers DH. A study of the association between Crohn's disease and psychiatric illness. Gastroenterology. 1984;**86**(2):324-330. Available from:

https://www.ncbi.nlm.nih.gov/pubmed/6690360

[256] Engel GL. Psychological factors in ulcerative colitis in man and gibbon. Gastroenterology. 1969;**57**(3):362-365. Available from: https://www.ncbi.nlm.nih.gov/pubmed/5808494

[257] Mawdsley JE, Macey MG, Feakins RM, Langmead L, Rampton DS. The effect of acute psychologic stress on systemic and rectal mucosal measures of inflammation in ulcerative colitis. Gastroenterology. 2006;**131**(2):410-419. DOI: 10.1053/j.gastro.2006.05.017

[258] Angelopoulos NV, Mantas C, Dalekos GN, Vasalos K, Tsianos EV. Psychiatric factors in patients with ulcerative colitis according to disease activity. European Psychiatry. 1996;**10**(2):87-99. Available from: https://www.researchgate.net/profile/Nikiforos-Angelopoulos/publication/289310216_Psychiatric_factors_in_patients_with_ulcerative_colitis_according_to_disease_activity/links/58a746fe4585150402f29fda/Psychiatric-factors-in-patients-with-ulcerative-colitis-according-to-disease-activity.pdf

[259] Marrie RA, Walld R, Bolton JM, Sareen J, Walker JR, Patten SB, et al. Physical comorbidities increase the risk of psychiatric comorbidity in immune-mediated inflammatory disease. General Hospital Psychiatry. 2018;**51**:71-78. DOI: 10.1016/j.genhosppsych.2018.01.003

[260] Araki M, Shinzaki S, Yamada T, Arimitsu S, Komori M, Shibukawa N, et al. Psychologic stress and disease activity in patients with inflammatory bowel disease: A multicenter cross-sectional study. PLoS ONE. 2020;**15**(5):e0233365. DOI: 10.1371/journal.pone.0233365

[261] Lix LM, Graff LA, Walker JR, Clara I, Rawsthorne P, Rogala L, et al. Longitudinal study of quality of life and psychological functioning for active, fluctuating, and inactive disease patterns in inflammatory bowel disease. Inflammatory Bowel Diseases. 2008;**14**(11):1575-1584. DOI: 10.1002/ibd.20511

[262] Bernstein CN. Psychological stress and depression: Risk factors for IBD? Digestive Diseases. 2016;**34**(1-2):58-63. DOI: 10.1159/000442929

[263] Graff LA, Walker JR, Bernstein CN. Depression and anxiety in inflammatory bowel disease: A review of comorbidity and management. Inflammatory Bowel Diseases. 2009;**15**(7):1105-1118. DOI: 10.1002/ibd.20873

[264] Walker JR, Ediger JP, Graff LA, Greenfeld JM, Clara I, Lix L, et al. The Manitoba IBD cohort study: A population-based study of the prevalence of lifetime and 12-month anxiety and mood disorders. The American Journal of Gastroenterology. 2008;**103**(8):1989-1997. DOI: 10.1111/j.1572-0241.2008.01980.x

[265] Guthrie E, Jackson J, Shaffer J, Thompson D, Tomenson B, Creed F. Psychological disorder and severity of inflammatory bowel disease predict health-related quality of life in ulcerative colitis and Crohn's disease. The American Journal of Gastroenterology. 2002;**97**(8):1994-1999. DOI: 10.1111/j.1572-0241.2002.05842.x

[266] Ursell LK, Metcalf JL, Parfrey LW, Knight R. Defining the human microbiome. Nutrition Reviews. 2012;**70**(Suppl. 1):S38-S44. DOI: 10.1111/j.1753-4887.2012.00493.x

[267] Chen X, D'Souza R, Hong S-T. The role of gut microbiota in the gut-brain axis: Current challenges and perspectives. Protein & Cell. 2013;**4**(6):403-414. DOI: 10.1007/s13238-013-3017-x

[268] Favier CF, de Vos WM, Akkermans ADL. Development of bacterial and bifidobacterial communities in feces of newborn babies. Anaerobe. 2003;**9**(5):219-229. DOI: 10.1016/j.anaerobe.2003.07.001

[269] Koenig JE, Spor A, Scalfone N, Fricker AD, Stombaugh J, Knight R, et al. Succession of microbial consortia in the developing infant gut microbiome. Proceedings of the National Academy of Sciences of the United States of America. 2011;**108**(Suppl. 1):4578-4585. DOI: 10.1073/pnas.1000081107

[270] Milani C, Duranti S, Bottacini F, Casey E, Turroni F, Mahony J, et al. The first microbial colonizers of the human gut: Composition, activities, and health implications of the infant gut microbiota. Microbiology and Molecular Biology Reviews. 2017;**81**(4):e00036-17. DOI: 10.1128/MMBR.00036-17

[271] Thursby E, Juge N. Introduction to the human gut microbiota. Biochemical Journal. 2017;**474**(11):1823-1836. DOI: 10.1042/BCJ20160510

[272] Fitzgibbon G, Mills KHG. The microbiota and immune-mediated diseases: Opportunities for therapeutic intervention. European Journal of Immunology. 2020;**50**(3):326-337. Available from: https://onlinelibrary.wiley.com/doi/10.1002/eji.201948322

[273] West CE, Renz H, Jenmalm MC, Kozyrskyj AL, Allen KJ, Vuillermin P, et al. The gut microbiota and inflammatory noncommunicable diseases: Associations and potentials for gut microbiota therapies. The Journal of Allergy and Clinical Immunology. 2015;**135**(1):3-13; quiz 14. DOI: 10.1016/j.jaci.2014.11.012

[274] Mehta S, Huey SL, McDonald D, Knight R, Finkelstein JL. Nutritional interventions and the gut microbiome in children. Annual Review of Nutrition. 2021;**41**:479-510. DOI: 10.1146/annurev-nutr-021020-025755

[275] Qureshi F, Edhi F, Ozga AT. The impact of delivery mode on the human microbiome: A global perspective. Mako: NSU Undergraduate Student Journal. 2022;**2022**(1):1. Available from: https://nsuworks.nova.edu/mako/vol2022/iss1/1/

[276] Cattell M, Lai S, Cerny R, Medeiros DM. A new mechanistic scenario for the origin and evolution of vertebrate cartilage. PLoS ONE. 2011;**6**(7):e22474. DOI: 10.1371/journal.pone.0022474

[277] Wang H-X, Wang Y-P. Gut Microbiota-brain axis. Chinese Medical Journal. 2016;**129**(19):2373-2380. Available from: https://journals.lww.com/00029330-201610050-00016

[278] Cryan JF, O'Mahony SM. The microbiome-gut-brain axis: From bowel to behavior. Neurogastroenterology and Motility. 2011;**23**(3):187-192. Available from: https://onlinelibrary.wiley.com/doi/10.1111/j.1365-2982.2010.01664.x

[279] Barbara G, Brummer RJ, Delzenne N, Doré J, Ghoeler L, Konsman J, et al. Investigating the crosstalk between the gut microbiota and the host: The gut-brain axis. Consensus Report Warsaw. 2007. Available from: http://citeseerx.ist.psu.edu/viewdoc/download?doi=10.1.1.470.9602&rep=rep1&type=pdf

[280] Kinlein SA, Karatsoreos IN. The hypothalamic-pituitary-adrenal axis

as a substrate for stress resilience: Interactions with the circadian clock. Frontiers in Neuroendocrinology. 2020;**56**:100819. DOI: 10.1016/j.yfrne.2019.100819

[281] Whirledge S, Cidlowski JA. Glucocorticoids, stress, and fertility. Minerva Endocrinologica. 2010;**35**(2):109-125. Available from: https://www.ncbi.nlm.nih.gov/pubmed/20595939

[282] Rea K, Dinan TG, Cryan JF. The microbiome: A key regulator of stress and neuroinflammation. Neurobiology of Stress. 2016;**4**:23-33. DOI: 10.1016/j.ynstr.2016.03.001

[283] Gądek-Michalska A, Tadeusz J, Rachwalska P, Bugajski J. Cytokines, prostaglandins and nitric oxide in the regulation of stress-response systems. Pharmacological Reports. 2013;**65**(6):1655-1662. DOI: 10.1016/s1734-1140(13)71527-5

[284] Biesmans S, Bouwknecht JA, Ver Donck L, Langlois X, Acton PD, De Haes P, et al. Peripheral administration of tumor necrosis factor-alpha induces neuroinflammation and sickness but not depressive-like behavior in mice. BioMed Research International. 2015;**2015**:716920. DOI: 10.1155/2015/716920

[285] Ohland CL, Kish L, Bell H, Thiesen A, Hotte N, Pankiv E, et al. Effects of Lactobacillus helveticus on murine behavior are dependent on diet and genotype and correlate with alterations in the gut microbiome. Psychoneuroendocrinology. 2013;**38**(9):1738-1747. DOI: 10.1016/j.psyneuen.2013.02.008

[286] Clapp M, Aurora N, Herrera L, Bhatia M, Wilen E, Wakefield S. Gut microbiota's effect on mental health: The gut-brain axis. Clinics and Practice. 2017;**7**(4):987. DOI: 10.4081/cp.2017.987

[287] Günther C, Rothhammer V, Karow M, Neurath MF, Winner B. The gut-brain axis in inflammatory bowel disease—current and future perspectives. International Journal of Molecular Sciences. 2021;**22**(16):8870. Available from: https://www.mdpi.com/1422-0067/22/16/8870

[288] Severance EG, Tveiten D, Lindström LH, Yolken RH, Reichelt KL. The gut microbiota and the emergence of autoimmunity: Relevance to major psychiatric disorders. Current Pharmaceutical Design. 2016;**22**(40):6076-6086. DOI: 10.2174/1381612822666160914183804

[289] Ancona A, Petito C, Iavarone I, Petito V, Galasso L, Leonetti A, et al. The gut–brain axis in irritable bowel syndrome and inflammatory bowel disease. Digestive and Liver Disease. 2021;**53**(3):298-305. Available from: https://www.sciencedirect.com/science/article/pii/S159086582031046X

[290] Sinagra E, Utzeri E, Morreale GC, Fabbri C, Pace F, Anderloni A. Microbiota-gut-brain axis and its affect inflammatory bowel disease: Pathophysiological concepts and insights for clinicians. World Journal of Clinical Cases. 2020;**8**(6):1013-1025. Available from: https://www.wjgnet.com/2307-8960/full/v8/i6/1013.htm

[291] Moloney RD, Johnson AC, O'Mahony SM, Dinan TG, Greenwood-Van Meerveld B, Cryan JF. Stress and the microbiota-gut-brain axis in visceral pain: Relevance to irritable bowel syndrome. CNS Neuroscience & Therapeutics. 2016;**22**(2):102-117. Available from: https://onlinelibrary.wiley.com/doi/10.1111/cns.12490

[292] Bavelaar BM. The microbiome-gut-brain-axis in coeliac disease: Mechanisms for depression and anxiety co-morbidity. studenttheses.uu.nl; 2013. Available from: https://studenttheses.uu.nl/handle/20.500.12932/14597

Chapter 13

Autoimmune Diseases of the GI Tract Part II: Emergence of Diagnostic Tools and Treatments

Mahmoud Khatib A.A. Al-Ruweidi, Nada Khater, Haya Rashid Alkaabi, Maram Hasan, Mohammed Murtaza and Huseyin C. Yalcin

Abstract

Autoimmune diseases (AD) have emerged as a pandemic in our modern societies, especially after the World War II. In part I, we have reviewed five main diseases and shed light on different aspects from introducing the concept of autoimmunity, the description of the disease's pathogenesis and the diagnosis, the role of antibodies as markers for the prediction of the disease, the link between the gut and brain through what is known as the gut–brain axis, and the relationship of this axis in GI autoimmune diseases. In this chapter, we review the role of antibodies as markers for the prediction of the disease, artificial intelligence in GI autoimmune diseases, the nutritional role and implications in the five GI autoimmune diseases, and finally the treatment of those diseases.

Keywords: achalasia, atrophic autoimmune Gastritis, celiac disease, eosinophilic esophagitis, inflammatory bowel diseases, Crohn disease, ulcerative colitis, autoantibodies, artificial intelligence, machine learning, immunological nutrition

1. Introduction

Autoimmune diseases can be defined as the inability of the human system to distinguish its own bodies from foreign bodies [1, 2]. The diagnosis of autoimmune diseases is not easy. However, with the emergence of the serological tool and our progress in understanding the science of the immunology, antibodies provide an excellent role in the prediction of GI autoimmune diseases. They serve as markers for the prediction or confirming the presence of an autoimmune disease. The emergence of artificial intelligence (AI) and the integration of machine learning (ML) algorithms in many applications and their incorporation into the health sector as well open the gate for improved diagnosis and management of the diseases. In ADs, they could be great asset as many of the diagnostic tests depend on imaging techniques that their interpretations could vary from one clinician to another. The treatment of GI autoimmune diseases could be variable from the need for elimination diets to surgical interventions depending on the case and the disease.

In the previous chapter, we provided an introductory background on autoimmune diseases, definition of pathophysiology and etiology of autoimmune diseases, a review of the five most common GI autoimmune diseases, the role of psychological association with GI Tract autoimmunity, and microbiome and AD: The Gut–Brain Axis. In this chapter, which is a continuation of the second chapter we discuss other aspects that include shading the light and in-depth review of the fascinating roles of antibodies as predictors for the GI autoimmune diseases, the role of dietary and nutritional implications, the use of artificial intelligence in diagnosis of GI autoimmune diseases, and the treatment of GI autoimmune diseases.

2. Antibodies as predictors of the GI autoimmune diseases

The presence of autoantibodies in patient's serum has been considered a common symbol of autoimmune diseases. Autoantibodies are produced by pathogenic B cells, to target individual's own tissues. Many have considered them a clinical marker of diseases that can diagnose and predict prognosis of the disease. However, one GI disease can have more than one autoantibody, and many other diseases share the same autoantibodies. Some autoantibodies are specific to specific diseases, and some are not [3, 4]. This section is intended to give an overview of the most common and important autoantibodies in GI autoimmune diseases.

2.1 Anti-parietal cell antibody (APCA)

Anti-parietal cell (APCA) is an autoantibody that targets H+/K+ ATPase, a heterodimer made of alpha- and beta-subunits. This enzyme is a proton pump located on parietal cells, that is involved in the production and release of high amount of hydrochloric acid [5]. Studies have shown that the isotypes of APCA immunoglobulins ate A, M, and G isotypes [6]. Many studies have associated APCA with autoimmune GI diseases, such as atrophic gastritis and Helicobacter pylori-associated atrophic gastritis. For example, H+/K+ ATPase has been considered a major antigen in H. pylori-associated antigastric autoimmunity [7]. Antibodies against this antigen are believed to have a crucial role in H. pylori-associated atrophic gastritis too. This was concluded by Ito et al., as the levels of APCA were significantly higher in patients with severe atrophy than in patients with mild atrophy ($P = 0.01$) [8]. Furthermore, H+/K+ ATPase is also a major antigen in autoimmune gastritis [9]. It is important to note that chronic gastritis, most commonly autoimmune gastritis and H. pylori gastritis, can result in atrophic gastritis [10]. To help identify atrophy, Claeys et al. state that APCA, which are closely associated with classical autoimmune gastritis, can be used as useful indicators for the atrophy of body mucosa in chronic H. pylori gastritis [7, 11].

Moreover, APCA can also predict one's risk for developing atrophic gastritis and its severity. For instance, Zhang et al. detected an overall APCA prevalence of 19.5%. They discovered that APCA prevalence was strongly associated with an approximately fourfold increased risk of chronic atrophic gastritis (CAG) (46.2% vs. 18.0%, adjusted OR = 3.8; 95% CI: 3.1–4.7). This striking association was even more increased with raising severity of chronic autoimmune gastritis (CAG) defined by PGI levels. As a result, they concluded that examining APCA levels might be a useful marker to be added when screening patients for CAG [11]. De Block et al. also conclude that individuals with positive anti-gastric parietal cell are at a higher risk for atrophic gastritis [12]. To summarize, occurrence of APCA can help predict the development of atrophic gastritis in the future.

2.2 Intrinsic factor antibodies (IFA)

Intrinsic factor antibodies are IgG autoantibodies that attack a 60-Kd intrinsic factor glycoprotein secreted by parietal cells that bind to vitamin B-12 and allow for its absorption. There are two types of those autoantibodies, the first is called Type I, which targets cobalamin binding sites and prevents the combination of IF and vitamin B-12. The other type is called Type 2, which targets ileal mucosa receptor and prevents IF-vitamin B-12 complex attachment to it [13, 14]. In addition to H^+/K^+ ATPase, intrinsic factor (IF) is also a crucial autoantigen in pernicious anemia [15]. IFA have been detected in 13 to 60% of patients with pernicious anemia [16–19] Type I IFA is found to be the predominant type in those cases. While Type II is only found in about half of those cases, Type II IFA is rarely detected in the absence of Type I IFA [14]. IFA has also been associated with autoimmune body gastritis. For instance, Lahner states that intrinsic factor autoantibodies are 100% specific for biopsy-proven autoimmune body gastritis. Moreover, they detected IFA in 27% of patients with ABG and none in healthy controls. Finally, Lahner et al. concluded that testing patients for IFA along with APCA can significantly increase the diagnostic accuracy for atrophic body gastritis and pernicious anemia [20].

2.3 Anti-transglutaminase (TGA)

Anti-transglutaminase (TGA) are autoantibodies targeting tissue transglutaminase (tTG) or transglutaminase 2, which is a 76-kD calcium-dependent ubiquitous enzyme released during inflammation that catalyzes the post-translational modification of proteins [21]. This ubiquitously expressed enzyme also plays a role as a GTPase, ATPase, and protein kinase [22]. This enzyme has been considered a specific marker for celiac disease (CD). Dieterich et al. were one of the first scientists to determine the role of tTG in CD [23]. Sabatino et al., further explain that tTG has at least two roles in CD one is being a deamidating enzyme to enhance the immunostimulatory effect of gluten, and the other as a target autoantigen in the immune response [21]. In a systematic review done by Ghatti et al., 11 studies detected intestinal transglutaminase 2 Immunoglobin A (IgA) deposits in 100% of adults with overt CD, while the prevalence in children ranged between 73.2 and 100% [24]. Similarly, in a study examining children, Borrelli et al. detected anti-TG2 IgA deposits in all 53 patients with confirmed CD and three out of three potential patients with CD. As a result, Borrelli et al. concluded that intestinal deposits of anti-TG2 appear early in the course of the disease and are of constant presence in patients with CD [25]. Furthermore, other studies detected TGA presence in serum of patients with CD. For example, Miller et al. detected TGA presence in 46 patients with untreated CD (sensitivity 100%) [26]. Moreover, in a study testing 37 patients with biopsy-confirmed CD, Damoiseaux et al. found that 86.5% have IgA antirecombinant human tissue transglutaminase antibodies (rh-tTGA) [27]. In addition, Tola et al. found significantly high levels of TGA in patients with CD [28]. TGA can be even found in the serum of asymptomatic individuals who later in life develop CD [29], which further emphasizes its importance in detecting CD. In fact, Rubio-Tapia et al. have found that elevated IgA anti-TGA has been associated with an increased mortality rate among men aged 50 years old. They also concluded that IgA anti-TGA could be used as a nonspecific marker of serious disease in older men [30]. There are few studies documenting IgA anti-TGA in Crohn's disease; however, there are other conflicting reports about anti-TGA IgG presence [31, 32]. Tola et al. have found significantly low positive values in IBD (Crohn's disease and ulcerative colitis (UC)). In addition, Tursi et al. also detected antitransglutaminase (anti-tTG) in 5 out of 27 (18.52%) patients with Crohn's disease [33]. While Shor et al. also detected positive IgG tTG in 4 out of 26 patients with UC, and 2 out of 194 in healthy controls (11.1% versus 1%; P = 0.018) [28]. As a result, TGA was not found to be useful in IBD; therefore, serological screening testing was only recommended if there is a relevant clinical suspicion of Crohn's [34]. While in IBD, Watanabe et al. detected significantly higher levels of antibodies against tissue transglutaminase in patients, which also correlated with disease severity [35]. IgA against the autoantigen tissue transglutaminase (tTG) is frequently associated with untreated Crohn's disease but disappears with gluten exclusion [23]. TGA has also been associated with Crohn's disease and its severity.

Moreover, Fevre et al. also detected anti-tissue transglutaminase antibodies (TTG Ab) in 23% of patients diagnosed with eosinophilic esophagitis (EoE) during the study. Shor et al. also detected positive IgG tTG in 4 out of 26 patients with ulcerative colitis, and 2 out of 194 in healthy controls (11.1% vs. 1%; P = 0.018) [28].

2.4 Anti-gliadin antibodies (AGA)

Anti-gliadin antibodies are antibodies that are targeted toward Gliadin, a protein found in Bread wheat, rye, and barley [36]. AGA IgA antibodies have been shown to be one of the hallmarks of CD. For instance, Jassim et al. tested AGA-IgA and AGA-IgG in 58 patients with celiac disease and 27 healthy control and found that both antibodies were significantly higher in the CD patients than in control [3]. In addition, Damoiseaux et al. found that 73% of 37 patients with biopsy-confirmed CD have IgA AGA in their serum [27]. Moreover, Lindqvist et al. have found that patients with psoriatic arthritis have an increased prevalence of high serum IgA AGA and of CD [37]. In fact, CD was commonly found in patients with isolated positive AGA; therefore, Taylor et al. recommended that all those patients should be referred to gastroscopy (OGD) and D2 biopsy to undergo further investigation [38]. Both AGA and anti-tTG antibodies are considered good serologic indicators of CD, and they can be even found in the serum of asymptomatic individuals who later in life develop CD [29, 39]. The sensitivities detected for tTG, AGA IgA, and AGA IgG are 90 to 98%, 80 to 90%, and 75 to 85%, respectively. While the specificities were found to be 95 to 97%, 85 to 95%, and 75 to 90%, respectively [40, 41]. Moving to Crohn's disease, Tursi et al. detected AGA in 8 out of the 27 patients with Crohn's disease (29.63%) [33]. Furthermore, Shor et al. detected high levels of AGA IgG in 17 out of 83 patients with Crohn's disease, and 20 out of 194 in healthy controls (20.5% vs. 10.3%; P = 0.023) [28].

2.5 Anti-endomysial antibodies (EMA)

The endomysium is a perivascular connective tissue that separates smooth muscle fibers from each other [42]. Dieterich et al. stated that tissue transglutaminase is the target antigen in endomysium in CD [23]. Detection of anti-endomysial antibodies (EMA) in blood has been used as the most specific test to diagnose CD. However, EMA lacks sensitivity, particularly in the earlier stages of disease exhibiting mild villous atrophy [38, 43]. On the other hand, Farrell et al. state that sensitivity of EMA

IgA is equal to or exceeds 90%, while the specificity approaches 100% in untreated patients with CD. Kanthi et al., similarly, mention that the sensitivity and specificity of IgA EMAs are found to be 85–98% and 97–100%, respectively. As a result of that, blood EMA testing is estimated to have a high positive predictive value [44]. Another characteristic of EMA includes that the antibodies' levels fall after following a gluten-free diet [45]. Similar to IgA AGA, IgA EMA antibody will also not be detected in IgA deficient CD patients [46]. As for using it, Keren et al. recommended testing for EMA to help select patients who would be qualified for a biopsy [47]. While others used it for screening and estimating the prevalence of CD [48]. Kanthi et al. stated that EMA IgG_1 have been used for diagnosing celiac disease, especially in IgA-deficient patients [44]. EMA has also been detected in other diseases. For example, Damoiseaux et al. detected IgA EMA presence in 86.5% of 37 patients with biopsy-confirmed CD [27]. Moving on to Crohn's disease, Tursi et al. only found anti-endomysial antibody (EMA) in 4 out of 27 patients with Crohn's disease (14.81%) [33].

2.6 Anti-Saccharomyces cerevisiae antibodies (ASCA)

Anti-S. cerevisiae antibodies (ASCA) are autoantibodies targeted toward the mannose residues on unicellular fungus *S. cerevisiae* (S. cerevisiae) [49]. Several studies associated those antibodies with GI autoimmune diseases. For instance, Shor et al. detected high levels of IgA ASCA in 16 out 83 patients with Crohn's disease patients, while only 2 of the 198 healthy controls had positive ASCA titers (19.3% vs. 1%; P = 0.000). In addition to IgA ASCA, the high titers of IgG ASCA were detected in 23 out of 83 patients with Crohn's disease, while only one healthy control of the 194 had a positive IgG ASCA titers (27.7% versus 0.5%; P = 0.000) [28]. Even in a pediatric population, El-Matary et al. detected a correlation between both ASCA IgA and IgG titers and clinical Crohn's disease activity [50]. Furthermore, Smids et al. detected IgA ASCA in 23% of Crohn's disease patients and only 3% of UC patients [51]. However, ASCA is not a specific marker for Crohn's disease, since it was also detected in patients with CD. Kotze et al. tested patients with Crohn's disease, and 3 groups of CD patients, including those at time of diagnosis, patients that follow gluten-free diet, and lastly, others who admit transgression in their gluten-free diet. Kotze et al. found statistically significant levels of ASCA IgA in patients with Crohn's disease, in addition to patients with CD at diagnosis and others that admit transgression in their gluten-free diet. Furthermore, ASCA IgG was also positive in Crohn's disease and in all groups of CD. They concluded in their study that ASCA detection is associated with the inflammation of small intestine [52]. Moreover, it was also detected in CD. For example, Damoiseaux et al. detected ASCA presence in 16 of 37 patients with biopsy-confirmed CD [27]. Also, Granito et al. detected IgA and/or IgG ASCA in 59% of 105 subjects with CD at the time of diagnosis. In their study, they did not find any significant correlation between ASCA positivity and severity of small intestinal mucosal damage. Furthermore, they tested 93% of revaluated coeliac patients again after they had followed a gluten-free diet and did not detect IgA ASCA. Instead, 83% of the subjects maintained their IgG ASCA reactivity [53]. ASCA can even help in predicting the development of CD in patients before they present with symptoms [53, 54]. Granito et al. called them the "potential/silent" CD and suggested diagnosing them with CD in case of positive serological markers (EmA and tTG) and typical HLA predisposing genotype (DQ8 or DQ2) [55]. In a study involving a Korean cohort, Choi et al. detected a positive rate of ASCA in 44.35% of patients with intestinal Behcet disease, compared to 8.8% in healthy control subjects [56]. Furthermore, Cheng et al. concluded in a metanalysis of 9 studies, that there is a strong correlation between ASCA and gastrointestinal Behcet disease, specially ASCA-IgG (OR = 5.50 (95% CI 2.58 to 11.55), p = 0.000) and ASCA-IgG + IgA (OR = 5.36 (95% CI 1.40 to 20.45), p = 0.014). The study also found that in gastrointestinal Behcet disease the positivity rate of ASCA was higher significantly than that in UC: IgA (OR = 2.13 (95% CI 1.30 to 3.50), p = 0.003); IgG + IgA (OR = 2.19 (95% CI 1.03 to 4.66), p = 0.042); IgG/IgA ((=2.03 (95% CI 1.30 to 3.17), p = 0.002). Moreover, the frequency of ASCA-IgG was found to be significantly higher in patients with Crohn's disease than in those with gastrointestinal Behcet disease (OR = 0.48 (95% CI 0.28 to 0.83), p = 0.009, [57]. This shows that ASCA plays a significant role in pathogenesis of autoimmune gastrointestinal diseases. ASCA have also been detected in other diseases. Shor et al. also detected IgG ASCA in Crohn's disease, Graves' disease, SLE, vasculitis, and cryoglobulinemia patients [28].

2.7 Perinuclear anti-neutrophil cytoplasmic antibodies p-ANCA

Perinuclear anti-neutrophil cytoplasmic antibodies (p-ANCA) are a subset of anti-neutrophil cytoplasmic antibodies (ANCA) that target the heterogeneous collection of antigens, such as myeloperoxidase (MPO), cathepsin-G, elastase, lactoferrin, and bactericidal/permeability-increasing protein. p-ANCA mostly recognizes MPO, followed by neutrophil elastase, lactoferrin, and other antigens [58]. Atypical p-ANCA binds to those antigens in neutrophil granules leading to the staining of rim of the neutrophil nuclei and intranuclear foci [19]. p-ANCA is thought to be more dominant in UC than in Crohn's patients. For instance, p-ANCA has been detected in 40–80% of patients with ulcerative colitis compared to 5–25% of patients with Crohn's disease [58]. In addition, Smids et al. detected p-ANCA in 45% of UC patients, and only 5% of Crohn's patients [51]. Moreover, Ruemmele et al. state that p-ANCA are 92% specific for detecting ulcerative colitis, as those autoantibodies were absent in all non-IBD controls [59]. Smids also confirms p-ANCA specificity to UC (p = 0.0001) [51]. In addition to IBD, Freeman states that p-ANCA can also be present in patients with histologically-defined celiac disease with or without concomitant lymphocytic colitis [60]. Damoiseaux et al. also confirm the presence of p-ANCA in celiac disease patients as they detected its presence in 8 of 37 patients with celiac disease (21.6%) [27].

2.8 Autoantibodies against epithelial cell adhesion molecules

In a study examining eosinophilic esophagitis (EoE) patients, Dellon detected higher levels of anti-DSG3 IgG4 in EoE patients' serum compared to healthy controls (p = 0.02). In addition to that marker, he also detected very high levels of Anti-NC16A IgG4 EoE patient's serum when compared with healthy controls (p < .001), which then led him to conclude that this marker is useful for diagnostic utility as a serum-based EoE biomarker [61].

2.9 IgE/IgG to food antigens

It has been shown that EoE patients' serum exhibited various IgE against food antigens. For instance, Roy-Ghanta et al. found that 82% of 23 patients with biopsy-proven EoE exhibited serum IgE targeting one or more food-associated allergens.

Most common food allergens were onion, wheat, carrot, and tomato [62, 63]. Moreover, Erwin et al. have also concluded that EoE-sensitized patients have higher IgE titers in comparison to nonsensitized patients (median, 150 vs. 13 IU/mL; $P < .001$) [64]. It is also common for EoE patients to have IgE targeting some milk proteins. For example, using ImmunoCAP assays for specific milk allergens, Erwin et al. have detected positive IgE antibodies in 31 out of 34 EoE patients. He then detected a strong correlation between IgE antibodies targeting Bos d 4 (α-lactalbumin) and Bos d 5 (β-lactoglobulin) and milk extract ($R = 0.89$ and $R = 0.76$ respectively; $p < 0.001$) [65]. In another example, Schuyler et al. also confirm the prevalence of antibodies against milk proteins in EoE patients. He found that 79% of 67 children diagnosed with EoE had cow milk (CM) sensitization (sIgE ≥0.10 IU/mL) compared with unselected controls, where only 22% of 101 had CM sensitization. When comparing specific IgG4 and total IgG4, both were significantly detected in EoE patients in comparison to unselected controls ($p < 0.001$ vs. $p < 0.01$, respectively). Just like Erwin et al., Schuyler et al. also found significantly high titer of antibodies against alpha-lactalbumin; however, the antibody was sIgG4, when compared to control ($p < 0.001$). He also detected another targeted protein in milk which was caseins ($p < 0.001$) [66]. Clayton also reports the presence of IgG4 targeting food in EoE patient's serum [67].

2.10 Aeroallergen-specific IgE

There can be a difference in aeroallergen-specific IgE serum levels between age groups in EoE patients. For instance, Erwin et al. have noticed that children have higher aeroallergen-specific IgE serum levels than adults. Regarding specificities, Erwin et al. have shown that prevalence of sensitization to one or more aeroallergen specificities was higher than that in children (93% vs. 65%), while the sensitization to each individual aeroallergens ranged from 12 to 61% [65].

2.11 Circulating antimyenteric autoantibodies (CAA)

CAA are circulating antibodies that target the myenteric neurons located in the GI tract. Several studies have associated those autoantibodies with the pathogenesis of achalasia disease. For instance, Storch et al. have detected IgG antibodies directed at Auerbach's plexus, also named myenteric plexus, in patients with achalasia with varying duration and stages of diseases (specificity 93%, sensitivity 64%, $p < 0.0001$) [68]. Furthermore, Verne et al. also detected them in 7 out of 18 achalasia patients. Those autoantibodies were found to stain most of the neurons found within plexi in the intestinal and esophageal sections, even nitric oxide synthase positive and negative neurons. While none of the controls exhibited neuronal staining [69]. Moreover, Ruiz-de-León et al. also confirmed CAA association with achalasia, as he found CAA in 54.3% of patients with achalasia and only 7.5% of healthy individuals ($P < 0.001$) [70]. When examining nuclear or cytoplasmic fluorescence patterns, Kallel-Sellami et al. found significantly high titers of CAA in patients with achalasia, in comparison to healthy controls (33% vs. 12%, $P = 0.03$ and 48% vs. 23%, $P = 0.001$ respectively) [71]. On the other hand, Kraichely et al. did not detect any specific myenteric neuronal antibody in all the 70 patients with primary achalasia he examined. Instead, they found significantly high levels of GAD65 autoantibody in patients with achalasia, which is an autoantibody found in other autoimmune diseases such as type 1 diabetes mellitus ($P < 0.0001$) [72].

3. Artificial intelligence (AI) in the diagnosis of GI autoimmune diseases

Artificial intelligence is a study of methods capable of imitating intelligent human behavior (e.g., making decisions under uncertain conditions) [73]. Machine learning (ML) is a subset of AI. The introduction of machine learning (ML) has revolutionized the image processing and analysis field in medicine. ML in computer science can be defined as the process by which computers can learn without being explicitly programmed. Machine learning is intended to assist in learning from data. There are many datasets available today, leading to an increase in ML demand [74]. The information extracted from the data can sometimes be difficult to interpret after viewing it. The use of ML can make machines more efficient at handling data. There are two famous models in ML which are unsupervised and supervised machines [75]. Supervised learning is an optimum choice for smaller volumes of data and clearly labeled data [76]. For large datasets, unsupervised learning generally results in better performance and results. If a large dataset is readily available and labeled, deep learning techniques are optimum for use [77]. The application of AI and ML in healthcare has a promising potential in providing medical solutions for the healthcare sector. One of the aspects that could be a valuable addition to the healthcare sector is the incorporation of AI into the diagnosis. In this section, we aim to shed the light on some of the recent applications of AI in GI autoimmune diseases.

3.1 AI in achalasia diagnostics

AI has not explored much of achalasia diagnosis. One of the scarce examples is the work of Carlson et al. where they used functional luminal imaging probe panometry as a method to detect achalasia subtypes using ML. Manometry was performed on 180 patients with achalasia's 3 subtypes. FLIP is a technique that is used to measure distensive pressures and distension-induced esophageal contractions. Correlation analysis, single tree, and random forest were adopted to develop classification trees to identify achalasia subtypes. Their decision tree model accurately identified spastic (type III) versus nonspastic (types I and II) achalasia with 90% and 78% accuracy, respectively. The train and test cohorts correctly identified achalasia subtypes I, II, and III with 71% and 55% accuracy, respectively [78]. In a recent conference proceeding, Jiang et al. reported an automated real-time esophagus achalasia detection method for esophagoscopy assistance through the use of convolutional neural network (CNN) to detect all achalasia frames in esophagoscopy videos. Since it is hard to distinguish achalasia features, they further introduced dense pooling connections and dilated convolutions in the CNN to better extract features from esophagoscopy frames. They reported a real-time achalasia detection system that achieved 0.872 accuracy and 0.943 AUC score on their dataset [79].

3.2 AI in AAG diagnostics

The atrophic gastritis can benefit from the applications of AI in the diagnosis as well. It is often hard to distinguish between the different types of gastritis. One of the most promising applications is the recent report by Franklin et al. that utilized a CNN machine learning model that can distinguish between cases of HPG and autoimmune gastritis with accuracy equal to GI pathologists [80]. This could be beneficial particularly in AAG since it is hard to diagnose pathologically depending on the expertise of the clinician.

3.3 AI in celiac disease diagnostics

Diagnosis of celiac disease (CD) is difficult because its symptoms are shared with many other diseases. However, AI can be used to further facilitate the diagnosis of CD. Joceli et al. proposed a web-based Clinical Decision-Support System (CDSS) using ML algorithms to identify CD. The database used for testing and training the algorithms consisted of clinical data of patients with 35 attributes of CD-related symptoms recorded per case. For the training set, a total of 178 cases were recorded out of which 46% were diagnosed with CD. For the testing set, a total of 38 cases were recorded out of which 37% were CD positive. The study used different variations of 13 algorithms equating the total number of models to 270. The algorithms were trained on the training set, and the best variation of each algorithm was used on the testing set. The selection criteria were the area under the curve of the receiver operating curve (AUC ROC). The results were compared with clinical diagnosis and the golden standard, and the results showed that the best algorithm was able to diagnose the CD cases with great accuracy. This preliminary work shows the prospective of using AI can be used to aid physicians in their diagnosis of diseases like CD [81].

3.4 AI in EoE diagnostics

The applications of AI in EoE have been on rise. One of the most recent applications by Guimarães et al. is the utilization of CNN networks in endoscopic images of EoE. Their study examined 484 real-world endoscopic images taken from 134 subjects within three distinct categories (normal, EoE, and candidiasis). In their results, they found that global accuracy (0.915 [95% confidence interval (CI) 0.880–0.940]), specificity (0.936 [95%CI 0.910–0.955]), and sensitivity (0.871 [95%CI 0.819–0.910) were all higher than for the endoscopists on the test set. The global area under the receiver operating characteristic curve was 0.966 [95%CI 0.954–0.975] [82]. One study by Dnaiel et al. applied machine learning to EoE biopsies and created a dataset for training a multilabel segmentation deep network. Their model was able to segment intact and notintact eosinophils with a mean intersection over union (mIoU) value of 0.93. This segmentation was able to quantify intact eosinophils with a mean absolute error of 0.611 eosinophils and to classify EoE disease activity with an accuracy of 98.5%. Their model achieved 94.8% accuracy, 94.3% sensitivity, and 95.14% specificity in detecting EoE disease activity when using whole slide images from the validation cohort [83]. EoE diagnosis could be flourished with the introduction of AI as more already ongoing research on it in the literature [84–87].

3.5 AI in IBD diagnostics

AI has also been explored widely in IBD diagnosis. The need for AI in identifying IBD and correctly identifying the type of Crohn's disease and ulcerative colitis has been pointed out by Suandram et al. The problem with diagnosing IBD through endoscopy is the subjectivity of the endoscopist in interpreting the results rather than the endoscopic result visualization. To aid in the decision-making, making AI-based applications exist, such as computer-aided diagnosis (CADx). The works reviewed by Suneha et al. have shown great accuracy in detecting and differentiating IBM diseases. Mossotto, 2017 was able to classify UC and Crohn's disease with an accuracy of 83.3% using pediatric data involving endoscopic images and histology [88]. Barash,

Study	Disease	Classification Technique	Predicting Classes	Accuracy (%)	Sensitivity (%)	Specificity (%)
(Carlson et al. 2021)	Achalasia	Decision Trees	Type I	55	9	86
			Type II		72	23
			Type III		64	97
(J. Zhang et al. 2021)	Atrophic Gastritis	Improve-DenseNet	AG/ Non-AG	98.63	95.42	93.87
(Y. Zhang et al. 2020)		DenseNet 121	Mild, Moderate, and Severe	94.24	94.58	94.01
(Tenório et al. 2011)	Celiac Disease	Bayesian Classifier (AODE-F1)	CD/ Non-CD	80	78	80
(Manandhar et al. 2021)	Inflammatory Bowl Diseases*	Decision Trees	IBD/ Non-IBD	0.72 ± 0.02	0.81 ± 0.04	0.63 ± 0.04
		Elastic Net		0.69 ± 0.02	0.77 ± 0.05	0.62 ± 0.06
		Neural Networks		0.63 ± 0.04	0.80 ± 0.22	0.46 ± 0.18
		Random Forrest		0.74 ± 0.02	0.84 ± 0.03	0.64 ± 0.04
		Support Vector Machines		0.67 ± 0.02	0.77 ± 0.06	0.58 ± 0.06
	Inflammatory Bowl Diseases*	Random Forrest	Crohn's/UC	0.83 ± 0.03	0.85 ± 0.04	0.80 ± 0.06

Table 1.
AI applications summary in GI tract autoimmune diseases.

2021 was able to diagnose CD ulcer severity with great accuracy based on capsule endoscopy images [89]. On the other hand, Gottlieb, 2020 used an endoscopy video to the grade the UC severity [90]. Takenaka, 2020 predicted the UC remission with 90% accuracy using endoscopic images and histology [91]. For detailed reviewing of AI in IBD, we refer the reader to in-scope reviews (**Table 1**) [92–101].

4. The role of dietary in GI autoimmune diseases: Nutritional implications

The value of the nutrition in the treatment and the prevention of the diseases has been known for thousands of years before the current modern medicine. The growing interest in the value of nutrition made it clearer that many of the diseases that have boomed in the modernism era are entangled with the poor nutrition and the lifestyle of the individuals. In this section, we aim to explore the role of the nutrition in the GI autoimmune diseases.

4.1 Role of dietary interventions in achalasia

Nutrition in patients with achalasia has often been overlooked. Achalasia is initially characterized by dysphagia when eating solid and liquid foods. Solid food tends to cause more dysphagia than liquids. Most patients modify their eating habits to ease the progress of the food bolus: eating more slowly or using certain maneuvers, such as raising the arms or arching the back [102].

The disease is extremely rare and has a high success rate in treatment. The clinicians usually recommend the patients to eat what they can tolerate and usually the patients resume the regular diet after the treatment. An adequate nutrition modification should be a part of the therapy. If the patient experiences swallowing difficulties, they may be advised to reduce their fiber intake as soluble fibers increase the viscosity of the bolus, decreasing its absorption, while insoluble fibers have a high water-binding capacity, increasing the bulk of the bolus. Low-fiber diets would be physiologically advantageous in situations where luminal narrowing is present, such as in achalasia due to high LES pressure. There is a possibility that some patients will have to switch to high-calorie/protein liquids if this is necessary for their condition. Patients with persistent vomiting might also benefit from supplementation with thiamine (and other vitamins and minerals). Achalasia patients who continue to have difficulty meeting their nutritional needs orally may need gastric access for enteral feeding, but this is rarely needed due to the effective treatment options available [103–105].

4.2 Role of dietary interventions in AAG

AAG patients are reported to have the malabsorption of food-bound vitamin B12 due to decreased IF production resulting in hematological, gastroenterological, and neuropsychiatric disorders. In addition, they are reported to have malabsorption of iron resulting in microcytic anemia. They are also reported to have a vitamin C deficiency that leads to decreased antioxidant defense, immunity, and protein synthesis. They are also reported to have calcium deficiency that could lead to osteopenia/osteoporosis. Furthermore, they are reported to have vitamin D deficiency that could lead to secondary hyperparathyroidism, osteopenia/osteoporosis, decreased immunity, and an increased risk of autoimmune disease development [106]. It is recommended that patients with AAG to follow an anti-inflammatory diet and avoid the food that causes inflammatory responses [107]. Some foods in particular such as garlic could be of beneficial use in the anti-inflammatory intake [108–111]. In addition, probiotics that can have positive influence on the gut microbiota have been shown to be good for the diet of AAG patients [112].

4.3 Role of dietary interventions in celiac disease (CD)

Gluten is considered an environmental trigger for CD. Unlike other autoimmune diseases, the progression and chronic dynamics of CD are reversible. The reconstruction of the mucosa is also achievable when accompanied by total gluten avoidance [113]. Hence, a strict gluten-free diet (GFD) results in intestinal and extraintestinal symptoms improvement, intestinal villi regrowth, and autoantibodies negativity. Furthermore, this diet reverses the complications of CD that includes malabsorption, osteopenia, osteoporosis, diarrhea, bloating, constipation, and abdominal pain [114]. Besides a GFD, lactose present in milk and most dairy products should be avoided at the early stages of treatment due to a brush border lactase deficiency that is a secondary result of the surface epithelial cells damage [113]. Another thing to consider is a diet low in fermentable, oligosaccharides, disaccharides, monosaccharides, and polyols (FODMAPs). Since irritable bowel syndrome (IBS) symptoms are prevalent in 38% of CD-treated patients, these symptoms persist even when they are following a strict GFD [115]. Lactose-free milk/yogurt, feta, cheddar, mozzarella, parmesan, brie, butter, and plant-based milk/yogurt are good alternatives that have low lactose content. A variety of dairy products that are low in lactose could provide CD patients with sufficient calcium. However, when choosing nondairy, lactose-free products that are made from soy, rice, and nuts it is crucial to find products that are supported with calcium since plant-based products are poor in calcium. CD individuals should aim for 200 – 300 mg of calcium/250 ml per serving [116].

Oats, rice, corn/quinoa/millet bread, sourdough, starch, corn tortilla, potato, soba/rice sticks/kelp/brown rice noodles, sago, samp, wonton wrapper, rice-based products, quinoa-based products, and quinoa/chickpea/sourdough pasta are all good substitutes that are gluten-free and would help CD patients to have a varied and balanced diet.

4.4 Role of dietary interventions in eosinophilic esophagitis

In most cases, but not all, EoE is triggered by food antigens. Hence, the nutrition plays an important part in both the pathogenesis and the treatment of the disease [117]. In pediatric and adult populations, food antigens are clearly antigenic triggers for EoE induction and exacerbation [118]. In 1995, Kelly and Sampson proposed that acid persistent esophageal eosinophilia can be caused by food antigen exposure in children [117]. Ever since the direction toward studying the role of food allergens in the pathogenesis and the treatment of EoE has been established. Some of the therapies that have proven the efficacy are the empiric elimination diets, such as the famous 6 food elimination diets (6-FED). These six foods are wheat, milk, egg, nuts, soy, fish, and eggs [119]. In animal studies, it was shown that accumulation of eosinophils in the murine esophagus occurred after the introduction of peanuts and eggs [120, 121]. Statistics show that 77% of patients with EoE have at least one positive skin-prick test (SPT) for at least 1 food allergy and up to 50% of adults have at least 1 positive test for food allergy [122, 123].

Currently, there are diets approaches that are used for EoE patients: 1) A crystalline amino acid-based elemental diet (ELED), 2) 6-FED, 3) 4-FED, 4) 2-FED) 5) Cow's milk elimination diet [124–126]. The amino acid diet is useful as it can eliminate all possible allergens and it has shown improvement in the symptoms in many cases, and the diet can last for 6 weeks. Initially, the 6-FED was studied in pediatric patients from Chicago in 2006, in which six food groups responsible for most IgE-mediated food reactions were eliminated for 6 weeks [127]. All studies have consistently shown that nuts and fish/seafood rarely trigger EoE in response to a 6FED, but cow's milk is by far the most common cause of EoE, followed by wheat/gluten, egg, and, to a lesser extent, wheat/gluten [118, 127–132]. The 4-FED is based on the elimination of the most common food triggers in EoE (animal milk, gluten-containing cereals, eggs, and legumes). Cow's milk (85%), egg (35%), wheat (33%), and soy (19%) were the most common food triggers. The 2-FED is based on the elimination of milk and gluten [126]. After reintroduction of individual foods, cow's milk was found to be the only trigger food in 55% of pediatric responders [133]. Therefore, the 1-FED or the cow's milk elimination diet could be recommended for some patients. One of the clinical practices in the dietary therapy is that a clinician could start with a 1-FED diet, if no response is observed the clinician could upgrade to 2-FED, 4-FED, or 6-FED. Patients could have more than food allergens; therefore, 6-FED is considered the most efficient diet (**Figure 1**) [125].

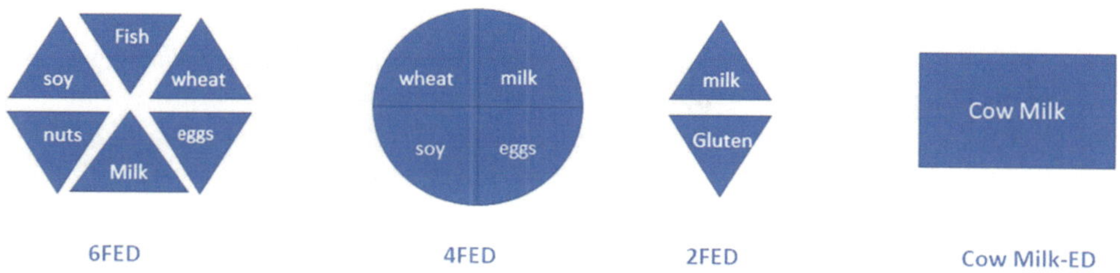

Figure 1.
The various types of food elimination diets.

4.5 Role of dietary interventions on inflammatory bowel disease (IBD)

The triggers of IBD include internal (enteric microflora) and external (food) triggers [134]. Overconsumption of sugar and refined carbohydrates was associated with the manifestation of Crohn's disease. Furthermore, excess intake of sugar over the years could alter the intestinal bacterial flora and general milieu, which could damage the mucosa or alter bile acid composition. These alterations could be the result of infective agents or sugar fermentation [135]. A balanced diet that includes fruits, vegetables, meat, olive oil, and fish (blue fish particularly) should be prescribed to IBD patients. Insoluble fiber might have negative effects in case of major intestinal stenosis coexisting with IBD. However, insoluble fiber intake should not be restricted in IBD patients. Moreover, dairy products are a crucial part of IBD nutrition intervention due to their calcium content. Products that contain lactose could be avoided if the patient had lactose intolerance and substituted with plant-based products that contain enough calcium. Supplementation with calcium and vitamin D3 might be required along with systemic steroids treatment as well as other treatments that have greater local effects such as budesonide or beclomethasone. Furthermore, iron and folic acid deficiencies should be closely monitored due to their huge prevalence in IBD patients. Deficiencies in iron or folic acid contribute to anemia in this population and could be easily treated orally or intravenously [134].

5. Treatments of GI autoimmune diseases

5.1 Achalasia

Esophagectomy is only necessary for 5% of patients with end-stage achalasia. Among the options for treating achalasia are botulinum toxin injection, pneumatic dilation, laparoscopic Heller myotomy, and peroral endoscopic myotomy (POEM). Botulinum toxin injections are one of the first-line treatment options in achalasia. The injection reduces the lower esophageal sphincter (LES) pressure by inhibiting the release of acetylcholine from nerve endings [136]. The injection is extremely safe and rarely causes any adverse reactions. The injection is, however, limited in its durability, which lasts only for a few months [137–143]. Another common treatment option is the pneumatic dilation. Under fluoroscopic guidance, the balloon dilates the LES fibers through intraluminal dilation and can be either 30, 35, or 40 mm in diameter. If no success is achieved, the clinician will go for a bigger balloon size. The success rate as per Eckardt score is achieved in 84% of the patients [144]. Another common treatment is the laparoscopic Heller myotomy. This treatment was based on surgical myotomy to disrupt the LES fibers through an incision but now it has been minimally invasive laparoscopic myotomy with a partial fundoplication. Clinical success is not purely determined by Eckardt's scores. The primary outcome measure was improvement of dysphagia, which was treated as a dichotomous variable. Overall, 87.7% of studies reported improvement in dysphagia through this treatment [144]. POEM is the last common treatment in achalasia, and this treatment was only established 12 years ago [145]. The clinical success in POEM was 98% [144].

5.2 Autoimmune atrophic gastritis (AAG)

In the early stage of AAG, due to the reduced gastric acid secretion and intrinsic factors the clinician should focus on preventing the deficiency of B12, iron, and folate as the development of anemia could be prevented with supplementation of these nutritions. In case of the presence of pernicious anemia already, the clinician should consider B12 repletion, cyanocobalamin, and iron supplements to restore hemoglobin function. Also, clinician should note that AAG is usually associated with autoimmune diseases, such as autoimmune thyroid disease, type 1 diabetes mellitus, and Addison disease [146–156].

5.3 Celiac disease

A lifelong strict GFD can be considered the only treatment for celiac disease [157]. For patients who have refractory type celiac disease, they might need a pharmacological intervention besides the strict GFD diet. The use of drugs that work on proteolytic destruction of gluten peptides, inhibition of intestinal permeability to prevent gluten absorption, inhibition of TG2, or modulation of the immune response to gluten to prevent T cell activation is a promising option [158, 159]. Currently, the most promising treatment is the vaccine Nexvax2, which is an adjuvant-free mixture of tripeptides immunodominant epitopes for gluten-specific CD4-positive T cells. However, it is still in the preliminary stages [160–162].

5.4 EoE

Overall, EoE is treated with three main categories: drugs, diet, and dilation [163–165]. The diet therapy has been discussed in a previous section. Pharmacological treatment includes topical corticosteroids, such as fluticasone or budesonide, swallowed rather than inhaled, for an initial duration of 8 weeks. It has been shown that the patients' symptoms have improved as decreased esophageal eosinophilia was apparent, and were generally well-tolerated by patients [166–172]. Proton pump inhibitors (PPI) are usually given to the patients of EoE since the patients usually suffer from regurgitation and acid reflux. The response to PPI is hugely variable between 30 to 70% [173]. PPI-responsive and PPI-resistant EoE have yet to be identified. In patients with PPI-responsive EoE, expression of the potassium channel gene, KCNJ2, is lower. CYP2C19 rapid metabolizers and allergy patients are more likely to lose EoE control despite continued PPI treatment [174]. Since the long-term use of corticosteroids can result in harmful effects, immunomodulators, such as 6-mercaptopurine and azathioprine, are often used for the treatment of the patients. They might have a role in inducing and maintaining long-term clinical and histologic remission in EoE in limited cases but their side effects can be discouraging [175, 176]. Monoclonal antibodies have been investigated in the last few years against EoE including some famous drugs including mepolizumab (anti-IL-5), reslizumab (anti-IL-5), QAX576 (anti-IL-13), omalizumab (anti-immunoglobulin-E), and infliximab (anti-TNF-α) [177]. IL-5 produced by Th2 lymphocytes has a critical role in eosinophil activation. Animal studies have shown that overexpression of IL-5 can induce EoE [178, 179]. IL-5 receptors, which are mainly expressed on the surface of eosinophils, are blocked by mepolizumab, a monoclonal antibody against IL-5 [180]. The use of mepolizumab seems promising in decreasing the number of eosinophils and reducing the dependency on corticosteroids but more clinical studies need to be conducted [181–183]. Another humanized anti-IL-5 mAb called reslizumab prevents IL-5 from binding to its receptor. The available trials show an improvement in eosinophil count but not in the symptoms and the drug was generally safe [184–186]. In the pathogenesis of EoE, IL3 plays a multifunctional role. An anti-IL3 therapy could be efficient in EoE one of the most famous anti-IL13 drugs is QAX576. Patients tolerated QAX576 well. Patients decreased by 60.0% and sustained for 6 months on the QAX576, which is an anti-IL3 drug. Unfortunately, the primary endpoint was not reached. A trend for improved symptoms was observed particularly dysphagia. Six months after treatment, QAX576 helped to improve expression of esophageal transcripts related to EoE, such as eotaxin-3, periostin, and markers of mast cells and barrier function [187]. Since mast cells are involved in the pathogenesis of EoE, targeting them directly could be an efficient treatment for EoE. Malizumab is a monoclonal anti-IgE antibody that prevents mast cell activation by binding to IgE [188]. However, in most of the trials, the response was poor in patients or reoccurrence of symptoms presented after a short time [189–191]. TNF-α and IFN-γ are found in esophageal mucosal biopsy of the EoE patients. A potent inhibitor of TNF-α is infliximab that is a chimeric IgG1mAb. Infliximab was not shown to be of no benefit for EoE patients [192].

Dilation is also sometimes used in the treatment of EoE. The most common use of esophageal dilation is in adults with EoE and strictures. Conservatively applied, this approach is safe and has a low complication rate. Dilation treats structural alterations in EoE. Although esophageal dilation is well tolerated by patients and can provide long-term symptomatic relief, it does not improve histologic changes [193–195].

5.5 IBD

IBD can have a wide range of treatments. In Crohn's disease, treatments include immunomodulators, corticosteroids, and monoclonal antibodies. 5-aminosalicylates is the most commonly prescribed for symptoms management in mild and moderate disease [196]. Corticosteroids are efficient, prednisone can be efficient in the course of treatment [197]. Budesonide (Entocort EC) may be preferred for diseases affecting the ileum and/or proximal colon since it is delivered specifically to those areas [196]. Immunomodulators, such as thiopurines and methotrexate are the most effective immunomodulators used in Crohn's disease [198]. Anti-TNF agents, anti-integrin agents, and anti-interleukin-12/23p40 antibody therapy are considered the most efficient in treating Crohn's disease. The continuation of any of them in the treatment plan depends on the remission success [199]. In moderate- to high-risk patients, anti-TNF agents, such as certolizumab pegol (Cimzia) and adalimumab (Humira) induce and maintain remission, or patients with inadequate responses to corticosteroids or immunomodulators [200]. In anti-integrin agents, vedolizumab is the most favorable drug because of its specificity to leukocyte trafficking in the gut and has demonstrated effectiveness in achieving clinical response, remission, and corticosteroid-free remission [201]. In, anti-interleukin-12/23p40, ustekinumab is promising for Crohn's disease as it was recently approved by FDA [202]. In Crohn's disease, 57% of the

Disease	Treatment
Achalasia	Esophagectomy for end-stage patients botulinum toxin injection, pneumatic dilation, laparoscopic Heller myotomy, and POEM
AAG	B12, Iron, and Folate nutritional supplementation therapy
CD	Gluten-Free Diet (GFD) Drugs that can destruct gluten peptides, inhibition of intestinal permeability to prevent gluten absorption, inhibition of TG2, or modulation of the immune response to gluten Nexvax2 vaccine
EoE	Diet: FED diets and amino acid formula diets Dilation Drugs: PPI, glucocorticoids, anti-IL5, Anti IgE, anti IL13, and anti-TNF-α
Crohn's Disease	Drugs: Glucocorticoids, ant-TNF, anti-integrin agents, and anti-IL12/23p40
UC	Does not differ much from Crohn's treatment

Table 2.
Common treatments for GI autoimmune.

patients might need surgical intervention to treat fistulas, abscesses, perforation, obstruction, strictures, uncontrolled bleeding, dysplasia, malignancy, and perianal disease [196, 203].

In UC, the treatment options do not differ much from Crohn's Disease. The mainstay of therapy for mild-to-moderate UC is sulfasalazine and other 5-ASA agents [204]. Corticosteroids are also efficient in UC patients and it's usually given to the cases of severe symptoms. Prednisone is the most used corticosteroid. Immunomodulators are also used such as their usage in Crohn's Disease. Azathioprine and 6-mercaptopurine (6-MP) are purine analogs that are the commonly most used in the treatment [205]. Also, monoclonal antibodies used in UC such as infliximab has proven their efficiency [206]. In addition, vedolizumab has proven to be efficient as well in UC [207]. Surgical treatment is an option as well in UC. This is generally considered a last resort when all other options have failed. The most common type of surgery is a subtotal or total colectomy with a temporary stoma [208]. Also, laparoscopic surgeries are a safe option in UC (**Table 2**) [209].

6. Conclusions

To conclude, GI autoimmune diseases can be compromising the patient's life. Nevertheless, the exploration of more diagnostic options, such as the antibodies, the growing applications of artificial intelligence in autoimmune diseases diagnosis, understanding the interaction of nutrition whether in the pathogenesis or the management, and the efficient treatment plans can help for better diagnosis, management, and treatment of GI autoimmune diseases, which is a sub-category of autoimmune diseases that are considered a pandemic in our modern societies.

Acknowledgement

The publication of this chapter is funded by Qatar National Library.

Conflict of interest

The authors declare no conflict of interest.

Author details

Mahmoud Khatib A.A. Al-Ruweidi[1,2], Nada Khater[3], Haya Rashid Alkaabi[4], Maram Hasan[1], Mohammed Murtaza[5] and Huseyin C. Yalcin[1*]

1 Biomedical Research Center, Qatar University, Doha, Qatar

2 Pharmaceutical Sciences Department, College of Pharmacy, QU Health, Qatar University, Doha, Qatar

3 Department of Sport Science, College of Arts and Sciences, Qatar University, Doha, Qatar

4 Weill Cornell Medical College Qatar, Doha, Qatar

5 Department of Electrical Engineering, Qatar University, Doha, Qatar

*Address all correspondence to: hyalcin@qu.edu.qa

References

[1] Singh K, Chang C, Gershwin ME. IgA deficiency and autoimmunity. Autoimmunity Reviews [Internet]. 2014;**13**(2):163-177. DOI: 10.1016/j.autrev.2013.10.005

[2] Sinha AA, Lopez MT, McDevitt HO. Autoimmune diseases: the failure of self tolerance. Science [Internet]. 1990;**248**(4961):1380-1388. DOI: 10.1126/science.1972595

[3] Xiao ZX, Miller JS, Zheng SG. An updated advance of autoantibodies in autoimmune diseases. Autoimmunity Reviews [Internet]. 2021;**20**(2):102743. DOI: 10.1016/j.autrev.2020.102743

[4] Eggert M, Zettl UK, Neeck G. Autoantibodies in autoimmune diseases. Current Pharmaceutical Design [Internet]. 2010;**16**(14):1634-1643. DOI: 10.2174/138161210791164144

[5] Ganser AL, Forte JG. K+-stimulated ATPase in purified microsomes of bullfrog oxyntic cells. Biochimica et Biophysica Acta (BBA) -Biomembranes [Internet]. 1973;**307**(1):169-180. Available from: https://www.sciencedirect.com/science/article/pii/0005273673900357

[6] Callaghan JM, Khan MA, Alderuccio F, van Driel IR, Gleeson PA, Toh BH. Alpha and beta subunits of the gastric H+/K(+)-ATPase are concordantly targeted by parietal cell autoantibodies associated with autoimmune gastritis. Autoimmunity [Internet]. 1993;**16**(4):289-295. DOI: 10.3109/08916939309014648

[7] Claeys D, Faller G, Appelmelk BJ, Negrini R, Kirchner T. The gastric H+,K+-ATPase is a major autoantigen in chronic Helicobacter pylori gastritis with body mucosa atrophy. Gastroenterology [Internet]. 1998;**115**(2):340-347. DOI: 10.1016/s0016-5085(98)70200-8

[8] Ito M, Haruma K, Kaya S, Kamada T, Kim S, Sasaki A, et al. Role of anti-parietal cell antibody in Helicobacter pylori-associated atrophic gastritis: Evaluation in a country of high prevalence of atrophic gastritis. Scandivian Journal of Gastroenterology [Internet]. 2002;**37**(3):287-293. DOI: 10.1080/003655202317284183

[9] Karlsson FA, Burman P, Lööf L, Mårdh S. Major parietal cell antigen in autoimmune gastritis with pernicious anemia is the acid-producing H+,K+-adenosine triphosphatase of the stomach. The Journal of Clinical Investigation [Internet]. 1988;**81**(2):475-479. DOI: 10.1172/JCI113344

[10] Güreşçi S. What is gastritis? What is gastropathy? How is it classified? Turkish Journal of Gastroenterology [Internet]. 2014;**25**:233-247. Available from: https://www.academia.edu/download/66562103/17dc7b5045bab5b9ae5726bba1c856632cd3.pdf

[11] Zhang Y, Weck MN, Schöttker B, Rothenbacher D, Brenner H. Gastric parietal cell antibodies, Helicobacter pylori infection, and chronic atrophic gastritis: Evidence from a large population-based study in Germany. Cancer Epidemiology, Biomarkers & Prevention [Internet]. 2013;**22**(5):821-826. DOI: 10.1158/1055-9965.EPI-12-1343

[12] De Block CE, Van Campenhout CM, De Leeuw IH, Keenoy BM, Martin M, Van Hoof V, et al. Soluble transferrin receptor level: a new marker of iron deficiency anemia, a common manifestation of gastric autoimmunity in

type 1 diabetes. Diabetes Care [Internet]. 2000;**23**(9):1384-1388. DOI: 10.2337/diacare.23.9.1384

[13] Conn DA. Detection of type I and type II antibodies to intrinsic factor. Medical Laboratory Sciences [Internet]. 1986;**43**(2):148-151. Available from: https://www.ncbi.nlm.nih.gov/pubmed/3736366

[14] McPherson RA, Pincus MR, Henry JB. Henry's clinical diagnosis and management by laboratory methods. 21st ed. Philadelphia: Saunders Elsevier; 2007

[15] Troilo A, Grassi A, Petrone L, Cianchi F, Benagiano M, Bella CD, et al. Intrinsic factor recognition promotes T helper 17/T helper 1 autoimmune gastric inflammation in patients with pernicious anemia. Oncotarget [Internet]. 2019;**10**(30):2921-2929. Available from: https://www.oncotarget.com/lookup/doi/10.18632/oncotarget.26874

[16] Morawiec-Szymonik E, Foltyn W, Marek B, Głogowska-Szeląg J, Kos-Kudła B, Kajdaniuk D. Antibodies involved in the development of pernicious anemia and other autoimmune diseases. Polish Archives of Internal Medicine [Internet]. 2020;**130**(1):31-37. DOI: 10.20452/pamw.15094

[17] Toh BH, van Driel IR, Gleeson PA. Pernicious anemia. The New England Journal of Medicine [Internet]. 1997;**337**(20):1441-1448. DOI: 10.1056/NEJM199711133372007

[18] Bizzaro N, Antico A. Diagnosis and classification of pernicious anemia. Autoimmunity Reviews [Internet]. 2014;**13**(4-5):565-568. DOI: 10.1016/j.autrev.2014.01.042

[19] Notsu T, Adachi K, Mishiro T, Fujihara H, Toda T, Takaki S, et al. Prevalence of autoimmune gastritis in individuals undergoing medical checkups in Japan. Internal Medicine [Internet]. 2019;**58**(13):1817-1823. DOI: 10.2169/internalmedicine.2292-18

[20] Lahner E, Norman GL, Severi C, Encabo S, Shums Z, Vannella L, et al. Reassessment of intrinsic factor and parietal cell autoantibodies in atrophic gastritis with respect to cobalamin deficiency. American Journal of Gastroenterology [Internet]. 2009;**104**(8):2071-2079. DOI: 10.1038/ajg.2009.231

[21] Di Sabatino A, Vanoli A, Giuffrida P, Luinetti O, Solcia E, Corazza GR. The function of tissue transglutaminase in celiac disease. Autoimmunity Reviews [Internet]. 2012;**11**(10):746-753. DOI: 10.1016/j.autrev.2012.01.007

[22] Wang Z, Griffin M. TG2, a novel extracellular protein with multiple functions. Amino Acids [Internet]. 2012;**42**(2-3):939-949. DOI: 10.1007/s00726-011-1008-x

[23] Dieterich W, Ehnis T, Bauer M, Donner P, Volta U, Riecken EO, et al. Identification of tissue transglutaminase as the autoantigen of celiac disease. Nature Medicine [Internet]. 1997;**3**(7):797-801. DOI: 10.1038/nm0797-797

[24] Gatti S, Rossi M, Alfonsi S, Mandolesi A, Cobellis G, Catassi C. Beyond the intestinal Celiac mucosa: Diagnostic role of anti-TG2 deposits, a systematic review. Frontiers in Medicine [Internet]. 2014;**1**:9. DOI: 10.3389/fmed.2014.00009

[25] Borrelli M, Maglio M, Korponay-Szabó IR, Vass V, Mearin ML, Meijer C, et al. Intestinal anti-transglutaminase 2 immunoglobulin A deposits in children

at risk for coeliac disease (CD): Data from the PreventCD study. Clinical & Experimental Immunology [Internet]. 2018;**191**(3):311-317. DOI: 10.1111/cei.13078

[26] Miller A, Paspaliaris W, Elliott PR, d'Apice A. Anti-transglutaminase antibodies and coeliac disease. Australian and New Zealand Journal of Medicine [Internet]. 1999;**29**(2):239-242. Available from: https://onlinelibrary.wiley.com/doi/10.1111/j.1445-5994.1999.tb00690.x

[27] Damoiseaux JGMC, Bouten B, Linders AMLW, Austen J, Roozendaal C, Russel MGVM, et al. Diagnostic value of anti-Saccharomyces cerevisiae and Antineutrophil cytoplasmic antibodies for Inflammatory Bowel disease: High prevalence in patients with Celiac disease. Journal of Clinical and Immunology [Internet]. 2002;**22**(5):281-288. DOI: 10.1023/A:1019926121972

[28] Shor DB-A, Orbach H, Boaz M, Altman A, Anaya J-M, Bizzaro N, et al. Gastrointestinal-associated autoantibodies in different autoimmune diseases. American Journal of Clinical Experimental Immunology [Internet]. 2012;**1**(1):49-55. Available from: https://www.ncbi.nlm.nih.gov/pubmed/23885314

[29] Hoffenberg EJ, MacKenzie T, Barriga KJ, Eisenbarth GS, Bao F, Haas JE, et al. A prospective study of the incidence of childhood celiac disease. The Journal of Pediatrics [Internet]. 2003 Sep;**143**(3):308-314. DOI: 10.1067/s0022-3476(03)00282-8

[30] Rubio-Tapia A, Ludvigsson JF, Choung RS, Brantner TL, Rajkumar SV, Landgren O, et al. Increased mortality among men aged 50 years old or above with elevated IgA anti-transglutaminase antibodies: NHANES III. BMC Gastroenterology [Internet]. 2016;**16**(1):136. DOI: 10.1186/s12876-016-0547-8

[31] Farrace MG, Picarelli A, Di Tola M, Sabbatella L, Marchione OP, Ippolito G, et al. Presence of anti-'tissue' transglutaminase antibodies in inflammatory intestinal diseases: An apoptosis-associated event? Cell Death & Differentiation [Internet]. 2001;**8**(7):767-770. Available from: https://www.nature.com/articles/4400880

[32] Sjöber K, Eriksson S, Tenngart B, Roth EB, Leffler H, Stenberg P. Factor XIII and tissue transglutaminase antibodies in coeliac and inflammatory bowel disease. Autoimmunity [Internet]. 2002;**35**(5):357-364. DOI: 10.1080/73-0891693021000005402

[33] Tursi A, Giorgetti GM, Brandimarte G, Elisei W. High prevalence of celiac disease among patients affected by Crohn's disease. Inflammatory Bowel Diseases [Internet]. 2005;**11**(7):662-666. DOI: 10.1097/01.mib.0000164195.75207.1e

[34] Bizzaro N, Villalta D, Tonutti E, Doria A, Tampoia M, Bassetti D, et al. IgA and IgG tissue transglutaminase antibody prevalence and clinical significance in connective tissue diseases, inflammatory bowel disease, and primary biliary cirrhosis. Digestive Diseases and Sciences [Internet]. 2003;**48**(12):2360-2365. DOI: 10.1023/B:DDAS.0000007875.72256.e8

[35] Watanabe C, Komoto S, Hokari R, Kurihara C, Okada Y, Hozumi H, et al. Prevalence of serum celiac antibody in patients with IBD in Japan. Journal of Gastroenterology [Internet]. 2014;**49**(5):825-834. DOI: 10.1007/s00535-013-0838-6

[36] van Heel DA, West J. Recent advances in coeliac disease. Gut [Internet].

2006;**55**(7):1037-1046. Available from: https://gut.bmj.com/content/55/7/1037.short

[37] Lindqvist U, Rudsander A, Boström A, Nilsson B, Michaëlsson G. IgA antibodies to gliadin and coeliac disease in psoriatic arthritis. Rheumatology [Internet]. 2002;**41**(1):31-37. DOI: 10.1093/rheumatology/41.1.31

[38] Rostami K, Kerckhaert J, Tiemessen R, von Blomberg BM, Meijer JW, Mulder CJ. Sensitivity of antiendomysium and antigliadin antibodies in untreated celiac disease: Disappointing in clinical practice. American Journal of Gastroenterology [Internet]. 1999;**94**(4):888-894. DOI: 10.1111/j.1572-0241.1999.983_f.x

[39] Mäki M, Mustalahti K, Kokkonen J, Kulmala P, Haapalahti M, Karttunen T, et al. Prevalence of Celiac disease among children in Finland. The New England Journal of Medicine [Internet]. 2003;**348**(25):2517-2524. DOI: 10.1056/NEJMoa021687

[40] Kelly CP, Feighery CF, Gallagher RB, Gibney MJ, Weir DG. Mucosal and systemic IgA anti-gliadin antibody in celiac disease. Contrasting patterns of response in serum, saliva, and intestinal secretions. Digestive Diseases and Sciences [Internet]. 1991;**36**(6):743-751. Available from: http://link.springer.com/10.1007/BF01311231

[41] Swallow K, Wild G, Sargur R, Sanders DS, Aziz I, Hopper AD, et al. Quality not quantity for transglutaminase antibody 2: the performance of an endomysial and tissue transglutaminase test in screening coeliac disease remains stable over time. Clinical & Experimental Immunology [Internet]. 2013;**171**(1):100-106. Available from: https://academic.oup.com/cei/article-abstract/171/1/100/6420920

[42] Stecco C. In: Hammer WI, editor. Functional atlas of the human fascial system; 2015

[43] Abrams JA, Diamond B, Rotterdam H, Green PHR. Seronegative celiac disease: Increased prevalence with lesser degrees of villous atrophy. Digestive Diseases and Sciences [Internet]. 2004;**49**(4):546-550. DOI: 10.1023/b:ddas.0000026296.02308.00

[44] Kanthi Y, Persley KM. Endomysial and related antibodies. In: Johnson LR, editor. Encyclopedia of Gastroenterology [Internet]. New York: Elsevier; 2004. pp. 681-682. Available from: https://www.sciencedirect.com/science/article/pii/B012386860200215X

[45] Farrell RJ, Kelly CP. Celiac disease and refractory celiac disease. In: Sleisenger and Fordtran's Gastrointestinal and Liver Disease. Amsterdam, Netherlands: Elsevier; 2010. pp. 1797-1820

[46] Abdulkarim AS, Murray JA. Celiac disease. In: Johnson LR, editor. Encyclopedia of Gastroenterology [Internet]. New York: Elsevier; 2004. pp. 278-283. Available from: https://www.sciencedirect.com/science/article/pii/B0123868602001106

[47] Keren DF. Chapter 9.15 - autoimmune disease. In: Wild D, editor. The Immunoassay Handbook (Fourth Edition) [Internet]. Oxford: Elsevier; 2013. pp. 869-890. Available from: https://www.sciencedirect.com/science/article/pii/B9780080970370000695

[48] Pizzorno JE, Murray MT, Joiner-Bey H. 16 - Celiac disease. In: Pizzorno JE, Murray MT, Joiner-Bey H, editors. The Clinician's Handbook of Natural Medicine (Third Edition) [Internet]. Edinburgh:

Churchill Livingstone; 2016. pp. 188-193. Available from: https://www.sciencedirect.com/science/article/pii/B9780702055140000257

[49] Lewis JD. The utility of biomarkers in the diagnosis and Therapy of Inflammatory Bowel disease. Gastroenterology [Internet]. 2011;**140**(6):1817-1826.e2. Available from: https://www.sciencedirect.com/science/article/pii/S0016508511001636

[50] El-Matary W, Dupuis K, Sokoro A. Anti-saccharomyces cerevisiae antibody titres correlate well with disease activity in children with Crohn's disease. Acta Paediatrica [Internet]. 2015;**104**(8):827-830. Available from: https://onlinelibrary.wiley.com/doi/10.1111/apa.13026

[51] Smids C, Horjus Talabur Horje CS, Groenen MJM, van Koolwijk EHM, Wahab PJ, van Lochem EG. The value of serum antibodies in differentiating inflammatory bowel disease, predicting disease activity and disease course in the newly diagnosed patient. Scandivian Journal of Gastroenterology [Internet]. 2017;**52**(10):1104-1112. doi:10.1080/00365521.2017.1344875

[52] Kotze LM d S, Nisihara RM, Utiyama SR d R, Kotze PG, Theiss PM, Olandoski M. Antibodies anti-saccharomyces cerevisiae (ASCA) do not differentiate Crohn's disease from celiac disease. Arquivos de Gastroenterology [Internet]. 2010;**47**(3):242-245. DOI: 10.1590/s0004-28032010000300006

[53] Granito A, Zauli D, Muratori P, Muratori L, Grassi A, Bortolotti R, et al. Anti-saccharomyces cerevisiae and perinuclear anti-neutrophil cytoplasmic antibodies in coeliac disease before and after gluten-free diet. Alimentary Pharmacology and Therapeutics [Internet]. 2005;**21**(7):881-887. Available from: https://onlinelibrary.wiley.com/doi/10.1111/j.1365-2036.2005.02417.x

[54] Israeli E, Grotto I, Gilburd B, Balicer RD, Goldin E, Wiik A, et al. Anti-saccharomyces cerevisiae and antineutrophil cytoplasmic antibodies as predictors of inflammatory bowel disease. Gut [Internet]. 2005;**54**(9):1232-1236. DOI: 10.1136/gut.2004.060228

[55] Granito A, Muratori L, Muratori P, Guidi M, Lenzi M, Bianchi FB, et al. Anti-saccharomyces cerevisiae antibodies (ASCA) in coeliac disease. Gut [Internet]. 2006;**55**(2):296. Available from: https://www.ncbi.nlm.nih.gov/pubmed/16407392

[56] Choi CH, Kim TI, Kim BC, Shin SJ, Lee SK, Kim WH, et al. Anti-Saccharomyces cerevisiae antibody in intestinal Behçet's disease patients: Relation to clinical course. Dis Colon Rectum [Internet]. 2006;**49**(12):1849-1859. DOI:10.1007/s10350-006-0706-z

[57] Cheng L, Li L, Liu C, Yan S, Li Y. Meta-analysis of anti-saccharomyces cerevisiae antibodies as diagnostic markers of Behçet's disease with gastrointestinal involvement. BMJ Open [Internet]. 2020;**10**(10):e033880. doi:10.1136/bmjopen-2019-033880

[58] Beniwal P, Harrell L. The status of diagnostic markers for inflammatory bowel disease. Current Gastroenterology Reports [Internet]. 2010;**12**(6):479-484. doi:10.1007/s11894-010-0145-9

[59] Ruemmele FM, Targan SR, Levy G, Dubinsky M, Braun J, Seidman EG. Diagnostic accuracy of serological assays in pediatric inflammatory bowel disease. Gastroenterology [Internet]. 1998;**115**(4):822-829. DOI: 10.1016/s0016-5085(98)70252-5

[60] Freeman HJ. Perinuclear antineutrophil cytoplasmic antibodies

in collagenous or lymphocytic colitis with or without celiac disease. Canadian Journal of Gastroenterology [Internet]. 1997;**11**(5):417-420. DOI: 10.1155/1997/210856

[61] Dellon ES, Lin L, Beitia R, Moran TP, Qian Y. Serum autoantibodies against epithelial cell adhesion molecules as disease biomarkers of eosinophilic esophagitis. Clinical & Experimental Allergy [Internet]. 2018;**48**(3):343-346. DOI: 10.1111/cea.13070

[62] Roy-Ghanta S, Larosa DF, Katzka DA. Atopic characteristics of adult patients with eosinophilic esophagitis. Clinical Gastroenterology and Hepatology [Internet]. 2008;**6**(5):531-535. DOI: 10.1016/j.cgh.2007.12.045

[63] Khan S, Guo X, Liu T, Iqbal M, Jiang K, Zhu L, et al. An update on eosinophilic esophagitis: Etiological factors, coexisting Diseases, and complications. Digestion [Internet]. 2021;**102**(3):342-356. Available from: https://www.karger.com/DOI/10.1159/000508191

[64] Erwin EA, James HR, Gutekunst HM, Russo JM, Kelleher KJ, Platts-Mills TAE. Serum IgE measurement and detection of food allergy in pediatric patients with eosinophilic esophagitis. Annals of Allergy, Asthma & Immunology [Internet]. 2010;**104**(6):496-502. doi:10.1016/j.anai.2010.03.018

[65] Erwin EA, Tripathi A, Ogbogu PU, Commins SP, Slack MA, Cho CB, et al. IgE antibody detection and component analysis in patients with eosinophilic esophagitis. Journal of Allergy and Clinical Immunology Practice [Internet]. 2015;**3**(6):896-904.e3. DOI: 10.1016/j.jaip.2015.05.015

[66] Schuyler AJ, Erwin EA, Oken E, Rifas-Shiman S, Lidholm J, Wilson J, et al. The diagnostic utility of serum assays for total IgG4 and specific IgG4 antibodies to cow's milk proteins in children with eosinophilic esophagitis: Comparison with an unselected birth cohort. Journal of Allergy and Clinical Immunology [Internet]. 2017;**139**(2):AB48. DOI: 10.1016/j.jaci.2016.12.108

[67] Clayton F, Fang JC, Gleich GJ, Lucendo AJ, Olalla JM, Vinson LA, et al. Eosinophilic esophagitis in adults is associated with IgG4 and not mediated by IgE. Gastroenterology [Internet]. 2014;**147**(3):602-609. DOI: 10.1053/j.gastro.2014.05.036

[68] Storch WB, Eckardt VF, Wienbeck M, Eberl T, Auer PG, Hecker A, et al. Autoantibodies to Auerbach's plexus in achalasia. Cellular and Molecular Biology [Internet]. 1995;**41**(8):1033-1038. Available from: https://www.ncbi.nlm.nih.gov/pubmed/8747084

[69] Verne GN, Sallustio JE, Eaker EY. Anti-myenteric neuronal antibodies in patients with achalasia. A prospective study. Digestive Diseases and Scienecs [Internet]. 1997;**42**(2):307-313. DOI: 10.1023/a:1018857617115

[70] Ruiz-de-León A, Mendoza J, Sevilla-Mantilla C, Fernández AM, Pérez-de-la-Serna J, Gónzalez VA, et al. Myenteric antiplexus antibodies and class II HLA in achalasia. Digestive Diseases and Sciences [Internet]. 2002;**47**(1):15-19. DOI: 10.1023/a:1013242831900

[71] Kallel-Sellami M, Karoui S, Romdhane H, Laadhar L, Serghini M, Boubaker J, et al. Circulating antimyenteric autoantibodies in Tunisian patients with idiopathic achalasia. Diseases of the Esophagus [Internet]. 2013;**26**(8):782-787. DOI: 10.1111/j.1442-2050.2012.01398.x

[72] Kraichely RE, Farrugia G, Pittock SJ, Castell DO, Lennon VA. Neural autoantibody profile of primary achalasia. Digestive Diseases and Sciences [Internet]. 2010;**55**(2):307-311. DOI: 10.1007/s10620-009-0838-9

[73] Stafford IS, Kellermann M, Mossotto E, Beattie RM, MacArthur BD, Ennis S. A systematic review of the applications of artificial intelligence and machine learning in autoimmune diseases. NPJ Digital Medicine [Internet]. 2020;**3**:30. DOI: 10.1038/s41746-020-0229-3

[74] Kocheturov A, Pardalos PM, Karakitsiou A. Massive datasets and machine learning for computational biomedicine: Trends and challenges. Annals of Operations Research [Internet]. 2019;**276**(1):5-34. DOI: 10.1007/s10479-018-2891-2

[75] Berry MW, Mohamed A, Yap BW. In: Berry MW, Mohamed A, Yap BW, editors. Supervised and Unsupervised Learning for Data Science [Internet]. Cham: Springer; 2020. Available from: https://link.springer.com/10.1007/978-3-030-22475-2

[76] Singh A, Thakur N, Sharma A. A review of supervised machine learning algorithms. In: 2016 3rd International Conference on Computing for Sustainable Global Development (INDIACom) [Internet]. IEEE; 2016. pp. 1310-1315. Available from: https://ieeexplore.ieee.org/abstract/document/7724478/

[77] Mahesh B. Machine learning algorithms-A review. International Journal of Science and Research (IJSR) [Internet] [Internet]. 2020;**9**:381-386. Available from: https://www.researchgate.net/profile/Batta-Mahesh/publication/344717762_Machine_Learning_Algorithms_-A_Review/links/5f8b2365299bf1b53e2d243a/Machine-Learning-Algorithms-A-Review.pdf

[78] Carlson DA, Kou W, Rooney KP, Baumann AJ, Donnan E, Triggs JR, et al. Achalasia subtypes can be identified with functional luminal imaging probe (FLIP) panometry using a supervised machine learning process. Neurogastroenterology & Motility [Internet]. 2021;**33**(3):e13932. DOI:10.1111/nmo.13932

[79] Jiang K, Oda M, Shiwaku H, Misawa M, Mori K. Real-time esophagus achalasia detection method for esophagoscopy assistance. In: Medical Imaging 2022: Computer-Aided Diagnosis [Internet]. SPIE; 2022. pp. 26-31. Available from: https://www.spiedigitallibrary.org/conference-proceedings-of-spie/12033/1203304/Real-time-esophagus-achalasia-detection-method-for-esophagoscopy-assistance/10.1117/12.2613289.short

[80] Franklin MM, Schultz FA, Tafoya MA, Kerwin AA, Broehm CJ, Fischer EG, et al. A deep learning convolutional neural network can differentiate between Helicobacter pylori gastritis and autoimmune gastritis with results comparable to gastrointestinal pathologists. Archives of Pathology & Laboratory Medicine [Internet]. 2022;146(1):117-122. doi:10.5858/arpa.2020-0520-OA

[81] Tenório JM, Hummel AD, Cohrs FM, Sdepanian VL, Pisa IT, de Fátima Marin H. Artificial intelligence techniques applied to the development of a decision–support system for diagnosing celiac disease. International Journal of Medical Information [Internet]. 2011;**80**(11):793-802. Available from: https://www.sciencedirect.com/science/article/pii/S1386505611001651

[82] Guimarães P, Keller A, Fehlmann T, Lammert F, Casper M. Deep

learning-based detection of eosinophilic esophagitis. Endoscopy [Internet]. 2022;**54**(3):299-304. DOI: 10.1055/a-1520-8116

[83] Daniel N, Larey A, Aknin E, Osswald GA, Caldwell JM, Rochman M, et al. PECNet: A deep multi-label segmentation network for eosinophilic esophagitis biopsy diagnostics [Internet]. arXiv [cs.CV]. 2021. Available from: http://arxiv.org/abs/2103.02015

[84] Czyzewski T, Daniel N, Rochman M, Caldwell JM, Osswald GA, Collins MH, et al. Machine learning approach for biopsy-based identification of eosinophilic esophagitis reveals importance of global features. IEEE Open Journal of Engineering in Medicine and Biology [Internet]. 2021;**2**:218-223. DOI: 10.1109/ojemb.2021.3089552

[85] Adorno W 3rd, Catalano A, Ehsan L, Vitzhum von Eckstaedt H, Barnes B, McGowan E, et al. Advancing eosinophilic esophagitis diagnosis and phenotype assessment with deep learning computer vision. Biomed Eng Syst Technol Int Jt Conf BIOSTEC Revis Sel Pap [Internet]. 2021;**2021**:44-55. DOI: 10.5220/0010241900440055

[86] Sallis BF, Erkert L, Moñino-Romero S, Acar U, Wu R, Konnikova L, et al. An algorithm for the classification of mRNA patterns in eosinophilic esophagitis: Integration of machine learning. Journal of Allergy and Clinical Immunology [Internet]. 2018;**141**(4):1354-1364.e9. DOI: 10.1016/j.jaci.2017.11.027

[87] Shi Y, Liu Q, Xu J, Asad Z, Cui C, Correa H, et al. Eosinophilic esophagitis multi-label feature recognition on whole slide imaging using transfer learning. In: Medical Imaging 2022: Digital and Computational Pathology [Internet]. SPIE; 2022. pp. 277-284. Available from: https://www.spiedigitallibrary.org/conference-proceedings-of-spie/12039/1203919/Eosinophilic-esophagitis-multi-label-feature-recognition-on-whole-slide-imaging/10.1117/12.2611521.short

[88] Mossotto E, Ashton JJ, Coelho T, Beattie RM, MacArthur BD, Ennis S. Classification of paediatric inflammatory bowel disease using machine learning. Scientific Reports [Internet]. 2017;7(1):2427. DOI: 10.1038/s41598-017-02606-2

[89] Barash Y, Azaria L, Soffer S, Margalit Yehuda R, Shlomi O, Ben-Horin S, et al. Ulcer severity grading in video capsule images of patients with Crohn's disease: An ordinal neural network solution. Gastrointestinal Endoscopy [Internet]. 2021;**93**(1):187-192. DOI: 10.1016/j.gie.2020.05.066

[90] Gottlieb K, Requa J, Karnes W, Chandra Gudivada R, Shen J, Rael E, et al. Central Reading of ulcerative colitis clinical trial videos using neural networks. Gastroenterology [Internet]. 2021;**160**(3):710-719.e2. DOI: 10.1053/j.gastro.2020.10.024

[91] Takenaka K, Ohtsuka K, Fujii T, Negi M, Suzuki K, Shimizu H, et al. Development and validation of a deep neural network for accurate evaluation of endoscopic images from patients with ulcerative colitis. Gastroenterology [Internet]. 2020;**158**(8):2150-2157. DOI: 10.1053/j.gastro.2020.02.012

[92] Chen D, Fulmer C, Gordon IO, Syed S, Stidham RW, Vande Casteele N, et al. Application of artificial intelligence to clinical practice in inflammatory bowel disease--What the clinician needs to know. Journal of Crohn's and Colitis [Internet]. 2022;**16**(3):460-471. Available from: https://

academic.oup.com/ecco-jcc/article-abstract/16/3/460/6374854

[93] Cohen-Mekelburg S, Berry S, Stidham RW, Zhu J, Waljee AK. Clinical applications of artificial intelligence and machine learning-based methods in inflammatory bowel disease. Journal of Gastroenterology and Hepatology [Internet]. 2021;**36**(2):279-285. Available from: https://onlinelibrary.wiley.com/doi/10.1111/jgh.15405

[94] Kohli A, Holzwanger EA, Levy AN. Emerging use of artificial intelligence in inflammatory bowel disease. World Journal of Gastroenterology [Internet]. 2020;**26**(44):6923-6928. Available from: https://www.wjgnet.com/1007-9327/full/v26/i44/6923.htm

[95] Lin Y, Wang G, Yu J, Sung JJY. Artificial intelligence and metagenomics in intestinal diseases. Journal of Gastroenterology and Hepatology [Internet]. 2021;**36**(4):841-847. DOI: 10.1111/jgh.15501

[96] Manandhar I, Alimadadi A, Aryal S, Munroe PB, Joe B, Cheng X. Gut microbiome-based supervised machine learning for clinical diagnosis of inflammatory bowel diseases. American Journal of Physiology Gastrointestinal and Liver Physiology [Internet]. 2021;**320**(3):G328-G337. DOI: 10.1152/ajpgi.00360.2020

[97] Matalka II, Al-Omari FA, Salama RM, Mohtaseb AH. A novel approach for quantitative assessment of mucosal damage in inflammatory bowel disease. Diagnostic Pathology [Internet]. 2013;**8**:156. DOI: 10.1186/1746-1596-8-156

[98] Pinton P. Computational models in inflammatory bowel disease. Clinical and Translational Science [Internet]. 2022;**15**(4):824-830. DOI: 10.1111/cts.13228

[99] Sundaram S, Choden T, Mattar MC, Desai S, Desai M. Artificial intelligence in inflammatory bowel disease endoscopy: Current landscape and the road ahead. Therapeutic Advances in Gastrointestinal Endoscopy [Internet]. 2021;**14**:26317745211017810. DOI: 10.1177/26317745211017809

[100] Yang Y, Li Y-X, Yao R-Q, Du X-H, Ren C. Artificial intelligence in small intestinal diseases: Application and prospects. World Journal of Gastroenterology [Internet]. 2021;**27**(25):3734-3747. DOI: 10.3748/wjg.v27.i25.3734

[101] Zand A, Stokes Z, Sharma A, van Deen WK, Hommes D. Artificial intelligence for inflammatory bowel diseases (IBD); Accurately predicting adverse outcomes using machine learning. Digestive Diseases and Sciences [Internet]. 2022. DOI: 10.1007/s10620-022-07506-8

[102] Vaezi MF, Felix VN, Penagini R, Mauro A, de Moura EGH, Pu LZCT, et al. Achalasia: From diagnosis to management. Annals of the New York Academy of Sciences [Internet]. 2016;**1381**(1):34-44. DOI: 10.1111/nyas.13176

[103] Patel DA, Vaezi MF. Achalasia and nutrition: Is it simple physics or biology. Practical Gastroenterology [Internet]. 2016;**40**(11):42-48. Available from: https://med.virginia.edu/ginutrition/wp-content/uploads/sites/199/2014/06/Parrish-November-16-2.pdf

[104] Ghoshal UC, Thakur PK, Misra A. Frequency and factors associated with malnutrition among patients with achalasia and effect of pneumatic dilation. JGH Open [Internet]. 2019;**3**(6):468-473. DOI: 10.1002/jgh3.12191

[105] Zifodya J, Kim HP, Silver HJ, Slaughter JC, Higginbotham T, Vaezi MF.

Tu1199 nutritional status of patients with untreated achalasia. Gastroenterology. 2015;**148**(4):S-819-S-820. Available from: https://www.researchgate.net/publication/276143517_Tu1199_Nutritional_Status_of_Patients_With_Untreated_Achalasia

[106] Cavalcoli F, Zilli A, Conte D, Massironi S. Micronutrient deficiencies in patients with chronic atrophic autoimmune gastritis: A review. World Journal of Gastroenterology [Internet]. 2017;**23**(4):563-572. DOI: 10.3748/wjg.v23.i4.563

[107] Hołubiuk Ł, Imiela J. Diet and Helicobacter pylori infection. Prz Gastroenterol [Internet]. 2016;**11**(3):150-154. DOI: 10.5114/pg.2016.61487

[108] Cellini L, Di Campli E, Masulli M, Di Bartolomeo S, Allocati N. Inhibition of Helicobacter pylori by garlic extract (Allium sativum). FEMS Immunology and Medical Microbiology [Internet]. 1996;**13**(4):273-277. DOI: 10.1111/j.1574-695X.1996.tb00251.x

[109] Wang Y-C. Medicinal plant activity on Helicobacter pylori related diseases. World Journal of Gastroenterology [Internet]. 2014;**20**(30):10368-10382. DOI: 10.3748/wjg.v20.i30.10368

[110] Li Z, Ying X, Shan F, Ji J. The association of garlic with Helicobacter pylori infection and gastric cancer risk: A systematic review and meta-analysis. Helicobacter [Internet]. 2018;23(5):e12532. doi:10.1111/hel.12532

[111] Bonifácio BV, dos Santos Ramos MA, da Silva PB, Bauab TM. Antimicrobial activity of natural products against Helicobacter pylori: A review. Annals of Clinical Microbiology and Antimicrobials [Internet]. 2014;**13**:54. DOI: 10.1186/s12941-014-0054-0

[112] Ji J, Yang H. Using probiotics as supplementation for Helicobacter pylori antibiotic therapy. International Journal of Molecular Sciences [Internet]. 2020;**21**(3):1136. DOI: 10.3390/ijms21031136

[113] Parzanese I, Qehajaj D, Patrinicola F, Aralica M, Chiriva-Internati M, Stifter S, et al. Celiac disease: From pathophysiology to treatment. World Journal of Gastrointestinal Pathophysiology [Internet]. 2017;**8**(2):27-38. DOI: 10.4291/wjgp.v8.i2.27

[114] Caio G, Volta U, Sapone A, Leffler DA, De Giorgio R, Catassi C, et al. Celiac disease: A comprehensive current review. BMC Medicine [Internet]. 2019;**17**(1):142. DOI: 10.1186/s12916-019-1380-z

[115] Sainsbury A, Sanders DS, Ford AC. Prevalence of irritable Bowel syndrome–type symptoms in patients with celiac disease: A meta-analysis. Clinical Gastroenterology and Hepatology [Internet]. 2013;**11**(4):359-365.e1. Available from: https://www.sciencedirect.com/science/article/pii/S1542356512014917

[116] https://www.tandfonline.com/doi/full/10.1080/07315724.2000.10718086?casa_token=d-bip_kvctgAAAAA%3AFKW4Rhsq50MGljvSE6qotou2w1Hs8B1z-tnboF6e2BNocDb5pYTjcMO551MWU2wtVXRnPWFTz_OUog

[117] Kelly KJ, Lazenby AJ, Rowe PC, Yardley JH, Perman JA, Sampson HA. Eosinophilic esophagitis attributed to gastroesophageal reflux: Improvement with an amino acid-based formula. Gastroenterology [Internet]. 1995;**109**(5):1503-1512. DOI: 10.1016/0016-5085(95)90637-1

[118] Gonsalves N, Yang G-Y, Doerfler B, Ritz S, Ditto AM, Hirano I. Elimination

diet effectively treats eosinophilic esophagitis in adults; food reintroduction identifies causative factors. Gastroenterology [Internet], 2012;**142**(7):1451-9.e1. DOI: 10.1053/j.gastro.2012.03.001

[119] Cotton CC, Eluri S, Wolf WA, Dellon ES. Six-food elimination diet and topical steroids are effective for eosinophilic esophagitis: A meta-regression. Digestive Diseases and Sciences [Internet]. 2017;**62**(9):2408-2420. DOI: 10.1007/s10620-017-4642-7

[120] Rajavelu P, Rayapudi M, Moffitt M, Mishra A, Mishra A. Significance of para-esophageal lymph nodes in food or aeroallergen-induced iNKT cell-mediated experimental eosinophilic esophagitis. American Jpurnal of Physiology Gastrointestinal and Liver Physiology [Internet]. 2012;**302**(7):G645-G654. DOI: 10.1152/ajpgi.00223.2011

[121] Rubinstein E, Cho JY, Rosenthal P, Chao J, Miller M, Pham A, et al. Siglec-F inhibition reduces esophageal eosinophilia and angiogenesis in a mouse model of eosinophilic esophagitis. Journal of Pediatric and Gastroenterology and Nutrition [Internet]. 2011;**53**(4):409-416. Available from: https://journals.lww.com/00005176-201110000-00014

[122] Spergel JM, Brown-Whitehorn TF, Cianferoni A, Shuker M, Wang M-L, Verma R, et al. Identification of causative foods in children with eosinophilic esophagitis treated with an elimination diet. Journal of Allergy and Clinical Immunology [Internet]. 2012;**130**(2):461-7.e5. DOI: 10.1016/j.jaci.2012.05.021

[123] Penfield JD, Lang DM, Goldblum JR, Lopez R, Falk GW. The role of allergy evaluation in adults with eosinophilic esophagitis. Journal of Clinical and Gastroenterology [Internet]. 2010;**44**(1):22-27. DOI: 10.1097/MCG.0b013e3181a1bee5

[124] Wechsler JB, Schwartz S, Amsden K, Kagalwalla AF. Elimination diets in the management of eosinophilic esophagitis. Journal of Asthma and Allergy [Internet]. 2014;7:85-94. Available from: doi:10.2147/JAA.S47243

[125] Molina-Infante J, Lucendo AJ. Dietary therapy for eosinophilic esophagitis. Journal of Allergy and Clinical Immunology [Internet]. 2018;**142**(1):41-47. DOI: 10.1016/j.jaci.2018.02.028

[126] De Vlieger L, Smolders L, Nuyttens L, Verelst S, Breynaert C, Vanuytsel T, et al. A clinical perspective on the dietary therapies for pediatric eosinophilic esophagitis: The gap Between research and daily practice. Frontiers in Immunology [Internet]. 2021;**12**:677859. DOI: 10.3389/fimmu.2021.677859

[127] Kagalwalla AF, Sentongo TA, Ritz S, Hess T, Nelson SP, Emerick KM, et al. Effect of six-food elimination diet on clinical and histologic outcomes in eosinophilic esophagitis. Clinical Gastroenterology and Hepatology [Internet]. 2006;**4**(9):1097-1102. DOI: 10.1016/j.cgh.2006.05.026

[128] Lucendo AJ, Arias Á, González-Cervera J, Yagüe-Compadre JL, Guagnozzi D, Angueira T, et al. Empiric 6-food elimination diet induced and maintained prolonged remission in patients with adult eosinophilic esophagitis: A prospective study on the food cause of the disease. Journal of Allergy and Clinical Immunology [Internet]. 2013;**131**(3):797-804. Available from: https://www.sciencedirect.com/science/article/pii/S0091674912026449

[129] Molina-Infante J, Arias A, Barrio J, Rodríguez-Sánchez J, Sanchez-Cazalilla M, Lucendo AJ. Four-food group elimination diet for adult eosinophilic esophagitis: A prospective multicenter study. Journal of Allergy and Clinical Immunology [Internet]. 2014;**134**(5):1093-9.e1. DOI: 10.1016/j.jaci.2014.07.023

[130] Molina-Infante J, Arias Á, Alcedo J, Garcia-Romero R, Casabona-Frances S, Prieto-Garcia A, et al. Step-up empiric elimination diet for pediatric and adult eosinophilic esophagitis: The 2-4-6 study. Journal of Allergy and Clinical and Immunology [Internet]. 2018;**141**(4):1365-1372. DOI: 10.1016/j.jaci.2017.08.038

[131] Kagalwalla AF, Shah A, Li BUK, Sentongo TA, Ritz S, Manuel-Rubio M, et al. Identification of specific foods responsible for inflammation in children with eosinophilic esophagitis successfully treated with empiric elimination diet. Journal of Pediatric Gastroenterology and Nutrition [Internet]. 2011;**53**(2):145-149. DOI: 10.1097/MPG.0b013e31821cf503

[132] Rodríguez-Sánchez J, Gómez Torrijos E, López Viedma B, de la Santa BE, Martín Dávila F, García Rodríguez C, et al. Efficacy of IgE-targeted vs empiric six-food elimination diets for adult eosinophilic oesophagitis. Allergy [Internet]. 2014;**69**(7):936-942. Available from: https://onlinelibrary.wiley.com/doi/10.1111/all.12420

[133] Kagalwalla AF, Wechsler JB, Amsden K, Schwartz S, Makhija M, Olive A, et al. Efficacy of a 4-food elimination diet for children with eosinophilic esophagitis. Clinical Gastroenterology and Hepatology [Internet]. 2017;**15**(11):1698-1707.e7. DOI: 10.1016/j.cgh.2017.05.048

[134] Lucendo AJ, De Rezende LC. Importance of nutrition in inflammatory bowel disease. World Journal of Gastroenterology [Internet]. 2009;**15**(17):2081-2088. DOI: 10.3748/wjg.15.2081

[135] Martini GA, Brandes JW. Increased consumption of refined carbohydrates in patients with Crohn's disease. Klin Wochenschr [Internet]. 1976;**54**(8):367-371. DOI: 10.1007/BF01469792

[136] Hoogerwerf WA, Pasricha PJ. Pharmacologic therapy in treating achalasia. Gastrointestinal Endoscopy Clinics of North America [Internet]. 2001;**11**(2):311-324. Available from: https://www.ncbi.nlm.nih.gov/pubmed/11319064

[137] Gordon JM, Eaker EY. Prospective study of esophageal botulinum toxin injection in high-risk achalasia patients. American Journal of Gastroenterology [Internet]. 1997;**92**(10):1812-1817. Available from: https://www.ncbi.nlm.nih.gov/pubmed/9382042

[138] Annese V, Basciani M, Borrelli O, Leandro G, Simone P, Andriulli A. Intrasphincteric injection of botulinum toxin is effective in long-term treatment of esophageal achalasia. Muscle Nerve [Internet]. 1998;**21**(11):1540-1542. DOI: 10.1002/(sici)1097-4598(199811)21:11<1540::aid-mus27>3.0.co;2-2

[139] D'Onofrio V, Annese V, Miletto P, Leandro G, Marasco A, Sodano P, et al. Long-term follow-up of achalasic patients treated with botulinum toxin. Diseases of the Esophagus [Internet]. 2000;**13**(2):96-101. DOI: 10.1046/j.1442-2050.2000.00094.x

[140] Forootan Pishbijari H, Mortazavi Tabatabaei S, Jangodaz M. Injection of botulinum toxin in the treatment

of achalasia: A randomized, double blind, controlled clinical trial. Medical Journal of The Islamic Republic of Iran [Internet]. 2001;**15**:123-127. Available from: https://www.scopus.com/record/display.uri?eid=2-s2.0-85076576005&origin=inward&featureToggles=FEATURE_NEW_DOC_DETAILS_EXPORT:1

[141] Dughera L, Battaglia E, Maggio D, Cassolino P, Mioli PR, Morelli A, et al. Botulinum toxin treatment of oesophageal achalasia in the old old and oldest old: A 1-year follow-up study. Drugs Aging [Internet]. 2005;**22**(9):779-783. DOI: 10.2165/00002512-200522090-00006

[142] Bassotti G, D'Onofrio V, Battaglia E, Fiorella S, Dughera L, Iaquinto G, et al. Treatment with botulinum toxin of octo-nonagerians with oesophageal achalasia: A two-year follow-up study. Alimentary Pharmacology and Therapeutics [Internet]. 2006;**23**(11):1615-1619. DOI: 10.1111/j.1365-2036.2006.02907.x

[143] Oude Nijhuis RAB, Prins LI, Mostafavi N, van Etten-Jamaludin FS, Smout AJPM, Bredenoord AJ. Factors associated with Achalasia treatment outcomes: Systematic review and meta-analysis. Clinical Gastroenterology and Hepatology [Internet]. 2020;**18**(7):1442-1453. Available from: https://www.sciencedirect.com/science/article/pii/S1542356519311073

[144] Khashab MA, Vela MF, Thosani N, Agrawal D, Buxbaum JL, Abbas Fehmi SM, et al. ASGE guideline on the management of achalasia. Gastrointestinal Endoscopy [Internet]. 2020;**91**(2):213-227.e6. DOI: 10.1016/j.gie.2019.04.231

[145] Inoue H, Minami H, Kobayashi Y, Sato Y, Kaga M, Suzuki M, et al. Peroral endoscopic myotomy (POEM) for esophageal achalasia. Endoscopy [Internet]. 2010;**42**(4):265-271. DOI: 10.1055/s-0029-1244080

[146] Minalyan A, Benhammou JN, Artashesyan A, Lewis MS, Pisegna JR. Autoimmune atrophic gastritis: Current perspectives. Clinical and Experimental Gastroenterology [Internet]. 2017;**10**:19-27. DOI: 10.2147/CEG.S109123

[147] Varis K, Ihamäki T, Härkönen M, Samloff IM, Siurala M. Gastric morphology, function, and immunology in first-degree relatives of probands with pernicious anemia and controls. Scandivian Journal of Gastroenterology [Internet]. 1979;**14**(2):129-139. DOI: 10.3109/00365527909179858

[148] Ottesen M, Feldt-Rasmussen U, Andersen J, Hippe E, Schouboe A. Thyroid function and autoimmunity in pernicious anemia before and during cyanocobalamin treatment. Journal of Endocrinological Investigation [Internet]. Feb 1995;**18**(2):91-97. DOI: 10.1007/BF03349707

[149] Centanni M, Marignani M, Gargano L, Corleto VD, Casini A, Delle Fave G, et al. Atrophic body gastritis in patients with autoimmune thyroid disease: An underdiagnosed association. Archives of Internal Medicine [Internet]. 1999;**159**(15):1726-1730. DOI: 10.1001/archinte.159.15.1726

[150] Silveira PA, Wilson WE, Esteban LM, Jordan MA, Hawke CG, van Driel IR, et al. Identification of the Gasa3 and Gasa4 autoimmune gastritis susceptibility genes using congenic mice and partitioned, segregative and interaction analyses. Immunogenetics [Internet]. 2001;**53**(9):741-750. DOI: 10.1007/s00251-001-0391-y

[151] Lam-Tse W-K, Batstra MR, Koeleman BPC, Roep BO, Bruining MGJ, Aanstoot HJ, et al. The association

between autoimmune thyroiditis, autoimmune gastritis and type 1 diabetes. Pediatric Endocrinology Reviews [Internet]. 2003;**1**(1):22-37. Available from: https://www.ncbi.nlm.nih.gov/pubmed/16437010

[152] De Block CEM, De Leeuw IH, Bogers JJPM, Pelckmans PA, Ieven MM, Van Marck EAE, et al. Autoimmune gastropathy in type 1 diabetic patients with parietal cell antibodies: Histological and clinical findings. Diabetes Care [Internet]. 2003;**26**(1):82-88. DOI: 10.2337/diacare.26.1.82

[153] Alonso N, Granada ML, Salinas I, Lucas AM, Reverter JL, Juncà J, et al. Serum pepsinogen I: An early marker of pernicious anemia in patients with type 1 diabetes. The Journal of Clinical Endocrinology & Metabolism [Internet]. 2005;**90**(9):5254-5258. DOI: 10.1210/jc.2005-0580

[154] De Block CEM, De Leeuw IH, Van Gaal LF. Autoimmune gastritis in type 1 diabetes: A clinically oriented review. The Journal of Clinical Endocrinology & Metabolism [Internet]. 2008;**93**(2):363-371. DOI: 10.1210/jc.2007-2134

[155] Magris R, De Re V, Maiero S, Fornasarig M, Guarnieri G, Caggiari L, et al. Low pepsinogen I/II ratio and high Gastrin-17 levels typify chronic atrophic autoimmune Gastritis patients with gastric neuroendocrine Tumors. Clinical and Translational Gastroenterology [Internet]. 2020;**11**(9):e00238. DOI: 10.14309/ctg.0000000000000238

[156] Rodriguez-Castro KI, Franceschi M, Miraglia C, Russo M, Nouvenne A, Leandro G, et al. Autoimmune diseases in autoimmune atrophic gastritis. Acta Biomedica [Internet]. 2018;**89**(8-S):100-103. Available from: https://www.mattioli1885journals.com/index.php/actabiomedica/article/view/7919

[157] Bascuñán KA, Vespa MC, Araya M. Celiac disease: Understanding the gluten-free diet. European Journal of Nutrition [Internet]. 2017;**56**(2):449-459. DOI: 10.1007/s00394-016-1238-5

[158] Elli L, Roncoroni L, Hils M, Pasternack R, Barisani D, Terrani C, et al. Immunological effects of transglutaminase-treated gluten in coeliac disease. Human Immunology [Internet]. 2012;**73**(10):992-997. DOI: 10.1016/j.humimm.2012.07.318

[159] Lundin KEA, Sollid LM. Advances in coeliac disease. Current Opinions in Gastroenterology [Internet]. 2014;**30**(2):154-162. DOI: 10.1097/MOG.0000000000000041

[160] Goel G, King T, Daveson AJ, Andrews JM, Krishnarajah J, Krause R, et al. Epitope-specific immunotherapy targeting CD4-positive T cells in coeliac disease: Two randomised, double-blind, placebo-controlled phase 1 studies. The Lancet Gastroenterology & Hepatology [Internet]. 2017;**2**(7):479-493. Available from: https://www.sciencedirect.com/science/article/pii/S2468125317301103

[161] Di Sabatino A, Lenti MV, Corazza GR, Gianfrani C. Vaccine immunotherapy for celiac disease. Frontiers in Medicine [Internet]. 2018;**5**:187. DOI: 10.3389/fmed.2018.00187

[162] Yoosuf S, Therrien A, Leffler DA. Chapter 7 - non-dietary therapies for celiac disease. In: Schieptti A, Sanders D, editors. Coeliac Disease and Gluten-Related Disorders [Internet]. Amsterdam, Netherlands: Academic Press; 2022. pp. 111-160. Available from: https://www.sciencedirect.com/science/article/pii/B9780128215715000118

[163] Liacouras CA, Furuta GT, Hirano I, Atkins D, Attwood SE, Bonis PA, et al. Eosinophilic esophagitis: Updated consensus recommendations for children and adults. Journal of Allergy and Clinical Immunology [Internet]. 2011;**128**(1):3-20.e6. DOI: 10.1016/j.jaci.2011.02.040

[164] Straumann A. Treatment of eosinophilic esophagitis: Diet, drugs, or dilation? Gastroenterology [Internet]. 2012;**142**(7):1409-1411. DOI: 10.1053/j.gastro.2012.04.039

[165] Dellon ES. Diagnosis and management of eosinophilic esophagitis. Clinical Gastroenterology and Hepatology [Internet]. 2012;**10**(10):1066-1078. DOI: 10.1016/j.cgh.2012.06.003

[166] Teitelbaum JE, Fox VL, Twarog FJ, Nurko S, Antonioli D, Gleich G, et al. Eosinophilic esophagitis in children: Immunopathological analysis and response to fluticasone propionate. Gastroenterology [Internet]. 2002;**122**(5):1216-1225. DOI: 10.1053/gast.2002.32998

[167] Arora AS, Perrault J, Smyrk TC. Topical corticosteroid treatment of dysphagia due to eosinophilic esophagitis in adults. Mayo Clinic Proceedings [Internet]. 2003;**78**(7):830-835. DOI: 10.4065/78.7.830

[168] Konikoff MR, Noel RJ, Blanchard C, Kirby C, Jameson SC, Buckmeier BK, et al. A randomized, double-blind, placebo-controlled trial of fluticasone propionate for pediatric eosinophilic esophagitis. Gastroenterology [Internet]. 2006;**131**(5):1381-1391. DOI: 10.1053/j.gastro.2006.08.033

[169] Schaefer ET, Fitzgerald JF, Molleston JP, Croffie JM, Pfefferkorn MD, Corkins MR, et al. Comparison of oral prednisone and topical fluticasone in the treatment of eosinophilic esophagitis: A randomized trial in children. Clinical Gastroenterology and Hepatology [Internet]. 2008;**6**(2):165-173. DOI: 10.1016/j.cgh.2007.11.008

[170] Straumann A, Conus S, Degen L, Frei C, Bussmann C, Beglinger C, et al. Long-term budesonide maintenance treatment is partially effective for patients with eosinophilic esophagitis. Clinical Gastroenterology and Hepatology [Internet]. 2011;**9**(5):400-9.e1. DOI: 10.1016/j.cgh.2011.01.017

[171] Peterson KA, Thomas KL, Hilden K, Emerson LL, Wills JC, Fang JC. Comparison of esomeprazole to aerosolized, swallowed fluticasone for eosinophilic esophagitis. Digestive Diseases and Sciences [Internet]. 2010;**55**(5):1313-1319. DOI: 10.1007/s10620-009-0859-4

[172] Aceves SS, Bastian JF, Newbury RO, Dohil R. Oral viscous budesonide: a potential new therapy for eosinophilic esophagitis in children. American Journal of Gastroenterology [Internet]. 2007;**102**(10):2271-2279. DOI: 10.1111/j.1572-0241.2007.01379.x

[173] Dellon ES, Liacouras CA, Molina-Infante J, Furuta GT, Spergel JM, Zevit N, et al. Updated international consensus diagnostic criteria for eosinophilic esophagitis: Proceedings of the AGREE conference. Gastroenterology [Internet]. 2018;**155**(4):1022-1033.e10. DOI: 10.1053/j.gastro.2018.07.009

[174] Molina-Infante J, Zamorano J. Distinguishing eosinophilic esophagitis from gastroesophageal reflux disease upon PPI refractoriness: What about PPI-responsive esophageal eosinophilia? Digestion [Internet]. 2012;**85**(3):210. DOI: 10.1159/000335932

[175] Netzer P, Gschossmann JM, Straumann A, Sendensky A, Weimann R, Schoepfer AM. Corticosteroid-dependent eosinophilic oesophagitis: Azathioprine and 6-mercaptopurine can induce and maintain long-term remission. European Journal of Gastroenterology & Hepatology [Internet]. 2007;**19**(10):865-869. DOI: 10.1097/MEG.0b013e32825a6ab4

[176] Mottet C, Schoepfer AM, Juillerat P, Cosnes J, Froehlich F, Kessler-Brondolo V, et al. Experts opinion on the practical use of azathioprine and 6-mercaptopurine in inflammatory bowel disease. Inflammatory Bowel Diseases [Internet]. 2016;**22**(11):2733-2747. DOI: 10.1097/MIB.0000000000000923

[177] Eskian M, Khorasanizadeh M, Assa'ad AH, Rezaei N. Monoclonal antibodies for treatment of eosinophilic esophagitis. Clinical Reviews in Allergy Immunology [Internet]. 2018;**55**(1):88-98. DOI: 10.1007/s12016-017-8659-7

[178] Mavi P, Rajavelu P, Rayapudi M, Paul RJ, Mishra A. Esophageal functional impairments in experimental eosinophilic esophagitis. American Journal of Physiology-Gastrointestinal and Liver Physiology [Internet]. 2012;**302**(11):G1347-G1355. DOI: 10.1152/ajpgi.00013.2012

[179] Mishra A, Hogan SP, Brandt EB, Rothenberg ME. IL-5 promotes eosinophil trafficking to the esophagus. Journal of Immunology [Internet]. 2002;**168**(5):2464-2469. DOI: 10.4049/jimmunol.168.5.2464

[180] Stein ML, Collins MH, Villanueva JM, Kushner JP, Putnam PE, Buckmeier BK, et al. Anti–IL-5 (mepolizumab) therapy for eosinophilic esophagitis. Journal of Allergy and Clinical Immunology [Internet]. 2006;**118**(6):1312-1319. DOI: 10.1016/j.jaci.2006.09.007

[181] Haldar P, Brightling CE, Hargadon B, Gupta S, Monteiro W, Sousa A, et al. Mepolizumab and exacerbations of refractory eosinophilic asthma. The New England Journal of Medicine [Internet]. 2009;**360**(10):973-984. DOI: 10.1056/NEJMoa0808991

[182] Rothenberg ME, Klion AD, Roufosse FE, Kahn JE, Weller PF, Simon H-U, et al. Treatment of patients with the hypereosinophilic syndrome with Mepolizumab. The New England Journal of Medicine [Internet]. 2008;**358**(12):1215-1228. DOI: 10.1056/NEJMoa070812

[183] Straumann A, Conus S, Grzonka P, Kita H, Kephart G, Bussmann C, et al. Anti-interleukin-5 antibody treatment (mepolizumab) in active eosinophilic oesophagitis: A randomised, placebo-controlled, double-blind trial. Gut [Internet]. 2010;**59**(1):21-30. DOI: 10.1136/gut.2009.178558

[184] Spergel JM, Rothenberg ME, Collins MH, Furuta GT, Markowitz JE, Fuchs G, et al. Reslizumab in children and adolescents with eosinophilic esophagitis: Results of a double-blind, randomized, placebo-controlled trial. Journal of Allergy and Clinical Immunology [Internet]. 2012;**129**(2):456-463.e3. Available from: http://www.jacionline.org/article/S0091674911019063/abstract

[185] Walsh GM. Reslizumab for pediatric eosinophilic esophagitis. Immunotherapy [Internet]. 2010;**2**(4):461-465. DOI: 10.2217/imt.10.41

[186] Reslizumab MA. First global approval. Drugs [Internet]. 2016;**76**(8):907-911. DOI: 10.1007/s40265-016-0583-2

[187] Rothenberg ME, Wen T, Greenberg A, Alpan O, Enav B, Hirano I, et al. Intravenous anti-IL-13 mAb QAX576 for the treatment of eosinophilic esophagitis. Journal of Allergy and Clinical Immunology [Internet]. 2015;**135**(2):500-507. DOI: 10.1016/j.jaci.2014.07.049

[188] Normansell R, Walker S, Milan SJ, Walters EH, Nair P. Omalizumab for asthma in adults and children. Cochrane Database of Systematic Reviews 1. 2014

[189] Straumann A, Bauer M, Fischer B, Blaser K, Simon HU. Idiopathic eosinophilic esophagitis is associated with a T(H)2-type allergic inflammatory response. Journal of Allergy and Clinical Immunology [Internet]. 2001;**108**(6):954-961. DOI: 10.1067/mai.2001.119917

[190] Foroughi S, Foster B, Kim N, Bernardino LB, Scott LM, Hamilton RG, et al. Anti-IgE treatment of eosinophil-associated gastrointestinal disorders. Journal of Allergy and Clinical Immunology [Internet]. 2007;**120**(3):594-601. DOI: 10.1016/j.jaci.2007.06.015

[191] Loizou D, Enav B, Komlodi-Pasztor E, Hider P, Kim-Chang J, Noonan L, et al. A pilot study of omalizumab in eosinophilic esophagitis. PLoS One [Internet]. 2015;**10**(3):e0113483. DOI: 10.1371/journal.pone.0113483

[192] Straumann A, Bussmann C, Conus S, Beglinger C, Simon H-U. Anti–TNF-α (infliximab) therapy for severe adult eosinophilic esophagitis. Journal of Allergy and Clinical Immunology [Internet]. 2008;**122**(2):425-427. DOI: 10.1016/j.jaci.2008.06.012

[193] Cohen MS, Kaufman AB, Palazzo JP, Nevin D, Dimarino AJ Jr, Cohen S. An audit of endoscopic complications in adult eosinophilic esophagitis. Clinical Gastroenterology and Hepatology [Internet]. 2007;**5**(10):1149-1153. DOI: 10.1016/j.cgh.2007.05.017

[194] Schoepfer AM, Gonsalves N, Bussmann C, Conus S, Simon H-U, Straumann A, et al. Esophageal dilation in eosinophilic esophagitis: Effectiveness, safety, and impact on the underlying inflammation. American Journal of Gastroenterology [Internet]. 2010;**105**(5):1062-1070. DOI: 10.1038/ajg.2009.657

[195] Gonsalves NP, Aceves SS. Diagnosis and treatment of eosinophilic esophagitis. Journal of Allergy and Clinical Immunology [Internet]. 2020;**145**(1):1-7. DOI: 10.1016/j.jaci.2019.11.011

[196] Lichtenstein GR, Loftus EV, Isaacs KL, Regueiro MD, Gerson LB, Sands BE. ACG clinical guideline: Management of Crohn's disease in adults. Official Journal of the American College of Gastroenterology | ACG [Internet]. 2018;**113**(4):481. Available from: https://journals.lww.com/ajg/fulltext/2018/04000/acg_clinical_guideline__management_of_crohn_s.10.aspx

[197] Veauthier B, Hornecker JR. Crohn's disease: Diagnosis and management. American Family Physician [Internet]. 2018;**98**(11):661-669. Available from: https://www.ncbi.nlm.nih.gov/pubmed/30485038

[198] Mantzaris GJ. Thiopurines and methotrexate use in IBD patients in a biologic era. Current Treatment Options in Gastroenterology [Internet]. 2017;**15**(1):84-104. DOI: 10.1007/s11938-017-0128-0

[199] Shah ED, Siegel CA, Chong K, Melmed GY. Patients with Crohn's disease are more likely to remain on biologics

than Immunomodulators: A meta-analysis of treatment durability. Digestive Diseases and Sciences [Internet]. 2015;**60**(8):2408-2418. DOI: 10.1007/s10620-015-3618-8

[200] Adegbola SO, Sahnan K, Warusavitarne J, Hart A, Tozer P. Anti-TNF Therapy in Crohn's disease. International Journal of Molecular Sciences [Internet]. 2018;**19**(8):2244. DOI: 10.3390/ijms19082244

[201] Bressler B. Use of Vedolizumab for the treatment of Crohn's disease. Gastroenterology & Hepatology [Internet]. 2019;**15**(4):204-206. Available from: https://www.ncbi.nlm.nih.gov/pubmed/31435197

[202] Davies SC, Nguyen TM, Parker CE, MacDonald JK, Jairath V, Khanna R. Anti-IL-12/23p40 antibodies for maintenance of remission in Crohn's disease. Cochrane Database of Systematic Reviews [Internet]. 2019;**12**:CD01280. DOI: 10.1002/14651858.CD012804.pub2

[203] Meima-van Praag EM, Buskens CJ, Hompes R, Bemelman WA. Surgical management of Crohn's disease: A state of the art review. International Journal of Colorectal Diseases [Internet]. 2021;**36**(6):1133-1145. DOI: 10.1007/s00384-021-03857-2

[204] Feagan BG, Macdonald JK. Oral 5-aminosalicylic acid for induction of remission in ulcerative colitis. Cochrane Database of Systematic Reviews [Internet]. 2012;**10**:CD000543. DOI: 10.1002/14651858.CD000543.pub3

[205] Chao Y-S, Loshak H. Biologics Versus Immunomodulators for the Treatment of Ulcerative Colitis: A Review of Comparative Clinical Effectiveness and Cost-Effectiveness [Internet]. Ottawa (ON): Canadian Agency for Drugs and Technologies in Health; 2019. Available from: https://www.ncbi.nlm.nih.gov/pubmed/31693324

[206] Schabert VF, Watson C, Gandra SR, Goodman S, Fox KM, Harrison DJ. Annual costs of tumor necrosis factor inhibitors using real-world data in a commercially insured population in the United States. Journal of Medical Economics [Internet]. 2012;**15**(2):264-275. DOI: 10.3111/13696998.2011.644645

[207] Feagan BG, Rutgeerts P, Sands BE, Hanauer S, Colombel J-F, Sandborn WJ, et al. Vedolizumab as induction and maintenance therapy for ulcerative colitis. The New England Journal of Medicine [Internet]. 2013;**369**(8):699-710. DOI: 10.1056/NEJMoa1215734

[208] Buchs NC, Bloemendaal ALA, Wood CPJ, Travis S, Mortensen NJ, Guy RJ, et al. Subtotal colectomy for ulcerative colitis: Lessons learned from a tertiary centre. Colorectal Diseases [Internet]. 2017;**19**(5):O153-O161. DOI: 10.1111/codi.13658

[209] Diamond IR, Gerstle JT, Kim PCW, Langer JC. Outcomes after laparoscopic surgery in children with inflammatory bowel disease. Surgical Endoscopy [Internet]. 2010;**24**(11):2796-2802. DOI: 10.1007/s00464-010-1050-x